Lecture Notes in Bioinformatics 12881

Subseries of Lecture Notes in Computer Science

More information about this subseries at http://www.springer.com/series/5381

Eugenio Cinquemani · Loïc Paulevé (Eds.)

Computational Methods in Systems Biology

19th International Conference, CMSB 2021
Bordeaux, France, September 22–24, 2021
Proceedings

Springer

Editors
Eugenio Cinquemani
Univ. Grenoble Alpes, Inria
Grenoble, France

Loïc Paulevé
Univ. Bordeaux, Bordeaux INP, CNRS,
LaBRI, UMR5800
Talence, France

ISSN 0302-9743 ISSN 1611-3349 (electronic)
Lecture Notes in Bioinformatics
ISBN 978-3-030-85632-8 ISBN 978-3-030-85633-5 (eBook)
https://doi.org/10.1007/978-3-030-85633-5

LNCS Sublibrary: SL8 – Bioinformatics

This Springer imprint is published by the registered company Springer Nature Switzerland AG
The registered company address is: Gewerbestrasse 11, 6330 Cham, Switzerland

Preface

This volume contains the papers presented at CMSB 2021, the 19th International Conference on Computational Methods in Systems Biology, held during September 22–24, 2021 in a hybrid format, allowing for in-person participation in Bordeaux, France, and for online participation.

The CMSB annual conference series, initiated in 2003, provides a unique discussion forum for computer scientists, biologists, mathematicians, engineers, and physicists interested in a system-level understanding of biological processes. Topics covered by the CMSB proceedings include formalisms for modeling biological processes; frameworks for model verification, validation, analysis, and simulation of biological systems; high-performance computational systems biology; model inference from experimental data; multi-scale modeling and analysis methods; computational approaches for synthetic biology; machine learning and data-driven approaches; microbial ecology modeling and analysis; methods and protocols coping with populations and their variability; and models, applications, and case studies in systems and synthetic biology.

There were a total of 54 submissions over the 4 conference tracks (regular papers, tool papers, highlight presentation proposals, and posters). Every regular paper was reviewed by at least three Program Committee members, whereas every tool paper was reviewed by two Program Committee members and two Tool Evaluation Committee members. The latter committee provided a thorough evaluation of the tool quality, in terms of usability, accessibility, reproducibility, and documentation. The committee decided to accept 13 of the 25 submitted regular papers and 5 of the 7 submitted tool papers for publication in this volume and presentation. The conference also included 4 of the 7 proposed highlight presentations and 16 posters not included in the proceedings. The program of CMSB was further enriched by five notorious invited speakers: Diego di Bernardo (TIGEM, Italy), Laurence Calzone (Institut Curie, Paris), Giulia Giodano (University of Trento, Italy), Yang-Yu Liu (Harvard Medical School, USA), and Ion Petre (University of Turku, Finland). Additional information about CMSB 2021 is available on the conference website at https://cmsb2021.labri.fr.

We are deeply grateful to the members of the Program Committee, Tool Evaluation Committee, and the external reviewers for their invaluable contribution to the reviewing process and the feedback they provided to the authors. Special thanks go to Auriane Dantès and Isabelle Garcia for taking care of the administrative aspects of the local organization, to Clémence Frioux, Misbah Razzaq, and Laurent Simon for their help in scientific and practical organizational matters, to Samuel Pastva as chair of the Tool Evaluation Committee, and to François Fages and all the members of the CMSB Steering Committee, for their advice on the organization and the running of the conference. We thank EasyChair for the support offered by its conference system in the reviewing process and the production of these proceedings, and Springer for publishing the CMSB proceedings in its Lecture Notes in Computer Science series. Finally, we are grateful to the Laboratoire Bordelais de Recherche en Informatique (LaBRI) for supporting and hosting CMSB 2021, as well as to CNRS, Université de Bordeaux and its Département

Santé publique, Bordeaux INP, Inria, ANR, and the SysNum cluster of excellence, for their financial support.

Last but not least, we immensely thank all authors, speakers, and contributors for making CMSB a first-class scientific event.

September 2021

Eugenio Cinquemani
Loïc Paulevé

Organization

Program Committee

Eugenio Cinquemani (Co-chair)	Inria, France
Loïc Paulevé (Co-chair)	CNRS/LaBRI, Bordeaux, France
Alessandro Abate	University of Oxford, UK
Claudio Altafini	Linköping University, Sweden
Paolo Ballarini	CentraleSupelec, France
Ezio Bartocci	TU Wien, Austria
Luca Bortolussi	University of Trieste, Italy
Luca Cardelli	University of Oxford, UK
Milan Ceska	Brno University of Technology, Czech Republic
Neil Dalchau	Microsoft, USA
François Fages	Inria, France
Karoline Faust	KU Leuven, Belgium
Jerome Feret	Inria, France
Christoph Flamm	University of Vienna, Austria
Clémence Frioux	Inria, France
Ashutosh Gupta	TIFR, India
Jan Hasenauer	University of Bonn, Germany
Monika Heiner	Brandenburg Technical University Cottbus-Senftenberg, Germany
Jane Hillston	The University of Edinburgh, UK
Ina Koch	Johann Wolfgang Goethe University Frankfurt am Main, Germany
Jan Kretinsky	Technical University of Munich, Germany
Jean Krivine	CNRS, France
Pedro T. Monteiro	INESC-ID/IST - Universidade de Lisboa, Portugal
Laura Nenzi	University of Trieste, Italy
Jun Pang	University of Luxembourg, Luxembourg
Nicola Paoletti	Royal Holloway, University of London, UK
Ion Petre	University of Turku, Finland
Tatjana Petrov	University of Konstanz, Germany
Carla Piazza	University of Udine, Italy
Ovidiu Radulescu	University of Montpellier 2, France
Andre Ribeiro	Tampere University, Finland
Maria Rodriguez Martinez	IBM, Zurich Research Laboratory, Switzerland
Olivier Roux	LS2N, École Centrale de Nantes, France
Guido Sanguinetti	The University of Edinburgh, UK
Heike Siebert	DFG Research Center Matheon, Freie Universität Berlin, Germany
Abhyudai Singh	University of Delaware, USA

Scott Smolka	Stony Brook Universtiy, USA
Carolyn Talcott	SRI International, USA
Adelinde Uhrmacher	Universität Rostock, Germany
Andrea Vandin	Sant'Anna School of Advanced Studies, Pisa, Italy
Verena Wolf	Saarland University, Germany
Christoph Zechner	Max Planck Institute of Molecular Cell Biology and Genetics, Germany
David Šafránek	Masaryk University, Czech Republic

Tool Evaluation Committee

Samuel Pastva (Chair)	Masaryk University, Czech Republic
Georgios Argyris	Technical University of Denmark, Copenhagen, Denmark
Candan Çelik	Comenius University in Bratislava, Slovakia
Laura Cifuentes Fontanals	Freie Universität Berlin/Max Planck Institute for Molecular Genetics, Germany
Aurélien Desoeuvres	University of Montpellier, France
Lukrécia Mertová	Masaryk University, Laboratory SYBILA, Czech Republic
Gareth Molyneux	University of Oxford, UK
Loïc Paulevé	CNRS/LaBRI, Bordeaux, France
Misbah Razzaq	Ecole Centrale de Nantes, France

Organization Committee

Auriane Dantès	CNRS, LaBRI, France
Clémence Frioux	Inria, France
Isabelle Garcia	CNRS, LaBRI, France
Misbah Razzaq	Inserm BPH, Bordeaux, France
Laurent Simon	Bordeaux INP, LaBRI, France

Steering Committee

Alessandro Abate (Guest)	University of Oxford, UK
Luca Bortolussi (Guest)	University of Trieste, Italy
Luca Cardelli	University of Oxford, UK
Eugenio Cinquemani (Guest)	Inria Grenoble-Rhône-Alpes, France
Finn Drablos	NTNU, Norway
François Fages	Inria Saclay île-de-France, France
David Harel	Weizmann Institute of Science, Israel
Monika Heiner	Brandenburg Technical University Cottbus-Senftenberg, Germany

Tommaso Mazza	IRCCS Casa Sollievo della Sofferenza Mendel, Italy
Satoru Miyano	University of Tokyo, Japan
Loïc Paulevé (Guest)	CNRS, LaBRI, France
Ion Petre	University of Turku, Finland
Tatjana Petrov (Guest)	University of Konstanz, Germany
Gordon Plotkin	The University of Edinburgh, UK
Corrado Priami	CoSBi/Microsoft Research, University of Trento, Italy
Guido Sanguinetti (Guest)	The University of Edinburgh, UK
Carolyn Talcott	SRI International, USA
Adelinde Uhrmacher	University of Rostock, Germany
Verena Wolf	Saarland University, Germany

Additional Reviewers

Ackermann, Jörg
Backeköhler, Michael
Cairoli, Francesca
Gilbert, David
Klarner, Hannes
Krüger, Thilo
Labarthe, Simon
Mizera, Andrzej
Molyneux, Gareth
Piho, Paul
Regolin, Enrico
Sinzger, Mark
Tonello, Elisa

Contents

Reducing Boolean Networks with Backward Boolean Equivalence

Georgios Argyris[1], Alberto Lluch Lafuente[1], Mirco Tribastone[2], Max Tschaikowski[3], and Andrea Vandin[1,4](✉)

[1] DTU Technical University of Denmark, Kongens Lyngby, Denmark
[2] IMT School for Advanced Studies Lucca, Lucca, Italy
[3] University of Aalborg, Aalborg, Denmark
[4] Sant'Anna School for Advanced Studies, Pisa, Italy
andrea.vandin@santannapisa.it

Abstract. Boolean Networks (BNs) are established models to qualitatively describe biological systems. The analysis of BNs might be infeasible for medium to large BNs due to the state-space explosion problem. We propose a novel reduction technique called *Backward Boolean Equivalence* (BBE), which preserves some properties of interest of BNs. In particular, reduced BNs provide a compact representation by grouping variables that, if initialized equally, are always updated equally. The resulting reduced state space is a subset of the original one, restricted to identical initialization of grouped variables. The corresponding trajectories of the original BN can be exactly restored. We show the effectiveness of BBE by performing a large-scale validation on the whole GINsim BN repository. In selected cases, we show how our method enables analyses that would be otherwise intractable. Our method complements, and can be combined with, other reduction methods found in the literature.

Keywords: Boolean network · State transition graph · Attractor analysis · Exact reduction · Ginsim repository

1 Introduction

Boolean Networks (BNs) are an established method to model biological systems [28]. A BN consists of Boolean variables (also called nodes) which represent the activation status of the components in the model. The variables are commonly depicted as nodes in a network with directed links which represent influences between them. However, a full descriptive mathematical model underlying a BN consists of a set of Boolean functions, the *update functions*, that govern the Boolean values of the variables. Two BNs are displayed on top of Fig. 1. The BN on the left has three variables x_1, x_2, and x_3, and the BN on the right has two variables $x_{1,2}$ and x_3. The dynamics (the state space) of a BN is encoded into a

Partially supported by the DFF project REDUCTO 9040-00224B, the Poul Due Jensen Foundation grant 883901, and the PRIN project SEDUCE 2017TWRCNB.

E. Cinquemani and L. Paulevé (Eds.): CMSB 2021, LNBI 12881, pp. 1–18, 2021.
https://doi.org/10.1007/978-3-030-85633-5_1

state transition graph (STG). The bottom part of Fig. 1 displays the STGs of the corresponding BNs. The boxes of the STG represent the BN *states*, i.e. vectors with one Boolean value per BN variable. A directed edge among two STG states represents the evolution of the system from the source state to the target one. The target state is obtained by synchronously applying all the update functions to the activation values of the source state. There exist BN variants with other update schema, e.g. asynchronous non-deterministic [47] or probabilistic [43]. Here we focus on the synchronous case. BNs where variables are *multivalued*, i.e. can take more than two values to express different levels of activation [46], are supported via the use of *booleanization* techniques [18], at the cost, however, of increasing the number of variables.

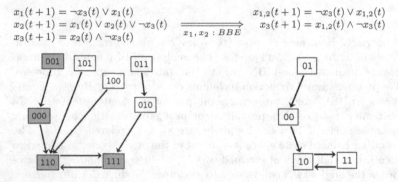

$$x_1(t+1) = \neg x_3(t) \vee x_1(t)$$
$$x_2(t+1) = x_1(t) \vee x_2(t) \vee \neg x_3(t)$$
$$x_3(t+1) = x_2(t) \wedge \neg x_3(t)$$

$\xrightarrow[x_1, x_2 \,:\, BBE]{}$

$$x_{1,2}(t+1) = \neg x_3(t) \vee x_{1,2}(t)$$
$$x_3(t+1) = x_{1,2}(t) \wedge \neg x_3(t)$$

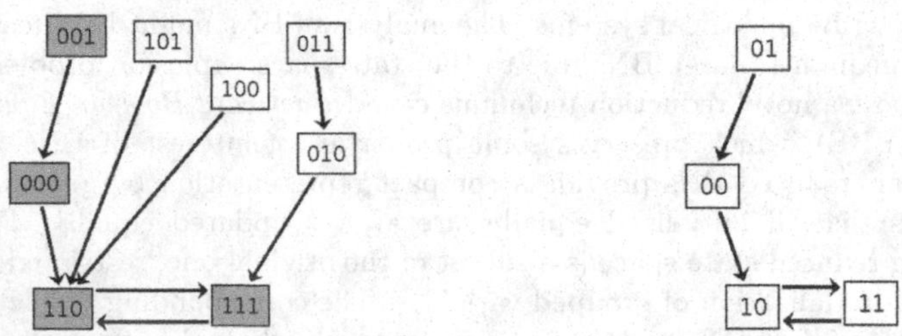

Fig. 1. A BN (top-left), its STG (bottom-left), the BBE-reduced BN (top-right) and its (reduced) STG (bottom-right).

BNs suffer from the state space explosion problem: there are exponentially many STG states with respect to the number of BN variables. This hampers BN analysis in practice, calling for reduction techniques for BNs. There exist manual or semi-automated ones based on domain knowledge. Such empirical reductions have several drawbacks: being semi-automated, they are error-prone, and do not scale. Popular examples are those based on the idea of *variable absorption*, proposed originally in [34,41,48]. The main idea is that certain BN variables can get *absorbed* by the update functions of their target variables by replacing all occurrences of the absorbed variables with their update functions. Other methods automatically remove *leaf* variables (variables with 0 outgoing links) or *frozen* variables (variables that stabilize after some iterations independently of the initial conditions) [4,39]. Several techniques [2,23] focus on reducing the STGs rather than the BN generating them. This requires to construct the original STG, thus still incurring the state space explosion problem.

Our research contributes a novel mathematically grounded method to automatically minimize BNs while exactly preserving behaviors of interest. We present Backward Boolean Equivalence (BBE), which collapses *backward Boolean equivalent* variables. The main intuition is that two BN variables are

BBE-equivalent if they maintain equal value in any state reachable from a state wherein they have the same value. In the STG in Fig. 1 (left), we note that for all states where x_1 and x_2 have same value (purple boxes), the update functions do not distinguish them. Notably, BBE is that it can be checked directly on the BN, without requiring to generate the STG. Indeed, as depicted in the middle of Fig. 1, x_1 and x_2 can be shown to be BBE-equivalent by inspecting their update functions: If x_1, x_2 have the same value in a state, i.e. $x_1(t) = x_2(t)$, then their update functions will not differentiate them since $x_2(t+1) = x_1(t) \vee x_2(t) \vee \neg x_3(t) = x_1(t) \vee x_1(t) \vee \neg x_3(t) = x_1(t) \vee \neg x_3(t) = x_1(t+1)$. We also present an iterative partition refinement algorithm [36] that computes the largest BBE of a BN. Furthermore, given a BBE, we obtain a *BBE-reduced* BN by collapsing all BBE-equivalent variables into one in the reduced BN. In Fig. 1, we collapsed x_1, x_2 into $x_{1,2}$. The reduced BN faithfully preserves part of the dynamics of the original BN: it exactly preserves all states and paths of the original STG where BBE-equivalent variables have same activation status. Figure 1 (right) shows the obtained BBE-reduced BN and its STG. We can see that the purple states of the original STG are preserved in the one of the reduced BN.

We implemented BBE in ERODE [10], a freely available tool for reducing biological systems. We built a toolchain that combines ERODE with several tools for the analysis, visualization and reduction of BNs, allowing us to apply BBE to all BNs from the GINsim repository (http://ginsim.org/models_repository). BBE led to reduction in 61 out of 85 considered models (70%), facilitating STG generation. For two models, we could obtain the STG of the reduced BN while it is not possible to generate the original STG due to its size. We further demonstrate the effectiveness of BBE in three case studies, focusing on their *asymptotic dynamics* by means of *attractors analysis*. Using BBE, we can identify the attractors of large BNs which would be otherwise intractable.

The article is organized as follows: Sect. 2 provides the basic definitions and the running example based on which we will explain the key concepts. In Sect. 3, we introduce BBE, present the algorithm for the automatic computation of maximal BBEs, and formalize how the STGs of the original and the reduced BN are related. In Sect. 4, we apply BBE to BNs from the literature. In Sect. 5 we discuss related works, while Sect. 6 concludes the paper.

2 Preliminaries

BNs can be represented visually using some graphical representation which, however, might not contain all the information about their dynamics [29]. An example is that of signed interaction (or regulatory) graphs adopted by the tool GinSim [31]. These representations are often paired with a more precise description containing either truth tables [39] or algebraic update functions [45]. In this paper we focus on such precise representation, and in particular on the latter. However, in order to better guide the reader in the case studies, wherein we manipulate BNs with a very large number of components, we also introduce signed interaction graphs.

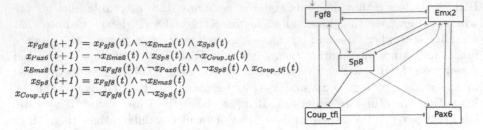

$$
\begin{aligned}
x_{Fgf8}(t+1) &= x_{Fgf8}(t) \wedge \neg x_{Emx2}(t) \wedge x_{Sp8}(t) \\
x_{Pax6}(t+1) &= \neg x_{Emx2}(t) \wedge x_{Sp8}(t) \wedge \neg x_{Coup\_tfi}(t) \\
x_{Emx2}(t+1) &= \neg x_{Fgf8}(t) \wedge \neg x_{Pax6}(t) \wedge \neg x_{Sp8}(t) \wedge x_{Coup\_tfi}(t) \\
x_{Sp8}(t+1) &= x_{Fgf8}(t) \wedge \neg x_{Emx2}(t) \\
x_{Coup\_tfi}(t+1) &= \neg x_{Fgf8}(t) \wedge \neg x_{Sp8}(t)
\end{aligned}
$$

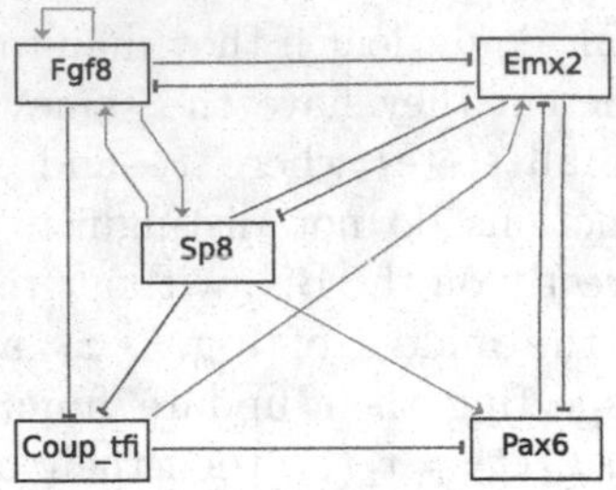

Fig. 2. (Left) the BN of cortical area development from [25]; (Right) its signed interaction graph.

We explain the concepts of current and next sections using the simple BN of Fig. 2 (left) taken from [25]. The model refers to the development of the outer part of the brain: the cerebral cortex. This part of the brain contains different areas with specialised functions. The BN is composed of five variables which represent the gradients that take part in its development: the morphogen *Fgf8* and four transcription factors, i.e., *Emx2, Pax6, Coup_tfi, Sp8.* During development, these genes are expressed in different concentrations across the surface of the cortex forming the different areas.

Figure 2 (right) displays the signed interaction graph that corresponds to the BN. The green arrows correspond to *activations* whereas the red arrows correspond to *inhibitions.* For example, the green arrow from *Sp8* to *Pax6* denotes that the former promotes the latter because variable x_{Sp8} appears (without negation) in the update function of x_{Pax6}, whereas the red arrow from *Pax6* to *Emx2* denotes that the former inhibits the latter because the negation of x_{Pax6} appears in the update function of x_{Emx2}.

We now give the formal definition of a BN:

Definition 1. *A BN is a pair* (X, F) *where* $X = \{x_1, ..., x_n\}$ *is a set of variables and* $F = \{f_{x_1}, ..., f_{x_n}\}$ *is a set of update functions, with* $f_{x_i} : \mathbb{B}^n \rightarrow \mathbb{B}$ *being the update function of variable* x_i.

A BN is often denoted as $X(t+1) = F(X, t)$, or just $X = F(X)$. In Fig. 2 we have $X = \{x_{Fgf8}, x_{Pax6}, x_{Emx2}, x_{Sp8}, x_{Coup\_tfi}\}$.

The *state* of a BN is an evaluation of the variables, denoted with the vector of values $\mathbf{s} = (s_{x_1}, \ldots, s_{x_n}) \in \mathbb{B}^n$. The variable x_i has the value s_{x_i}. When the update functions are applied synchronously, we have synchronous transitions between states, i.e. for $\mathbf{s}, \mathbf{t} \in \mathbb{B}^n$ we have $\mathbf{s} \rightarrow \mathbf{t}$ if $\mathbf{t} = F(\mathbf{s}) = (f_{x_1}(\mathbf{s}), \ldots, f_{x_n}(\mathbf{s}))$.

Suppose that the activation status of the variables x_{Fgf8}, x_{Emx2}, x_{Pax6}, x_{Sp8}, x_{Coup_tfi} is given by the state $\mathbf{s} = (1, 0, 1, 1, 1)$. After applying the update functions, we have $\mathbf{t} = F(\mathbf{s}) = (0, 0, 0, 0, 0)$.

The state space of a BN, called *State Transition Graph (STG)*, is the set of all possible states and state transitions.

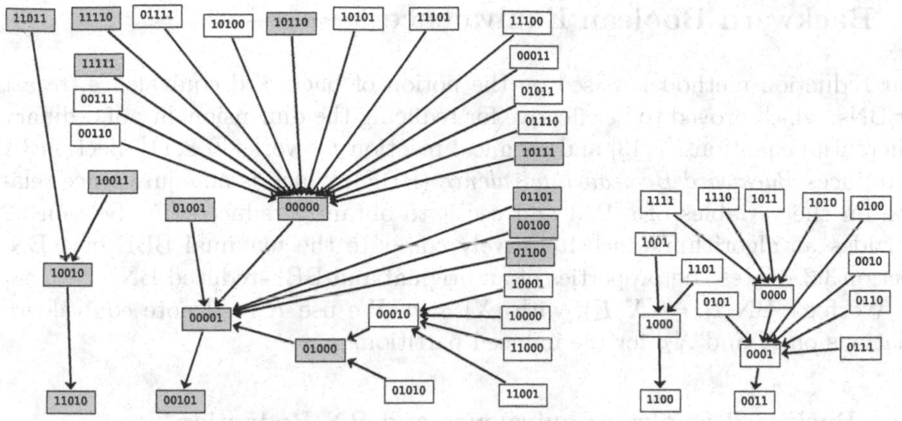

(a) STG of BN in Fig. 2 (b) STG of BN in Fig. 4

Fig. 3. The STGs of the BN of Fig. 2 and of its BBE-reduction in Fig. 4. We use CINsim's visual representation, where self-loops are implicit in nodes without outgoing edges.

Definition 2. *Let $B = (X, F)$ be a BN. We define the state transition graph of B, denoted with $STG(B)$, as a pair (S, T) with $S \subseteq \mathbb{B}^n$ being a set of vertices labelled with the states of B, and $T = \{\boldsymbol{s} \to \boldsymbol{t} \mid \boldsymbol{s} \in S, \boldsymbol{t} = F(\boldsymbol{s})\}$ a set of directed edges representing the transitions between states of B.*

We often use the notation $\mathbf{s} \to^+ \mathbf{t}$ for the transitive closure of the transition relation. The cardinality of the set of states is 2^n, which illustrates the state space explosion: we have exponentially many states on BN variables. Figure 3(a) displays the STG of the BN in Fig. 2.

Several BN properties are identified in STGs, e.g. attractors, basins of attraction, and transient trajectories [42]. Attractors are sets of states towards which a system tends to evolve and remain [27]. They are often associated with the interpretation of the underlying system; for example, Kauffman equated attractors with different cell types [20]. Hence, the main reduction methods that have been developed in the literature so far concentrate on how they affect the asymptotic dynamics i.e. the number of attractors and the distribution of their lengths. We define an attractor as follows:

Definition 3 ***(Attractor).*** *Let $B = (X, F)$ be a BN with $STG(B) = (S, T)$. We say that a set of states $A \subseteq S$ is an* attractor *iff*

1. $\forall \boldsymbol{s}, \boldsymbol{s}' \in A$, $\boldsymbol{s} \to^+ \boldsymbol{s}'$, and
2. $\forall \boldsymbol{s} \in A, \forall \boldsymbol{s}' \in S$, $\boldsymbol{s} \to^+ \boldsymbol{s}'$ implies $\boldsymbol{s}' \in A$.

Attractors are hence just absorbing strongly connected components in the STG. An attractor A such that $|A| = 1$ is called a *steady state* (also named *point attractor*). We also denote with $|A|$ the *length* of attractor A.

3 Backward Boolean Equivalence

Our reduction method is based on the notion of backward equivalence, recast for BNs, which proved to be effective for reducing the dimensionality of ordinary differential equations [9,13] and chemical reaction networks [6,8,11]. Section 3.1 introduces *Backward Boolean Equivalence* (BBE), which is an equivalence relation on the variables of a BN, and use it to obtain a reduced BN. Section 3.2 provides an algorithm which iteratively compute the maximal BBE of a BN. Section 3.3 relates the properties of an original and BBE-reduced BN.

We fix a BN $B = (X, F)$, with $|X| = n$. We use R to denote equivalence relations on X and X_R for the induced partition.

3.1 Backward Boolean Equivalence and BN Reduction

We first introduce the notion of *constant* state on an equivalence relation R.

Definition 4 ***(Constant State).*** *A state* $\mathbf{s} \in \mathbb{B}^n$ *is* constant *on* R *if and only if* $\forall(x_i, x_j) \in R$ *it holds that* $s_{x_i} = s_{x_j}$.

Consider our running example and an equivalence relation R given by the partition $X_R = \{\{x_{Sp8}, x_{Fgf8}\}, \{x_{Pax6}\}, \{x_{Emx2}\}, \{x_{Coup\_tfi}\}\}$. The states constant on R are colored in purple in Fig. 3. For example, the state $\mathbf{s} = (1, 0, 1, 1, 1)$ is constant on R because $s_{Sp8} = s_{Fgf8}$ (the first and fourth positions of $\mathbf{s}$, respectively). On the contrary, $(1, 0, 1, 0, 1)$ is not constant on R.

We now define *Backward Boolean Equivalence (BBE)*.

Definition 5 ***(Backward Boolean Equivalence).*** *Let* $B = (X, F)$ *be a BN,* X_R *a partition of the set* X *of variables, and* $C \in X_R$ *a class of the partition. A partition* X_R *is a* Backward Boolean Equivalence (BBE) *if and only if the following formula is valid:*

$$\Phi^{X_R} \equiv \left(\bigwedge_{\substack{C \in X_R \\ x, x' \in C}} \left(x = x' \right) \right) \longrightarrow \bigwedge_{\substack{C \in X_R \\ x, x' \in C}} \left(f_x(X) = f_{x'}(X) \right)$$

Φ^{X_R} says that if for all equivalence classes C the variables in C are equal, then the update functions of variables in the same equivalence class stay equal.

In other words, R is a BBE if and only if for all $\mathbf{s} \in \mathbb{B}^n$ constant on R it holds that $F(\mathbf{s})$ is constant on R. BBE is a relation where the update functions F preserve the "constant" property of states. The partition $X_R = \{\{x_{Sp8}, x_{Fgf8}\}, \{x_{Pax6}\}, \{x_{Emx2}\}, \{x_{Coup\_tfi}\}\}$ described above is indeed a BBE. This can be verified on the STG: all purple states (the constant ones) have outgoing transitions only towards purple states.

We now define the notion of BN reduced up to a BBE R. Each variable in the reduced BN represents one equivalence class in R. We denote by $f\{^a/_b\}$ the term arising by replacing each occurrence of b by a in the function f.

Definition 6. *The reduction of B up to R, denoted by B/R, is the BN* (X_R, F_R) *where* $F_R = \{f_{x_C} : C \in X_R\}$*, with* $f_{x_C} = f_{x_k}\{^{x_{C'}}/_{x_i} : \forall C' \in X_R, \forall x_i \in C'\}$ *for some* $x_k \in C$.

The definition above uses one variable per equivalence class, selects the update function of any variable in such class, and replaces all variables in it with a representative one per equivalence class. Figure 4 shows the reduction of the cortical area development BN. We selected the update function of x_{Sp8} as the update function of the class-variable $x_{\{Fgf8,Sp8\}}$, and replaced every occurrence of x_{Sp8} and x_{Fgf8} with $x_{\{Fgf8,Sp8\}}$. The STG of such reduced BN is given in Fig. 3(b).

$$
\begin{aligned}
x_{\{Fgf8,Sp8\}}(t+1) &= x_{\{Fgf8,Sp8\}}(t) \wedge \neg x_{\{Emx2\}}(t)\\
x_{\{Pax6\}}(t+1) &= \neg x_{\{Emx2\}}(t) \wedge x_{\{Fgf8,Sp8\}}(t) \wedge \neg x_{\{Coup\_tfi\}}(t)\\
x_{\{Emx2\}}(t+1) &= \neg x_{\{Fgf8,Sp8\}}(t) \wedge \neg x_{\{Pax6\}}(t) \wedge \neg x_{\{Fgf8,Sp8\}}(t) \wedge x_{\{Coup\_tfi\}}(t)\\
x_{\{Coup\_tfi\}}(t+1) &= \neg x_{\{Fgf8,Sp8\}}(t) \wedge \neg x_{\{Fgf8,Sp8\}}(t)
\end{aligned}
$$

Fig. 4. The BBE reducion of the cortical area development network of Fig. 2.

3.2 Computation of the Maximal BBE

A crucial aspect of BBE is that it can be checked directly on a BN without requiring the generation of the STG. This is feasible by encoding the logical formula of Definition 5 into a logical SATisfiability problem [3]. A SAT solver has the ability to check the validity of such a logical formula by checking for the unsatisfiability of its negation ($sat(\neg\Phi^{X_R})$). A partition X_R is a BBE if and only if $\mathsf{sat}(\neg\Phi^{X_R})$ returns "unsatifiable", otherwise a counterexample (a witness) is returned, consisting of variables assignments that falsify Φ^{X_R}. Using counterexamples, it is possible to develop a partition refinement algorithm that computes the largest BBE that refines an initial partition.

The partition refinement algorithm is shown in Algorithm 1. Its input are a BN and an initial partition of its variables X. A *default* initial partition that leads to the maximal reduction consists of one block only, containing all variables. In general, the modeller may specify a different initial partition if some variables should not be merged together, placing them in different blocks. The output of the algorithm is the largest partition that is a BBE and refines the initial one.

We now explain how the algorithm works for input the cortical area development BN and the initial partition $X_R = \{\{x_{Fgf8}, x_{Emx2}, x_{Pax6}, x_{Sp8}, x_{Coup\_tfi}\}\}$.

Iteration 1. The algorithm enters the *while* loop, and the solver checks if Φ^{X_R} is valid. X_R is not a BBE, therefore the algorithm enters the second branch of the *if* statement. The solver gives an example satisfying $\neg\Phi^{X_R}$: $s = (s_{x_{Fgf8}}, s_{x_{Pax6}}, s_{x_{Emx2}}, s_{x_{Sp8}}, s_{x_{Coup\_tfi}}) = (0, 0, 0, 0, 0)$. Since $t = F(s) = (0, 0, 0, 0, 1)$, the *for* loop partitions G into $X_{R_1} = \{\{x_{Fgf8}, x_{Pax6}, x_{Emx2}\ x_{Sp8}\}, \{x_{Coup\_tfi}\}\}$. The state $t = (0, 0, 0, 0, 1)$ is now constant on X_{R_1}.

Algorithm 1: Compute the maximal BBE that refines the initial partition X_R for a BN (X, F)

```
Result: maximal BBE H that refines X_R
H ← X_R;
while true do
    if Φ^H is valid then
        return H ;
    else
        s ← get a state that satisfy ¬Φ^H;
        H' ← ∅;
        for C ∈ H do
            C_0 = {x_i ∈ C : f_{x_i}(s) = 0};
            C_1 = {x_i ∈ C : f_{x_i}(s) = 1};
            H' = H' ∪ {C_1} ∪ {C_0};
        end
        H ← H' \ {∅};
    end
end
```

Iteration 2. The algorithm checks if $\Phi^{X_{R_1}}$ is valid (i.e. if X_{R_1} is a BBE). X_{R_1} is not a BBE. The algorithm gives a counterexample with $s = (0,0,0,0,1)$ and $t = F(s) = (0,0,1,0,1)$. The *for* loop refines X_{R_1} into $X_{R_2} = \{\{x_{Fgf8}, x_{Pax6}$ $x_{Sp8}\}, \{x_{Emx2}\}, \{x_{Coup\_tfi}\}\}$. X_{R_2} makes $t = (0,0,1,0,1)$ constant.

Iteration 3. The algorithm checks if G_2 is a BBE. The formula $\neg\Phi^{X_{R_2}}$ is satisfiable, so G_2 is not a BBE, and the solver provides an example with $s = (1,1,0,1,1)$ and $F(s) = (1,0,0,1,0)$. Hence, X_{R_2} is partitioned into $X_{R_3} = \{\{x_{Fgf8}, x_{Sp8}\}, \{x_{Pax6}\}\ \{x_{Emx2}\}, \{x_{Coup\_tfi}\}\}$.

Iteration 4. The SAT solver proves that $\Phi^{X_{R_3}}$ is valid.

The number of iterations needed to reach a BBE depends on the counterexamples that the SAT solver provides. As for all partition-refinement algorithms, it can be easily shown that the number of iterations is bound by the number of variables. Each iteration requires to solve a SAT problem which is known to be NP-complete, however we show in Sect. 4 that we can easily scale to the largest models present in popular BN repositories.

We first show that given an initial partition there exists exactly one *largest* BBE that refines it.[1]

After that, we prove that Algorithm 1 indeed provides the maximal BBE that refines the initial one.

Theorem 1. *Let $BN = (X, F)$ and X_R a partition. There exists a unique maximal BBE H that refines X_R.*

Theorem 2. *Algorithm 1 computes the maximal BBE partition refining X_R.*

[1] All proofs are given in the extended version of this paper [1].

3.3 Relating Dynamics of Original and Reduced BNs

Given a BN B and a BBE R, $STG(B/R)$ can be seen as the subgraph of $STG(B)$ composed of all states of $STG(B)$ that are constant on R and their transitions. Of course, those states are transformed in $STG(B/R)$ by "collapsing" BBE-equivalent variables in the state representation. This can be seen by comparing the STG of the our running example (left part of Fig. 3) and of its reduction (right part of Fig. 3). The states (and transitions) of the STG of the reduced BN correspond to the purple states of the original STG.

Let B be a BN with n variables, $S \subseteq \mathbb{B}^n$ be the states of its STG, and R a BBE for B. We use $S_{|R}$ to denote the subset of S composed by all and only the states constant on R. With $STG(B)_{|R}$ we denote the subgraph of $STG(B)$ containing $S_{|R}$ and its transitions. Formally $STG(B)_{|R} = (S_{|R}, T_{|R})$, where $T_{|R} = T \cap (S_{|R} \times S_{|R})$.

The following lemma formalizes a fundamental property of $STG(B)_{|R}$, namely that all attractors of B containing states constant on R are preserved in $STG(B)_{|R}$.

Lemma 1 ***(Constant attractors).*** *Let $B(X, F)$ be a BN, R be a BBE, and A an attractor. If $A \cap S_{|R} \neq \emptyset$ then $A \subseteq S_{|R}$.*

We now define the bijective mapping $m_R : S_{|R} \leftrightarrow S_R$ induced by a BBE R, where S_R are the states of $STG(B/R)$, as follows: $m_R(\mathbf{s}) = (v_{C_1}, \ldots, v_{C_{|X/R|}})$ where $v_{C_j} = s_{x_i}$ for some $x_i \in C_j$. In words m_R bijectively maps each state of $STG(B)_{|R}$ to their compact representation in $STG(B/R)$. Indeed, $STG(B)_{|R}$ and $STG(B/R)$ are isomorphic, with m_R defining their (bijective) relation. We can show this through the following lemma.

Lemma 2 ***(Reduction isomorphism).*** *Let $B(X, F)$ be a BN and R be a BBE. Then, it holds*

1. *For all states $\mathbf{s} \in S_{|R}$ it holds $F_R(m_R(\mathbf{s})) = m_R(F(\mathbf{s}))$.*
2. *For all states $\mathbf{s} \in S_R$ it holds $F(m_R^{-1}(\mathbf{s})) = m_R^{-1}(F_R(\mathbf{s}))$.*

The previous Lemma ensures that BBE does not generates spurious trajectories or attractors in the reduced system. We can now state the main result of our approach, namely that the BBE reduction of a BN for a BBE R exactly preserves all attractors that are constant on R up to renaming with m_R.

Theorem 3 ***(Constant attractor preservation).*** *Let $B(X, F)$ be a BN, R a BBE, and A an attractor. If $A \cap S_{|R} \neq \emptyset$ then $m_R(A)$ is an attractor for B/R.*

4 Application to BNs from the Literature

We hereby apply BBE to BNs from the GINsim repository. Section 4.1 validates BBE on all models from the repository, while Sect. 4.2 studies the runtime speedups brought by BBE on attractor-based analysis of selected case studies, showing cases for which BBE makes the analysis feasible. Section 4.3 compares

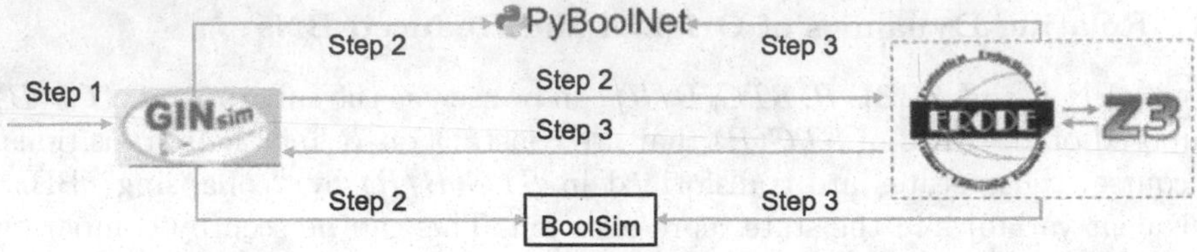

Fig. 5. BBE toolchain. (Step 1) We use GINsim [15] to access its model repository, and (Step 2) export it in the formats of the other tools in the toolchain to perform: STG generation (PyBoolNet [30]), attractor analysis (BoolSim [19]), and BBE reduction (ERODE [10]). (Step 3) We export the reduced models for analysis to PyBoolNet and BoolSim, or to GINsim.

BBE with the approach based on ODE encoding from [11], showing how such encoding leads to scalability issues and to the loss of reduction power.

The experiments have been made possible by a novel toolchain (Fig. 5) combining tools from the COLOMOTO initiative [33], and the reducer tool ERODE [10] which was extended here to support BBE-reduction. For Algorithm 1 we use the solver Z3 [17] which was already integrated in ERODE.

All experiments were conducted on a common laptop with an Intel Xeon(R) 2.80 GHz and 32 GB of RAM. We imposed an arbitrary timeout of 24 h for each task, after which we terminated the analysis. We refer to these cases as *time-out*, while we use *out-of-memory* if a tool terminated with a memory error.

4.1 Large Scale Validation of BBE on BNs

We validate BBE on real-world BNs in terms of the number of BNs that can be reduced and the average reduction ratio.

Configuration. We conducted our investigation on the whole GINsim model repository which contains 85 networks: 29 are Boolean, and 56 are multivalued. In multivalued networks (MNs), some variables have more than 2 activation statuses, e.g. $\{0, 1, 2\}$. These models are automatically *booleanized* [14,18] by GinSim when exporting in the input formats of the other tools in the tool-chain.

Most of the models in the repository have a specific structure [32] where a few variables are so-called *input variables*. These are variables whose update functions are either a stable function (e.g. $x(t+1) = 0$, $x(t+1) = 1$) or the identity function (e.g. $x(t+1) = x(t)$). These are named 'input' because their values are explicitly set by the modeler to perform experiments campaigns. We investigate two reduction scenarios relevant to input variables. In the first one, Algorithm 1 starts with initial partitions that lead to the *maximal reduction*, i.e. consisting of one block only. In the second scenario, we provide initial partitions that isolate inputs in singleton blocks. Therefore, we prevent their aggregation with other variables, and obtain reductions independent of the values of the input variables (we recall that BBE requires related variables to be initialized with same activation value). We call this case *input-distinguished (ID) reduction.*

Results. By using the maximal reduction setting, we obtained reductions on 61 of the 85 models, while we obtained ID reductions on 38 models. We summarize the reductions obtained for the two settings in Fig. 6, displaying the distribution of the reduction ratios $r_m = N_m/N$ and $r_i = N_i/N$, where N, N_m and N_i are the number of variables in the original BN, in the maximal BBE-reduction, and in the ID one, respectively. We also provide the average reduction ratios on the models, showing that it does not substantially change across Boolean or multivalued models. No reduction took more than 3 s.

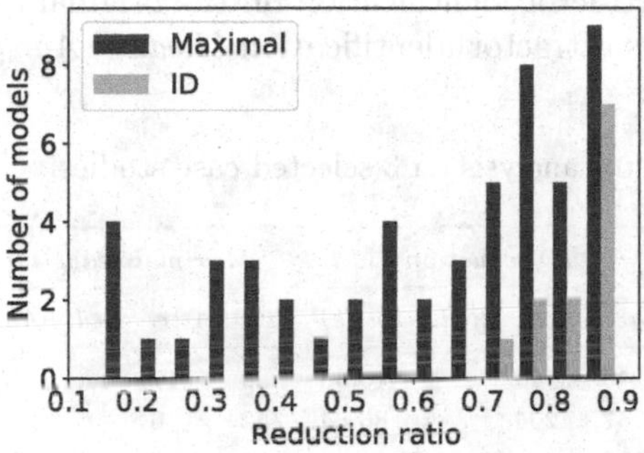

Average reduction ratios

	Maximal	*ID*
BNs	0.66	0.83
MNs	0.68	0.95
ALL	0.67	0.91

Fig. 6. (Left) Distribution of reduction ratios (reduced variables over original ones) on all models from the GINsim repository using the maximal and ID reduction strategy. Each bar counts the number of models with that reduction ratio, starting from 15% up to 90%, with step 5%. (Right) Average reduction ratios for Boolean, Multivalued and all models.

Interpretation. BBE reduced a large number of models (about 72%). In particular, this happened in 24 out of the 29 (83%) Boolean models and in 37 out of 56 (66%) multivalued networks. The average reduction ratio for the maximal and ID strategies are 0.67 and 0.91, respectively. For the former strategy, we get trivial reductions in 22 models wherein only input variables are related. In such trivial cases, the ID strategy does not lead to reduction. In other cases, the target variables of inputs (i.e. variables with incoming edges only from input variables considering the graphical representation of variables) appeared to be backward equivalent together with the input variables. This results in reductions with large equivalence classes consisting of input variables and their descendants. These are interesting reductions which get lost using the ID approach, as the input variables get isolated.

4.2 Attractor Analysis of Selected Case Studies

Hypothesis. We now investigate the fate of asymptotic dynamics after BBE-reduction, and test the computational efficiency in terms of time needed for attractor identification in the original and reduced models. We expect that BBE-reduction can be utilized to (i) gain fruitful insights into large BN models and (ii) to reduce the time needed for attractor identification.

Configuration. Our analysis focuses on three BNs from the GINsim repository. The first is the Mitogen-Activated Protein Kinases (MAPK) network [26] with 53 variables. The second refers to the survival signaling in large granular lymphocyte leukemia (T-LGL) [51] and contains 60 variables. The third is the merged Boolean model [40] of T-cell and Toll-like receptors (TCR-TLR5) which is the largest BN model in GINsim repository with 128 variables.

Results. The results of our analysis are summarized in Table 1 for the original, ID- and maximal-reduced BN. We present the number of variables (*size*) and of Attractors (*Attr.*), the time for attractor identification on the original model (*An. (s)*) and that for reduction plus attractor identification (*Red. + An. (s)*).

Table 1. Reduction and attractor analysis on 3 selected case studies.

	Original model			*ID reduction*			*Maximal reduction*		
	Size	*Attr.*	*An.(s)*	*Size*	*Attr.*	*Red.+An.(s)*	*Size*	*Attr.*	*Red.+An.(s)*
MAPK Network	53	40	16.50	46	40	15.33	39	17	3.49
T-LGL	60	264	123.43	57	264	86.84	52	6	3.49
TCR-TLR	128	—*Time Out*—		116	—*Time Out*—		95	2	31.29

Interpretation. ID reduction preserves all attractors reachable from any combination of activation values for inputs. This is an immediate consequence of 2, Theorem 3 and the fact that number of attractors in the original and the ID reduced BN is the same (see Table 1). Maximal reduction might discard some attractors. We also note that, despite the limited reduction in terms of obtained number of variables, we have important analysis speed-ups, up to two orders of magnitude. Furthermore, the largest model could not be analyzed, while it took just 30 s to analyze its maximal reduction identifying 2 attractors.

4.3 Comparison with ODE-Based Approach From [11]

As discussed, BBE is based on the backward equivalence notion firstly provided for ordinary differential equations (ODEs), chemical reaction networks, and Markov chains [9,11]. Notably, [11] shows how the notion for ODEs can be applied indirectly to BNs via an *odification* technique [49] to encode BNs as ODEs. Such odification transforms each BN variable into an ODE variable that takes values in the continuous interval [0,1]. The obtained ODEs preserve the attractors of the original BN because the equations of the two models coincide when all variables have value either 0 or 1. However, infinitely more states are added for the cases in which the variables do not have integer value.

Scalability. The technique from [11] has been proved able to handle models with millions of variables. Instead, the odification technique is particularly computationally intensive. Due to this, it failed on some models from the GINsim repository, including two from [22], namely *core_engine_budding_yeast_CC* and *coupled_budding_yeast_CC*, consisting of 39 and 50 variables, respectively. Instead, BBE could be applied in less than a second.

Reduction Power. Another example is the *TCR-TLR* model from the previous section. In this case, both the ODE-based and BBE techniques succeeded. However, BBE led to better reductions due to the added non-integer states in the ODEs. Intuitively, the ODE-based technique *counts* incoming influences from equivalence classes of nodes, while BBE only checks whether at least one of such influence is present or not. Figure 7 shows an excerpt of the graphical representation of the model by GINsim. We use background colors of nodes to denote BBE equivalence classes (white denotes singleton classes). We see a large equivalence class of magenta species, 3 of which (*IRAK4*, *IRAK1*, and *TAK1*) receive two influences by magenta species, while the others receive only one. This differentiates the species in the ODE-based technique, keeping only the top four in the *magenta* block, while all the others end up in singleton blocks. We compare the original equations of *MyD88* and *IRAK4* which have 1 and 2 incoming influences each.

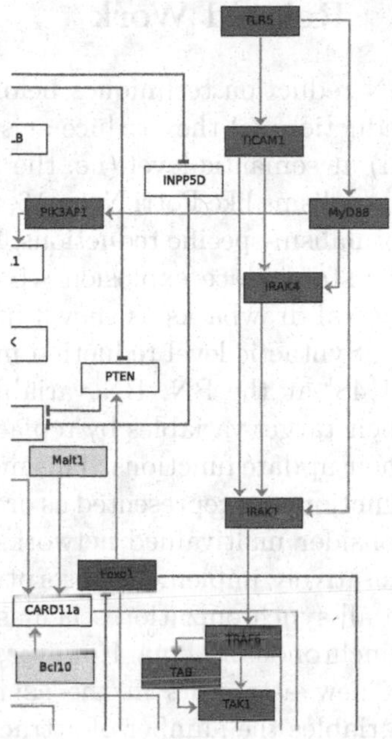

Fig. 7. Excerpt of GINsim's depict of TCR-TLR.

$$x_{MyD88}(t+1) = x_{TLR5}(t)$$
$$x_{IRAK4}(t+1) = (\neg x_{MyD88}(t) \wedge x_{TICAM1}(t)) \vee (x_{MyD88}(t))$$

We see that the two variables are BBE because their update functions depend only on the BBE-equivalent variables *TLR5* and *MyD88*, respectively. For *IRAK4*, the three variables in the update function are BBE. Therefore, they have same value allowing us to simplify the update function to just *MyD88*. The ODEs obtained for the 2 variables are, where $x'_{\_}$ denotes the derivative of $x_{\_}$:

$$x'_{MyD88} = x_{TLR5} - x_{MyD88}$$
$$x'_{IRAK4} = x_{MyD88} + x_{TICAM1} - x_{MyD88} \cdot x_{TICAM1} - x_{IRAK4}$$

Given that all variables appearing in the equations are backward equivalent, the two equations coincide with the original ones when all variables have values either 0 or 1. However, they differ for non-integer values. For example, in case all variables have value 0.5, we get 0 for the former, and 0.25 for the latter.

5 Related Work

BN reduction techniques belong to three families according to their domain of reduction: (i) they reduce at syntactic level (i.e. the BN [4,32,34,39,41,48,50]), (ii) at semantic level (i.e. the STG [2,23]), or (iii) they transform BNs to other formalisms like Petri Nets [16,44] and ordinary differential equations [49] offering formalism-specific reductions. However, (semantic) STG-reduction does not solve the state space explosion whereas the transformation to other formalisms has several drawbacks as shown in Sect. 4.3.

Syntactic level reduction methods usually perform variable absorption [4,34, 41,48] at the BN. BN variables can get absorbed by the update functions of their target variables by replacing all occurrences of the absorbed variables with their update functions. This method was first investigated in [34] wherein update functions are represented as ordinary multivalued decision diagrams. The authors consider multivalued networks with updates being applied asynchronously and iteratively implement absorption. The process, despite preserving steady states in all synchronization schemas [48], might lead to loss of cycle attractors in the synchronous schema. However, absorption of variables might lead to introduction of new attractors in the asynchronous case, i.e., by reducing the number of variables the number of attractors can stay the same or increase (attractors can split or new attractors can appear).

A similar study [48] presents a reduction procedure and proves that it preserves steady states. This procedure includes two steps. The first refers to the deletion of links between variables on their network structure. Deletion of pseudo-influences is feasible by simplifying the Boolean expressions in update functions. The second step of the procedure refers to the absorption of variables like in [34].

The difference between studies [48], [34] is that [48] exploits Boolean algebra instead of multivalued decision diagrams to explain absorption. Moreover, they refer only to Boolean networks, and do not consider any update schema. In studies [34,41,48], self-regulated BN variables (i.e. variables with a self-loop in the graphical representation) can not be selected for absorption. The inability to absorb self-regulated variables is inherent in the implementation of absorption in contrast to our method where the restrictions are encoded by the user at the initial partition and self-regulated variables can be merged with other variables.

In [41] the authors presented a two step reduction algorithm. The first step includes the absorption of input variables with stable function and the second step the absorption of single mediator variables (variables with one incoming and outgoing edge in the signed interaction graph). The first step of the algorithm in [41] is equally useful and compatible with the first step of [48]. Moreover, if we combine the first steps of [48] and [41], we may achieve interesting reductions which exactly preserve all asymptotic dynamics.

The first steps of [41,48] affect only a BN property called *stability*. Stability is the ability of a BN to end up to the same attractor when starting from slightly different initial conditions. In [4], the authors introduced the decimation procedure -a reduction procedure for synchronous BNs- to discuss how it affects stability. The crucial difference between decimation procedure and BBE-reduction is that the first was invented to study stability whereas the latter was invented to degrade state space explosion. The decimation procedure is summarized by the following four steps: (i) remove from every update functions the inputs that it does not depend on, (ii) find the constant value for variables with no inputs, (iii) propagate the constant values to other update functions and remove this variable from the system, and (iv) if a variable has become constant, repeat from step (i). The study also refers to leaf variables because their presence does not play any role in the asymptotic dynamics of a BN. However, both leaf and fixed-valued variables affect stability. Overall, the decimation procedure exactly preserves the asymptotic dynamics of the original model since it throws out only variables considered as asymptotically irrelevant.

6 Conclusion

We introduced an automatic reduction technique for synchronous Boolean Networks which preserves dynamics of interest. The modeller gets a reduced BN based on requirements expressed as an initial partition of variables. The reduced BN can recover a pure part of the original state space and its trajectories established by the reduction isomorphism. Notably, we draw connections between the STG of the original and that of the reduced BN through a rigorous mathematical framework. The dynamics preserved are those wherein collapsed variables have equal values.

We used our reduction technique to speed-up attractor identification. Despite that the length of the preserved attractors is consistent in the reduced model, some of them may get lost. In the future, we plan to study classes of initial partitions that preserve all attractors. We have shown the analysis speed-ups obtained for attractor identification as implemented in the tool BoolSim [24]. In the future we plan to perform a similar analysis on a recent attractor identification approach from [21].

Our method was implemented in ERODE [10], a freely available tool for reducing biological systems. Related *quantitative* techniques offered by ERODE have been recently validated on a large database of biological models [5,37,38]. In the future we plan to extend this analysis considering also BBE. We also plan to investigate whether BBE can be extended in order to be able to compare different models as done for its quantitative counterparts [7,12].

Our method could be combined with most of the existing methods found in literature. Our prototype toolchain consists of several tools from the COLOMOTO interoperability initiative. We aim to incorporate our toolchain into the COLOMOTO Interactive Notebook [35], a unified environment to edit, execute, share, and reproduce analyses of qualitative models of biological networks.

Multivalued BNs, i.e. whose variables can take more than two activation values, are currently supported only via a *booleanization* technique [14,18] that might hamper the interpretability of the reduced model. In future work we plan to generalize BBE to support directly multivalued networks.

References

1. Argyris, G., Lafuente, A.L., Tribastone, M., Tschaikowski, M., Vandin, A.: Reducing boolean networks with backward boolean equivalence - extended version (2021). https://arxiv.org/abs/2106.15476

2. Bérenguier, D., Chaouiya, C., Monteiro, P.T., Naldi, A., Remy, E., Thieffry, D., Tichit, L.: Dynamical modeling and analysis of large cellular regulatory networks. Chaos. Interdisc. J. Nonlinear Sci. **23**(2), 025114 (2013)
3. Biere, A., Biere, A., Heule, M., van Maaren, H., Walsh, T.: Handbook of Satisfiability: Volume 185 Frontiers in Artificial Intelligence and Applications. IOS Press, NLD (2009)
4. Bilke, S., Sjunnesson, F.: Stability of the Kauffman model. Phys. Rev. E **65**(1), 016129 (2001)
5. Cardelli, L., Perez-Verona, I.C., Tribastone, M., Tschaikowski, M., Vandin, A., Waizmann, T.: Exact maximal reduction of stochastic reaction networks by species lumping. Bioinformatics (2021). https://doi.org/10.1093/bioinformatics/btab081
6. Cardelli, L., Tribastone, M., Tschaikowski, M., Vandin, A.: Forward and backward bisimulations for chemical reaction networks. In: 26th International Conference on Concurrency Theory, CONCUR 2015, Madrid, Spain, 1–4 September 2015, pp. 226–239 (2015). https://doi.org/10.4230/LIPIcs.CONCUR.2015.226
7. Cardelli, L., Tribastone, M., Tschaikowski, M., Vandin, A.: Comparing chemical reaction networks: a categorical and algorithmic perspective. In: Proceedings of the 31st Annual ACM/IEEE Symposium on Logic in Computer Science, LICS 2016, New York, NY, USA, 5–8 July 2016, pp. 485–494 (2016). https://doi.org/10.1145/2933575.2935318
8. Cardelli, L., Tribastone, M., Tschaikowski, M., Vandin, A.: Efficient syntax-driven lumping of differential equations. In: Chechik, M., Raskin, J.-F. (eds.) TACAS 2016. LNCS, vol. 9636, pp. 93–111. Springer, Heidelberg (2016). https://doi.org/10.1007/978-3-662-49674-9_6
9. Cardelli, L., Tribastone, M., Tschaikowski, M., Vandin, A.: Symbolic computation of differential equivalences. In: Proceedings of the 43rd Annual ACM SIGPLAN-SIGACT Symposium on Principles of Programming Languages, POPL 2016, St. Petersburg, FL, USA, 20–22 January 2016, pp. 137–150 (2016). https://doi.org/10.1145/2837614.2837649
10. Cardelli, L., Tribastone, M., Tschaikowski, M., Vandin, A.: ERODE: a tool for the evaluation and reduction of ordinary differential equations. In: Legay, A., Margaria, T. (eds.) TACAS 2017. LNCS, vol. 10206, pp. 310–328. Springer, Heidelberg (2017). https://doi.org/10.1007/978-3-662-54580-5_19
11. Cardelli, L., Tribastone, M., Tschaikowski, M., Vandin, A.: Maximal aggregation of polynomial dynamical systems. Proc. Nat. Acad. Sci. **114**(38), 10029–10034 (2017)
12. Cardelli, L., Tribastone, M., Tschaikowski, M., Vandin, A.: Comparing chemical reaction networks: a categorical and algorithmic perspective. Theor. Comput. Sci. **765**, 47–66 (2019). https://doi.org/10.1016/j.tcs.2017.12.018
13. Cardelli, L., Tribastone, M., Tschaikowski, M., Vandin, A.: Symbolic computation of differential equivalences. Theor. Comput. Sci. **777**, 132–154 (2019). https://doi.org/10.1016/j.tcs.2019.03.018
14. Chaouiya, C., et al.: SBML qualitative models: a model representation format and infrastructure to foster interactions between qualitative modelling formalisms and tools. BMC Syst. Biol. **7**(1), 1–15 (2013)
15. Chaouiya, C., Naldi, A., Thieffry, D.: Logical modelling of gene regulatory networks with ginsim. In: van Helden, J., Toussaint, A., Thieffry, D. (eds.) Bacterial Molecular Networks, pp. 463–479. Springer, New York (2012). https://doi.org/10.1007/978-1-61779-361-5_23
16. Chaouiya, C., Remy, E., Thieffry, D.: Petri net modelling of biological regulatory networks. J. Discrete Algorithms **6**(2), 165–177 (2008)

17. de Moura, L., Bjørner, N.: Z3: an efficient SMT solver. In: Ramakrishnan, C.R., Rehof, J. (eds.) TACAS 2008. LNCS, vol. 4963, pp. 337–340. Springer, Heidelberg (2008). https://doi.org/10.1007/978-3-540-78800-3_24
18. Delaplace, F., Ivanov, S.: Bisimilar booleanization of multivalued networks. BioSystems **197**, 104205 (2020)
19. Di Cara, A., Garg, A., De Micheli, G., Xenarios, I., Mendoza, L.: Dynamic simulation of regulatory networks using squad. BMC Bioinformatics **8**(1), 462 (2007)
20. Drossel, B.: Random boolean networks. Rev. Nonlinear Dyn. Complex. **1**, 69–110 (2008)
21. Dubrova, E., Teslenko, M.: A sat-based algorithm for finding attractors in synchronous boolean networks. IEEE/ACM Trans. Comput. Biol. Bioinf. **8**(5), 1393–1399 (2011)
22. Fauré, A., Naldi, A., Lopez, F., Chaouiya, C., Ciliberto, A., Thieffry, D.: Modular logical modelling of the budding yeast cell cycle. Mol. BioSyst. **5**, 1787–96 (2009)
23. Figueiredo, D.: Relating bisimulations with attractors in boolean network models. In: Botón-Fernández, M., Martín-Vide, C., Santander-Jiménez, S., Vega-Rodríguez, M.A. (eds.) AlCoB 2016. LNCS, vol. 9702, pp. 17–25. Springer, Cham (2016). https://doi.org/10.1007/978-3-319-38827-4_2
24. Garg, A., Di Cara, A., Xenarios, I., Mendoza, L., De Micheli, G.: Synchronous versus asynchronous modeling of gene regulatory networks. Bioinformatics **24**(17), 1917–1925 (2008). https://doi.org/10.1093/bioinformatics/btn336
25. Giacomantonio, C.E., Goodhill, G.J.: A boolean model of the gene regulatory network underlying mammalian cortical area development. PLOS Comput. Biol. **6**(9), 1–13 (2010). https://doi.org/10.1371/journal.pcbi.1000936
26. Grieco, L., Calzone, L., Bernard-Pierrot, I., Radvanyi, F., Kahn-Perles, B., Thieffry, D.: Integrative modelling of the influence of MAPK network on cancer cell fate decision. PLoS Comput. Biol. **9**(10), e1003286 (2013)
27. Hopfensitz, M., Müssel, C., Maucher, M., Kestler, H.A.: Attractors in boolean networks: a tutorial. Comput. Stat. **28**(1), 19–36 (2013)
28. Kauffman, S.: Metabolic stability and epigenesis in randomly constructed genetic nets. J. Theor. Biol. **22**(3), 437–467 (1969)
29. Klamt, S., Haus, U.U., Theis, F.: Hypergraphs and cellular networks. PLoS Comput. Biol. **5**(5), e1000385 (2009)
30. Klarner, H., Streck, A., Siebert, H.: PyBoolNet: a Python package for the generation, analysis and visualization of boolean networks. Bioinformatics **33**(5), 770–772 (2017)
31. Naldi, A., Berenguier, D., Fauré, A., Lopez, F., Thieffry, D., Chaouiya, C.: Logical modelling of regulatory networks with GINsim 2.3. Biosystems **97**(2), 134–139 (2009)
32. Naldi, A., Monteiro, P.T., Chaouiya, C.: Efficient handling of large signalling-regulatory networks by focusing on their core control. In: Gilbert, D., Heiner, M. (eds.) CMSB 2012. LNCS, pp. 288–306. Springer, Heidelberg (2012). https://doi.org/10.1007/978-3-642-33636-2_17
33. Naldi, A., et al.: Cooperative development of logical modelling standards and tools with colomoto. Bioinformatics **31**(7), 1154–1159 (2015)
34. Naldi, A., Remy, E., Thieffry, D., Chaouiya, C.: Dynamically consistent reduction of logical regulatory graphs. Theor. Comput. Sci. **412**(21), 2207–2218 (2011)
35. Naldi, A., et al.: The colomoto interactive notebook: accessible and reproducible computational analyses for qualitative biological networks. Front. Physiol. **9**, 680 (2018) https://doi.org/10.3389/fphys.2018.00680

36. Paige, R., Tarjan, R.E.: Three partition refinement algorithms. SIAM J. Comput. **16**(6), 973–989 (1987)
37. Pérez-Verona, I.C., Tribastone, M., Vandin, A.: A large-scale assessment of exact model reduction in the biomodels repository. In: Computational Methods in Systems Biology - 17th International Conference, CMSB 2019, Trieste, Italy, 18–20 September 2019, Proceedings, pp. 248–265 (2019). https://doi.org/10.1007/978-3-030-31304-3_13
38. Perez-Verona, I.C., Tribastone, M., Vandin, A.: A large-scale assessment of exact lumping of quantitative models in the biomodels repository. Theor. Comput. Sci. (2021). https://doi.org/10.1016/j.tcs.2021.06.026. https://www.sciencedirect.com/science/article/pii/S0304397521003716
39. Richardson, K.A.: Simplifying boolean networks. Adv. Complex Syst. **8**(04), 365–381 (2005)
40. Rodríguez-Jorge, O., et al.: Cooperation between T cell receptor and toll-like receptor 5 signaling for CD4+ T cell activation. Sci. Signal. **12**(577), eaar3641 (2019)
41. Saadatpour, A., Albert, R., Reluga, T.C.: A reduction method for boolean network models proven to conserve attractors. SIAM J. Appl. Dyna. Syst. **12**(4), 1997–2011 (2013)
42. Schwab, J.D., Kühlwein, S.D., Ikonomi, N., Kühl, M., Kestler, H.A.: Concepts in boolean network modeling: what do they all mean? Comput. Struct. Biotechnol. J. **18**, 571–582 (2020). https://doi.org/10.1016/j.csbj.2020.03.001. http://www.sciencedirect.com/science/article/pii/S200103701930460X
43. Shmulevich, I., Dougherty, E.R., Kim, S., Zhang, W.: Probabilistic boolean networks: a rule-based uncertainty model for gene regulatory networks. Bioinformatics **18**(2), 261–274 (2002)
44. Steggles, L.J., Banks, R., Shaw, O., Wipat, A.: Qualitatively modelling and analysing genetic regulatory networks: a petri net approach. Bioinformatics **23**(3), 336–343 (2007)
45. Su, C., Pang, J.: Sequential control of boolean networks with temporary and permanent perturbations. arXiv preprint arXiv:2004.07184 (2020)
46. Thomas, R.: Regulatory networks seen as asynchronous automata: a logical description. J. Theor. Biol. **153**(1), 1–23 (1991)
47. Thomas, R.: Kinetic logic: a Boolean approach to the analysis of complex regulatory systems. In: Proceedings of the EMBO Course "Formal Analysis of Genetic Regulation", held in Brussels, 6–16 September 1977, vol. 29. Springer, Heidelberg (2013). https://doi.org/10.1007/978-3-642-49321-8
48. Veliz-Cuba, A.: Reduction of boolean network models. J. Theor. Biol. **289**, 167–172 (2011)
49. Wittmann, D.M., Krumsiek, J., Saez-Rodriguez, J., Lauffenburger, D.A., Klamt, S., Theis, F.J.: Transforming boolean models to continuous models: methodology and application to T-cell receptor signaling. BMC Syst. Biol. **3**(1), 98 (2009). https://doi.org/10.1186/1752-0509-3-98
50. Zañudo, J.G.T., Albert, R.: An effective network reduction approach to find the dynamical repertoire of discrete dynamic networks. Chaos Interdiscip. J. Nonlinear Sci. **23**(2), 025111 (2013). https://doi.org/10.1063/1.4809777
51. Zhang, R., et al.: Network model of survival signaling in large granular lymphocyte leukemia. Proc. Nat. Acad. Sci. **105**(42), 16308–16313 (2008)

Abstraction of Markov Population Dynamics via Generative Adversarial Nets

Francesca Cairoli[1(✉)], Ginevra Carbone[1], and Luca Bortolussi[1,2]

[1] Department of Mathematics and Geosciences, University of Trieste, Trieste, Italy
francesca.cairoli@phd.units.it

[2] Modeling and Simulation Group, Saarland University, Saarbrücken, Germany

Abstract. Markov Population Models are a widespread formalism used to model the dynamics of complex systems, with applications in Systems Biology and many other fields. The associated Markov stochastic process in continuous time is often analyzed by simulation, which can be costly for large or stiff systems, particularly when a massive number of simulations has to be performed (e.g. in a multi-scale model). A strategy to reduce computational load is to abstract the population model, replacing it with a simpler stochastic model, faster to simulate. Here we pursue this idea, building on previous works and constructing a generator capable of producing stochastic trajectories in continuous space and discrete time. This generator is learned automatically from simulations of the original model in a Generative Adversarial setting. Compared to previous works, which rely on deep neural networks and Dirichlet processes, we explore the use of state of the art generative models, which are flexible enough to learn a full trajectory rather than a single transition kernel.

1 Introduction

A wide range of complex systems can be modeled as a network of chemical reactions. Stochastic simulation is typically the only feasible analysis approach that scales in a computationally tractable manner with the increase in system size, as it avoids the explicit construction of the state space. The well known Gillespie Stochastic Simulation Algorithm [8] is widely used for simulating models, as it samples from the exact distribution over trajectories. This algorithm is effective to simulate systems of moderate complexity, but it does not scale well to systems with many species and reactions, large populations, or internal stiffness. In these scenarios, a more effective choice is to rely on approximate simulation algorithms such as tau-leaping [7] and hybrid simulation [14]. Nonetheless, when the number of simulations required is extremely large and possibly costly, e.g. when one needs to simulate a large population of heterogeneous cells in a multi-scale model of a tissue or to simulate many heterogeneous individuals in an population ecology scenario, all these methods become extremely computationally demanding, even for HPC facilities.

E. Cinquemani and L. Paulevé (Eds.): CMSB 2021, LNBI 12881, pp. 19–35, 2021.
https://doi.org/10.1007/978-3-030-85633-5_2

A viable approach to address such problem is model abstraction, which aims at reducing the underlying complexity of the model, and thus reduce its simulation cost. However, building effective model abstractions is difficult, requiring a lot of ingenuity and man power. Here we advocate the strategy of learning an abstraction from simulation data. Our strategy is to frame model abstraction as a supervised learning problem, and learn an abstract probabilistic model using state of the art deep learning. The probabilistic model should then be able to generate approximate trajectories efficiently and in constant time, i.e., independent on the complexity of the original system, thus sensibly reducing the simulation cost.

Related Work. The idea of using machine learning as a model abstraction tool to approximate and simplify the dynamics of a Markov Population Process has received some attention in recent years. In [5] the authors use a Mixture Density Network (MDN) [3] to approximate the transition kernel of the stochastic process. In [16] the authors extend the previous approach by introducing an automated search of the MDN architecture that better fit the data. In [4] the authors present a Bayesian model abstraction technique, based on Dirichlet Processes, that allows the quantification of the reconstruction uncertainty. In all cases, what is learned is an approximate transition kernel, i.e., the probabilistic distribution of a single simulation step.

In this paper we address a more general and more complex problem. Instead of learning an approximate transition kernel, we learn the distribution of an entire trajectory of fixed length. This latter problem is not solvable with any of the previously adopted approaches, and its major goal is to keep abstraction error under control. In fact, training the abstract model on a full trajectory, rather than on pairs of subsequent states, allows the abstract model to retain and capture more information about the dynamics of the Markov process.

Contributions. Our approach leverages Generative Adversarial Nets (GAN), which are one of the most strong and flexible techniques to learn probabilistic models. In fact, the GAN-based model abstraction technique is capable of learning a conditional distribution over the trajectory space, keeping into account the correlation, both spatial and temporal, among all the different species and conditioning both on initial states and model parameters. All the previous approaches focus on learning the distribution of the state of the system after a time Δt, the so called *transition kernel.* However, such approaches perform poorly when the time interval is small and the dynamics is transient, showing a clear propagation of the error as the approximate kernel is applied iteratively to form a trajectory. Furthermore, producing a full trajectory reduces even more the computational cost of simulating a large pool of trajectories for different initial settings.

Paper Structure. The paper is organized as follows: in Sect. 2 the relevant background notions are introduced, in Sect. 3 we describe in detail the abstraction procedure, Sect. 4 presents the case studies and the experimental evaluation. Conclusions are drawn in Sect. 5.

2 Background

2.1 Chemical Reaction Networks

Consider a system with n species evolving according to a stochastic model defined as a Chemical Reaction Network. Under the well-stirred assumption, the time evolution can be modelled as a Continuous Time Markov Chain (CTMC) on a discrete state space. The vector $\eta_t = (\eta_{t,1}, \ldots, \eta_{t,n}) \in S \subseteq \mathbb{N}^n$ denotes the state vector at time t, where $\eta_{t,i}$ is the number of individuals in species i at time t. The dynamics is encoded by a set of m reactions with parametric propensity functions that depends on the state of the system. Due to the memoryless property of CTMC, the probability of finding the system in state s at time t given that it was in state s_0 at time t_0 can be expressed as a system of ODEs known as Chemical Master Equation (CME). Since in general the CME is a system with countably many differential equations, its analytic or numeric solution is almost always unfeasible. An alternative computational approach is to generate trajectories using stochastic algorithms for simulation, like the well-known Gillespie's SSA [8] which produces statistically correct trajectories, i.e., sampled according to the stochastic process described by the CME.

2.2 Generative Adversarial Nets

Every dataset can be considered as a set of observations drawn from an unknown distribution $\mathbb{P}_r$. Generative models aim at learning a model that mimics this unknown distribution as closely as possible, i.e., learn a distribution $\mathbb{P}_{w_g}$ as similar as possible to $\mathbb{P}_r$, in order to then get samples from it that are new but look as if they could have belonged to the original dataset. Generative Adversarial Nets (GANs) [10] are deep learning-based generative models, that, given a dataset, are capable of generating new random but plausible examples.

Wasserstein GAN. In this work we consider the Wasserstein version of GAN (WGAN) [1,11] as it is known to be more stable and less sensitive to the choice of model architecture and hyperparameters compared to a traditional GAN. WGANs use the Wasserstein distance (also known as Earth-Mover's distance), rather than the Jensen Shannon divergence, to measure the difference between the model distribution $\mathbb{P}_{w_g}$ and the target distribution $\mathbb{P}_r$. Because of Kantorovich-Rubinstein duality [17] such distance can be computed as the supremum over all the 1-Lipschitz functions $f : S \to \mathbb{R}$:

$$W(\mathbb{P}_r, \mathbb{P}_{w_g}) = \sup_{\|f\|_L \leq 1} \left(\mathbb{E}_{x\sim\mathbb{P}_r}[f(x)] - \mathbb{E}_{x\sim\mathbb{P}_{w_g}}[f(x)] \right). \tag{1}$$

We approximate these functions f with a neural net C_{w_c} parametrized by weights w_c. To enforce the Lipschitz constraint we follow [11] and introduce a penalty over the norm of the gradients. It is known that a differentiable function is 1-Lipchitz if and only if it has gradients with norm at most 1 everywhere. The objective function, to be maximized w.r.t. w_c, becomes:

$$\mathcal{L}(w_c, w_g) := \mathbb{E}_{x\sim\mathbb{P}_r}[C_{w_c}(x)] - \mathbb{E}_{x\sim\mathbb{P}_{w_g}}[C_{w_c}(x)] - \lambda \mathbb{E}_{\hat{x}\sim\mathbb{P}_{\hat{x}}}(\|\nabla_{\hat{x}} C_{w_c}(\hat{x})\|_2 - 1)^2], \tag{2}$$

where λ is the penalty coefficient and $\mathbb{P}_{\hat{x}}$ is defined by sampling uniformly along straight lines between pairs of points sampled from $\mathbb{P}_r$ and $\mathbb{P}_{w_g}$. This is actually a softer constraint that however performs well in practice [11]. The C_{w_c} network is referred to as *critic* and it outputs different scores for real and fake samples, its objective function (Eq. (2)) provide an estimate of the Wasserstein distance among the two distributions. On the other hand, the distribution $\mathbb{P}_{w_g}$ is parametrized by w_g; we seek the parameters that make it as close as possible to $\mathbb{P}_r$. To achieve this, we consider a random variable Z with a fixed simple distribution $\mathbb{P}_Z$ and pass it through a parametric function, the *generator*, $G_{w_g} : Z \to S$ that generates samples following the distribution $\mathbb{P}_{w_g}$. Therefore, the WGAN architecture consists of two deep neural nets, a generator that proposes a distribution and a critic that estimate the distance between the proposed and the real (unknown) distribution. Using WGAN brings several important advantages compared to traditional GAN: it avoids the mode collapse problem, which makes WGAN more suitable for capturing stochastic dynamics, it drastically reduces the problem of vanishing gradients and it also have an objective function that correlates with the quality of generated samples, making the results easier to interpret.

Conditional GAN. Conditional Generative Adversarial Nets (cGAN) [13] are a type of GANs that involves the conditional generation of examples, i.e., the generator produces examples of a required type, e.g. examples that belong to a certain class, and thus they introduce control over the desired generated output. In our application, we want the generation of stochastic trajectories to be conditioned on some model parameters and on the initial state of the system.

Furthermore, dealing with inputs that are trajectories, i.e. sequences of fixed length, requires the use of convolutional neural networks (CNNs) [9] for both the generator and the critic. The architecture used in this work is thus a conditional Wasserstein Convolutional GAN with gradient penalty, it is going to be referred to as cWCGAN-GP.

3 GAN-Based Abstraction

3.1 Model Abstraction

The underlying idea is the following: given a stochastic process $\{\eta_t\}_{t\geq 0}$ with transition probabilities $\mathbb{P}_{s_0}(\eta_t = s) = \mathbb{P}(\eta_t = s \mid \eta_{t_0} = s_0)$, we aim at finding another stochastic process whose trajectories are faster to simulate but similar to the original ones. Time has to be discretized, meaning we fix an initial time t_0 and a time step Δt that suits our problem. We define $\tilde{\eta}_i := \eta_{t_0 + i\cdot\Delta t}$, $\forall i \in \mathbb{N}$. In addition, given a fixed time horizon H, we define time-bounded trajectories as $\tilde{\eta}_{[1,H]} = s_1 s_2 \cdots s_H \in S^H \subseteq \mathbb{N}^{H\times n}$. Given a state s_0 and a set of parameters θ, we can represent a trajectory of length H as a realization of a random variable over the state space S^H. The probability distribution for such random variable is given by the product of the transition probabilities at each time step: $\mathbb{P}_{s_0,\theta}(\tilde{\eta}_{[1,H]} = s_1 s_2 \cdots s_H) = \prod_{i=1}^{H} \mathbb{P}_{s_{i-1},\theta}(\tilde{\eta}_i = s_i)$. The CTMC, $\{\eta_t\}_{t\geq 0}$, is now expressed as a time-homogeneous Discrete Time Markov Chain $\{\tilde{\eta}_i\}_i$. An additional approximation has to be made: the abstract model takes values in $S' \subseteq \mathbb{R}^n_{\geq 0}$, a continuous space in which the state space $S \subseteq \mathbb{N}^n$ is embedded. In constructing the approximate probability distribution for trajectories we can decide to restrict our attention to arbitrary aspects of the process, rather than trying to preserve the full behavior. A *projection* π from S^H to an arbitrary space U^H can be used to reach this purpose, for instance, to monitor the number of molecules belonging to a certain subset of chemical species, i.e., $U \subseteq S$. Note that $\pi(\tilde{\eta}_{[0,H]})$ is a random variable over U^H. Such flexibility could be extremely helpful in capturing the dynamics of systems in which some species are not observable.

Abstraction Accuracy. Another important ingredient is a meaningful quantification of the error introduced by the abstraction procedure, i.e., the reconstruction accuracy. Such quantification must be based on a distance, d, among distributions. We choose the Wasserstein distance, together with the absolute and relative difference among means and variances of the histograms. Given a distribution over initial states s_0 and a distribution over parameters θ, we would like to measure the expected error at every time instant $t_i = t_0 + i \cdot \Delta t$ with $i \in \{1, \ldots, H\}$. Formally, we want to measure $\mathbb{E}_{s_0,\theta}\left[d(\pi(\eta_{[1,H]})\big|_i, \pi'(\eta'_{[1,H]})\big|_i)\right]$ where $\pi(\eta_{[1,H]})\big|_i$ denotes the i-th time components of the projected trajectory $\pi(\eta_{[1,H]}) \in U^H$. To estimate such quantity we use a well-known unbiased estimator, which is the average over the distances computed over a large sample set of initial settings. Computing the distance among SSA and abstract distributions at each time step quantifies how small the expected error is and, more importantly, how it evolves in time. As a matter of fact, it shows whether the error tends to propagate or not and how much each species contributes to the abstraction error. In practice, we compute $H \cdot n$ distances among distributions over $\mathbb{N}$ as we want to know how each species contributes in the reconstruction error.

3.2 Dataset Generation

Training Set. Choose a set of N_{train} initial settings and for each setting simulate k_{train} SSA trajectory of length H. The training set is composed of $N_{train} \cdot k_{train}$ pairs initial setting-trajectory, i.e. pairs $(\theta^i, s_0^i, \eta^{ij}_{[1,H]})$ for $i = 1, \ldots, N_{train}$ and $j = 1, \ldots, k_{train}$.

Test Set. Choose a set of N_{test} initial settings and for each setting simulate a large number, $k_{test} \gg k_{train}$, of SSA trajectory of length H. The test set is composed of $N_{test} \cdot k_{test}$ pairs initial setting-trajectory, i.e. pairs $(\theta^i, s_0^i, \eta^{ij}_{[1,H]})$ for $i = 1, \ldots, N_{test}$ and $j = 1, \ldots, k_{test}$.

Partial Observability. In case of partial observability, $U \subseteq S$, we fix an initial condition for species in U, and simulate a pool of trajectories each time sampling the initial value of species in $S \setminus U$. As a result, we are learning and abstract distribution that marginalizes over unobserved variables.

3.3 cWCGAN-GP Architecture

The critic C_{w_c} takes as input a batch of initial states, $s_0^1, \ldots, s_0^b$, a batch of parameters, $\theta_1, \ldots, \theta_b$, and a batch of subsequent trajectories, $\eta^1_{[1,H]}, \ldots, \eta^b_{[1,H]}$. For each $i \in \{1, \ldots, b\}$ the inputs, $\eta^i_{[1,H]}$, s_0^i and θ_i, are concatenated to form an input with dimension $b \times (H+1) \times (n+m)$. Formally, $C_{w_c} : S^{H+1} \times \Theta \to \mathbb{R}$. To enforce the Lipschitz property over C_{w_c} we add a gradient penalty term over $\mathbb{P}_{\hat{x}}$. Samples of $\mathbb{P}_{\hat{x}}$ are generated by sampling uniformly along straight lines connecting points coming from a batch of real trajectories and points coming from a batch of generated trajectories.

On the other hand, the generator G_{w_g} takes as input a batch of initial states, $s_0^1, \ldots, s_0^b$, a batch of parameters, $\theta_1, \ldots, \theta_b$, and a batch of random noise, $z^1, \ldots, z^b$, with dimension k, a user-defined hyper-parameter. For each $i \in \{1, \ldots, b\}$ the two inputs are, once again, concatenated to form an input with dimension $b \times (n + m + k)$. The generator outputs a batch of generated trajectories $\hat{\eta}^1_{[1,H]}, \ldots \hat{\eta}^b_{[1,H]}$. Formally, $G_{w_g} : S \times \Theta \times Z \to S^H$, such that $G_{w_g}(s_0, \theta, z) = \hat{\eta}_{[1,H]} = s_1 \cdots s_H$. See the pseudocode for the algorithm in Appendix E of [6].

3.4 Model Training

The cWCGAN-GP-based model abstraction framework consists in training two different CNNs. The loss function, introduced in Eq. (2), is a parametric function depending both on the generator weights w_g and the critic weights w_c. When training the critic, we keep the generator weights constant $\overline{w}_g$, and we maximize $\mathcal{L}(w_c, \overline{w}_g)$ w.r.t. w_c. Formally, we solve the problem

$$w_c^* = \underset{w_c}{\operatorname{argmax}} \Big\{ \mathcal{L}(w_c, \overline{w}_g) \Big\}.$$

On the other hand, in training the generator, we keep the critic weights constant $\overline{w}_c$, and we minimize $\mathcal{L}(\overline{w}_c, w_g)$ w.r.t. w_g. Formally, we solve the problem

$$w_g^* = \underset{w_g}{\operatorname{argmin}} \Big\{ \mathcal{L}(\overline{w}_c, w_g) \Big\} = \underset{w_g}{\operatorname{argmin}} \Big\{ -\mathbb{E}_{z,(s_0,\theta)} \Big[C_{\overline{w}_c}\big(G_{w_g}(z, s_0, \theta), s_0, \theta\big) \Big] \Big\}.$$

As mentioned in Sect. 2, the loss function derives from the Wasserstein distance between the real and generated distributions, see [1,11] for the mathematical details.

Intuitively, the generator generates a batch of samples, and these, along with real examples from the dataset, are provided to the critic, which is then updated to get better at estimating the distance between the real and the abstract distribution. The generator is then updated based on scores obtained by the generated samples from the critic. An important collateral advantage is that WGANs have a loss function that correlates with the quality of generated examples.

Training the cWCGAN-GP has a cost. Nonetheless, once it has been trained, its evaluation is extremely fast. Details about training and evaluation costs are discussed in Sect. 4.

Abstract Model Simulation. Once the training is over, we can discard the critic and focus only on the trained generator G. In order to generate an abstract trajectory starting from a state s_0^* with parameters θ^*, we just have to sample a value z from the random noise variable Z and evaluate the generator on the pair (s_0^*, θ^*, z). The output is a stochastic trajectory of length H: $G(s_0^*, \theta^*, z) = \hat{\eta}_{[1,H]}$. The stochasticity is provided by the random noise variable, de facto the generator acts as a distribution transformer that maps a simple random variable into a complex distribution. In order to generate a pool of p trajectories, we simply sample p different values from the random noise variable: $z_1, \ldots, z_p$. Therefore, the generation of a trajectory has a fixed computational cost.

4 Experimental Results

In this section we validate our GAN-based model abstraction procedure on the following case studies. More details are provided in Appendix A of [6].

- **SIR Model (Absorbing state).** The SIR epidemiological model describes the spread, in a population, of an infectious disease that grants immunity to those who recover from it. The population is divided in three mutually exclusive groups: susceptible (S), infected (I) and recovered (R). The possible reactions, given by the interaction of individuals are infection and recovery. An important feature is the presence of an absorbing states.
- **Ergodic SIRS Model.** A SIR model in which the population is not perfectly isolated, meaning there is always a chance of getting infected from some external individuals, and in which immunity is only temporary. As a consequence, this model has no absorbing state.
- **Genetic Toggle Switch Model (Bistability).** The toggle switch is a well-known bistable biological circuit consisting of two genes, G_1 and G_2, that mutually repress each other in the production of proteins P_1 and P_2 respectively. The system displays two stable equilibria.
- **Oscillator Model.** The circuit consists of three species A, B and C and three cyclic reactions: A converts B to itself, B converts C to itself, and C converts A to itself. The concentrations of the three species oscillates in time.
- **MAPK Model.** The mitogen-activated protein kinase cascade models the amplification of an output signal ($MAKP\_PP$) thorough a multi-level cascade with negative feedback which is ultra-sensitive to an input stimulus (V_1). The output signal shows either stable or oscillating behaviour, depending on the input signal.

In order to evaluate the performance of our abstraction procedure we consider two important measures: the accuracy of the abstract model, evaluated for each species at each time step of the time grid, and the computational gain compared to SSA simulation time.

Experimental Settings. The workflow can be divided in steps: (1) define a CRN model, (2) generate the synthetic datasets via SSA simulation, (3) learn the abstract model by training the cWCGAN-GP and, finally, (4) evaluate such abstraction. All the steps have been implemented in Python. In particular, CRN models are defined in the `.psc` format, CRN trajectories are simulated using Stochpy [12] (stochastic modeling in Python) and PyTorch [15] is used to craft the desired architecture for the cWCGAN-GP and to evaluate the latter on the test data. All the experiments were performed on a Intel Xeon Gold 6140 with 24 cores and a 128 GB RAM. The source code for all the experiments can be found at the following link: https://github.com/francescacairoli/WGAN_ModelAbstraction.

Datasets. For each case study with fixed parameters, the training set consists of 20K different SSA trajectories. In particular, $N_{train} = 2\text{K}$ and $k_{train} = 10$. The test set, instead, consists of 25 new initial settings and from each of these we simulate 2K trajectories, so to obtain an empirical approximation of the distribution targeted by model abstraction. When a parameter is allowed to vary, the training set consists of 50K SSA trajectories ($N_{train} = 1\text{K}$ and $k_{train} = 50$). We manually choose H and Δt so that the system is close to steady state at time $H \cdot \Delta t$, without spending there too many steps. The time interval should be small enough to capture the full transient behavior of the system. For systems with no steady state, such as the oscillating models, we choose H and Δt so to observe a full period of oscillation. The chosen values are the following: SIR: $\Delta t = 0.5$, $H = 16$; e-SIRS: $\Delta t = 0.1$, $H = 32$; Toggle Switch: $\Delta t = 0.1$, $H = 32$; Oscillator: $\Delta t = 1$, $H = 32$; MAPK: $\Delta t = 60$, $H = 32$.

Data Preparation. Data have been scaled to the interval $[-1, 1]$ to enhance the performance of the two CNNs and to avoid sensitivity to different scales in species counts. During the evaluation phase, the trajectories have been scaled back. Hence, results and errors are shown in the original scale.

4.1 cWCGAN-GP Architecture

The same architecture and the same set of hyper-parameters works well for all the analyzed case studies, showing great stability and usability of the proposed solution. The Wasserstein formulation of GANs, with gradient penalty, strongly contributes to such stability. Traditional GANs have been tested as well, but they do not have such strength. The details of the architecture follows the best practice suggestions provided in [11]. The critic network has two hidden one-dimensional convolutional layers, with $n+m$ channels, each containing 64 filters of size 4 and stride 2. We use a leaky-ReLU activation function with slope 0.2, we do layer normalization and at each layer we introduce a dropout with probability 0.2. An additional dense layer, with linear activation function, is used to connect the single output node, that contains the critic value. In order to enforce the Lipschitz constraint on the critic's model we add a gradient penalty term, as described in Sect. 2.2. On the other hand, the generator network takes as input the noise and the initial settings and it embeds the inputs in a larger space with N_{ch} channels (512 in our experiments) through a dense layer. Four one-dimensional convolutional transpose layers are then inserted, containing respectively 128, 256, 256 and 128 filters of size 4 with stride 2. Here we do batch normalization and use a leaky-ReLU activation function with slope 0.2. Finally, a traditional convolutional layer is introduced to reduce the number of output channels to n. The Adam algorithm [2] is used to optimize loss function of both the critic and the generator. The learning rate is set to 0.0001 and $\beta = \{0.5, 0.9\}$. The above settings are shared by all the case studies, the only exception is the more complex MAPK model for which a deeper cWCGAN-GP architecture is selected: a critic with five layers, each containing 256 filters of size 4 and stride 2, and a generator with five layers, containing respectively 128, 256, 512, 256 and 128 filters of size 4 with stride 2.

Training times depend on the dimension of the dataset, on the size of mini-batches, on the number of species, and on the architecture of the cWCGAN-GP. The latter has been kept constant for all the case studies. Batches of 256 samples have been used and the number of epochs varies from 200 to 500 depending on the complexity of the model. Moreover, each training iteration of the generator correspond to 5 iterations of the critic, to balance the power of the two player. The average time required for each training epoch is around one minute. Therefore, training the cWCGAN-GP model for 500 epochs takes around 8 h leveraging the GPU.

4.2 Results

Computational Gain. The time needed to generate abstract trajectories does not depend on the complexity of the original system. Moreover, as the cWCGAN-GP architecture is shared by all the case studies, the computational time required to generate abstract trajectories is the same for all the case studies. In particular, considering a noise variable of size 480, it takes around 1.75 milliseconds (ms) to simulate a single trajectory. However, when generating batches of at least 200 trajectories the overhead reduces and the time to generate a single trajectory stabilizes around 0.8 ms. The same does not hold for the SSA trajectories, whose computational costs depends on the complexity of the model and on the chosen reaction rates. In the case studies considered the time required to simulate a single trajectory varies from 0.04 to 0.22 s, but it easily increases for more complex models or for smaller reaction rates, whereas the cost of abstract simulation stays constant. Details about the computational gain for each model are presented in Table 1. Computations are performed exclusively on a single CPU processor, to perform a fair comparison. However, the evaluation of cWCGAN-GP can be further sped up using GPUs, especially for large batches of trajectories, but this would have introduced a bias in their favour. It is important to stress how GPU parallelization is extremely straightforward in PyTorch and how the time to generate a single trajectory decrease to 1.9×10^{-5} s when generating a batch of at least $2K$ trajectories (see last line of Table 1).

Table 1. Comparison of the average computational time required to simulate a single trajectory either via SSA (both direct or approximate methods) or via the cWCGAN-GP abstraction. 200 trajectories are needed to reduce the CPU (single processor) overhead, whereas, 2000 trajectories are required for the on GPU overhead.

Model	SIR	e-SIRS	Switch	Osc.	MAPK
SSA (direct)	0.043 s	0.047 s	0.041 s	0.042 s	0.224 s
- CPU (avg over 200):	0.024 s	0.024 s	0.020 s	0.021 s	0.211 s
SSA (τ-leaping)	0.054 s	0.052 s	0.044 s	0.042 s	0.26 s
- CPU (avg over 200):	0.018 s	0.028 s	0.024 s	0.021 s	0.24 s
cWCGAN-GP	0.00175 s	0.00175 s	0.00175 s	0.00175 s	0.00175 s
- CPU (avg over 200):	0.0008 s	0.0008 s	0.0008 s	0.0008 s	0.0008 s
- GPU (avg over 2000): $10^{-5}\times$	1.9 s	1.9 s	1.9 s	1.9 s	1.9 s

The training phase introduces a fixed overhead that affects the overall computational gain. For instance, the training phase of the MAPK model takes around 8 h, which is equivalent to the time needed to generate 140K SSA trajectories. It follows that, together with the trajectories needed to generate the training set, the cost of the training procedure is paid off when we simulate at least 200K trajectories. In a typical biological multi-scale scenario in which we seek to simulate the evolution in time of a tissue containing hundreds of thousands or millions of cells, simulating also some of their internal pathways, the number of trajectories needed for the training phase becomes negligible and the training time is soon paid off.

Measures of Performance. Results are presented as follows. For each model, we present a small batch of trajectories, both real and abstract. From the plots of such trajectories we can appreciate if the abstract trajectories are similar to real ones and if they capture the most important macroscopic behaviors. We also show the histograms of empirical distributions at time t_H for each species to quantify the behavior over all the 2K trajectories present in the test set (see Fig. 2–6). Additional plots are shown in Appendix B of [6] (Fig. 7–10).

Measuring Error Propagation. The reconstruction accuracy of the proposed abstraction procedure is performed on test sets consisting of 25 different initial settings. For each of these points 2K SSA trajectories represent the empirical approximation of the true distribution over S^H. From each of these initial settings we also simulate 2K abstract trajectories. Given a species $i \in \{1, \dots n\}$ and a time step $j \in \{1, \dots H\}$, we have the real one-dimensional distribution $\eta_{i,j}$ and the generated abstract distribution $\hat{\eta}_{i,j}$, where $\eta_{i,j}$ denotes the counts of species i at time t_j in a trajectory $\eta_{[1,H]}$. In order to quantify the reconstruction error, we compute five quantities: the Wasserstein distance among the two one-dimensional distributions, the absolute and relative difference among the two means and the absolute and relative difference among the two variances. By doing so, we are

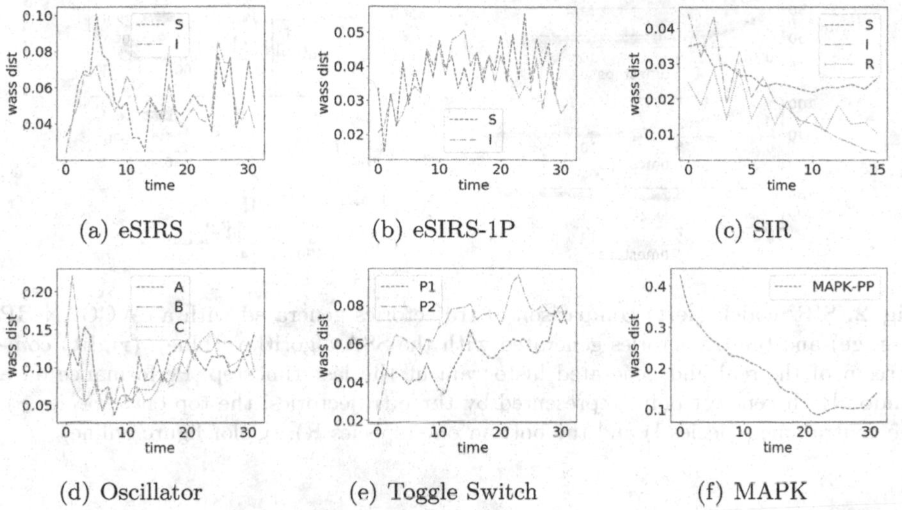

(a) eSIRS (b) eSIRS-1P (c) SIR

(d) Oscillator (e) Toggle Switch (f) MAPK

Fig. 1. Plots of the error over time for each model and each species. Errors are computed using the Wasserstein distance over the entire test set. Generated trajectories have been keep scaled to the interval $[-1, 1]$ so that the scale of the system does not affect the scale of the error measure.

capable of seeing whether the error propagates in time and whether some species are harder to reconstruct than others. The error plots for the Wasserstein distance are shown in Fig. 1. Plots of means and variances distances are provided in Appendix B of [6] (Fig. 11–14). In addition, for two-dimensional models, i.e. eSIRS, Toggle Switch and MAPK, we show the landscapes of these five measures of the reconstruction error at three different time steps: step t_1, step $t_{H/2}$ and step t_H (Fig. 15–17 in Appendix B). We observe that, in all the models, each species seems to contribute equally to the global error and, in general, the error stays constant w.r.t. time, i.e., it does not propagate. This was a major concern in previous methods, based on the abstraction of transition kernels. In fact, in order to simulate a trajectory of length H the abstract kernel has to be applied iteratively H times. As a consequence, this results in a propagation of the error introduced in the approximation of the transition kernel.

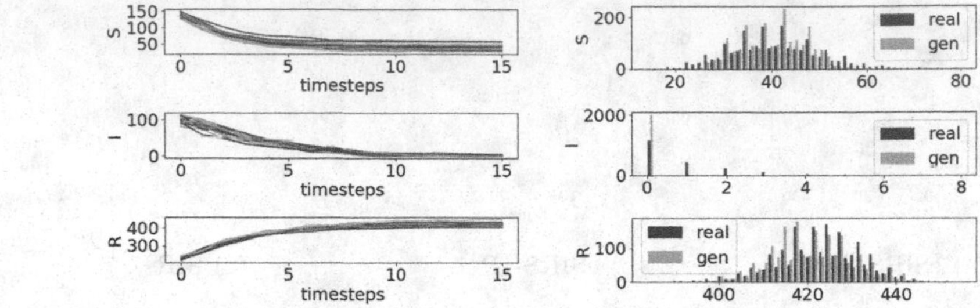

Fig. 2. SIR model: **(left)** comparison of trajectories generated with a cWCGAN-GP (orange) and the trajectories generated with the SSA algorithm (blue); **(right)** comparison of the real and generated histogram at the last timestep. Performance on a randomly chosen test point represented by three trajectories: the top one (species S), the central one (species I) and the bottom one (species R). (Color figure online)

SIR. The results for the SIR model are presented in Fig. 2 and Fig. 7 (Appendix B of [6]), which shows the performance on two, randomly chosen, test points. Each point is represented by three trajectories, the top one is for species S, the central one is for species I and the bottom one is for species R. The population size, given by $S + I + R$, is variable. The abstraction was trained on a dataset with fixed parameters, $\theta = \{3, 1\}$. Likewise, in the test set only the initial states are allowed to vary. We observe that our abstraction method is able to capture the absorbing nature of SIR trajectories. It is indeed very important that once state $I = 0$ or state $R = N$ are reached, the system should not escape from it. Abstract trajectories satisfy such property without requiring the imposition of any additional constraint. The empirical distributions, real and generated, at time t_H are almost indistinguishable.

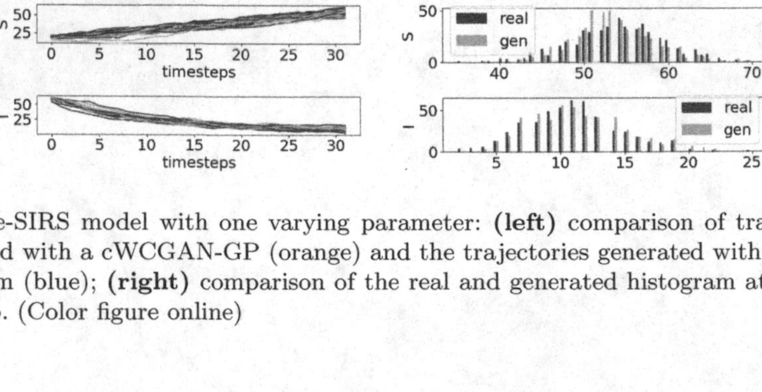

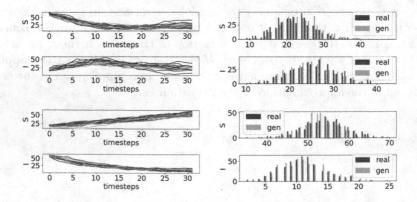

Fig. 3. e-SIRS model with one varying parameter: **(left)** comparison of trajectories generated with a cWCGAN-GP (orange) and the trajectories generated with the SSA algorithm (blue); **(right)** comparison of the real and generated histogram at the last timestep. (Color figure online)

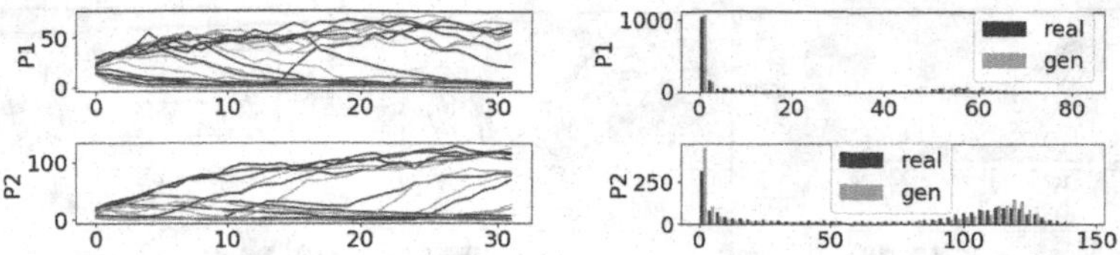

Fig. 4. Toggle Switch model: **(left)** comparison of trajectories generated with a cWCGAN-GP (orange) and the trajectories generated with the SSA algorithm (blue); **(right)** comparison of the real and generated histogram at the last timestep. Performance for a randomly chosen test point represented by a pair of trajectories: the top one (species P1) and the bottom one (species P2). (Color figure online)

e-SIRS. The e-SIRS model represents our baseline. We train two abstractions: in the first case the model is trained on a dataset with fixed parameters, $\theta = \{2.36, 1.67, 0.9, 0.64\}$, and in the second case we let parameter θ_1 vary as well. Results are very accurate in both scenarios. In the fixed-parameters case, Fig. 8 (Appendix B of [6]), the results are shown for two, randomly chosen, initial states. In the second case, Fig. 3, the results are shown on two, randomly chosen, pairs (s_0, θ_1). Each point is represented by a pair of trajectories, the top one is for species S and the bottom one is for species I. We performed a further analysis on the generalization capabilities of the abstraction learned on the dataset with one varying parameter, using larger test sets and computing mean and standard deviation of the distribution of Wasserstein distances over such sets. The mean stays around 0.04 with a tight standard deviation ranges from 0.01 to 0.05, showing little impact of the chosen conditional setting (see Fig. 18 in Appendix B of [6]).

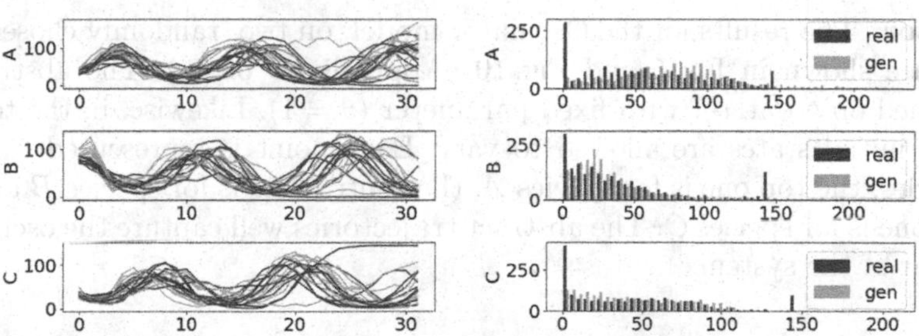

Fig. 5. Oscillator model: **(left)** comparison of trajectories generated with a cWCGAN-GP (orange) and the trajectories generated with the SSA algorithm (blue);**(right)** comparison of the real and generated histogram at the last timestep. Performance on a randomly chosen test point represented by three trajectories: the top one (species A), the central one (species B) and the bottom one (species C). (Color figure online)

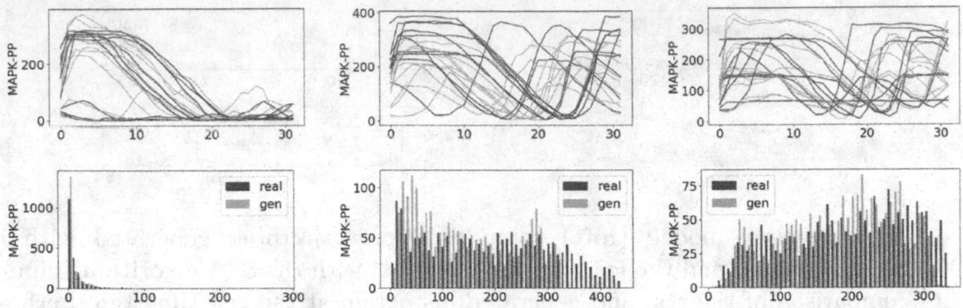

Fig. 6. MAPK model: **(top)** comparison of trajectories generated with a cWCGAN-GP (orange) and the trajectories generated with the SSA algorithm (blue);**(bottom)** comparison of the real and generated histogram at the last timestep. Performance on three, randomly chosen, test points. Each point is represented by the output species MAPK_PP. (Color figure online)

Toggle Switch. The results for the Toggle Switch model, on two, randomly chosen, test points, are shown in Fig. 4) and Fig. 9 (Appendix B of [6]). The abstraction was trained on a dataset with fixed symmetric parameters ($kp_i = 1, kb_i = 1, ku_i = 1, kd_i = 0.01$ for $i = 1, 2$). Likewise, in the test set only the initial states are allowed to vary. In this model, we tried to abstract only trajectories of the proteins $P1$ and $P2$, which are typically the observable species, ignoring the state of the genes. By doing so, we reduce the dimensionality of the problem but we also lose some information about the full state of the system. Nonetheless, the cWCGAN-GP abstraction is capable of capturing the bistable behaviour of such trajectories. In Fig. 4, each point is represented by two trajectories, the top one is for species $P1$, whereas the bottom one is for species $P2$.

Oscillator. The results for the Oscillator model, on two, randomly chosen, test points, are shown in Fig. 5 and Fig. 10 (Appendix B of [6]). The abstraction was trained on a dataset with fixed parameter ($\theta = 1$). Likewise, in the test set only the initial states are allowed to vary. Each point is represented by three trajectories, the top one is for species A, the central one is for species B and the bottom one is for species C. The abstract trajectories well capture the oscillating behaviour of the system.

MAPK. The results for the MAPK model, on three, randomly chosen, test points, are shown in Fig. 6. The abstraction was trained on a dataset considering only a varying V_1 parameter and the dynamics of species $MAPK\_PP$. This case study represents a complex scenario in which the abstract distribution should capture the marginalization over the other seven unobserved variables. Moreover, the emergent behaviour of the only observed variable, $MAPK\_PP$, is strongly influenced by the input parameter V_1 and further amplified by the multi-scale nature of the cascade: for some values of V_1 the system oscillates, whereas for others it stabilizes around an equilibrium. Results show that our abstraction technique is flexible enough to capture such sensitivity.

4.3 Discussion

Previous approaches to model abstraction, see Related work in Sect. 1, focus on approximating the transition kernel, meaning the distribution of possible next states after a time Δt, rather than learning the distribution of full trajectories of length H. The main reason for such choice is the limited scalability of the tool used for learning the abstraction. In fact, learning a distribution over $S^H \subseteq \mathbb{N}^{H \times n}$ with a Mixture Density Network is unfeasible even for small H. Moreover, in learning to approximate the transition kernel one must split the SSA trajectories of the dataset in pairs of subsequent states. By doing so, a lot of information about the temporal correlation among states is lost. Having a tool strong and stable enough to learn distributions over S^H allows us to preserve this information and make abstraction possible even for systems with a complex dynamics, which the abstraction of the transition kernel was failing to capture. For instance, we are now able to abstract the transient behaviour of multi-stable or oscillating systems. When attempting to abstract the transition kernel, either via MDN or via c-GAN, for such complex systems, we did not succeed in learning meaningful solutions. A collateral advantage in generating full trajectories, rather than single subsequent states, is that it introduces an additional computational speed-up in the time required to generate a large pool of trajectories of length H. For instance, if a cWGAN is used to approximate the transition kernel, it takes around 31 s to simulate the 50K trajectories of length 32 present in the test set. Our trajectory-based method takes only 3.4 s to generate the same number of trajectories. Furthermore, our cWCGAN-GP was trained with relative small datasets, which leaves room for further improvements where needed. An additional strength of our method is that one can train the abstract model only on species that are observable, reducing the complexity of the CRN model while preserving an accurate reconstruction for the species of interest. Once again, this was not possible with transition kernels and it may be extremely useful in real world applications.

In general, the cWCGAN-GP approximation does not provide any statistical guarantee about the reconstruction error. In addition, the set of observations used to learn the abstraction is rather small, typically 10 samples for each initial setting. Therefore, it is not surprising that the real and the abstract distributions are not indistinguishable from a statistical point of view, as shown in Appendix D of [6]. However, the abstract model is actually capable of capturing, from the little amount of information provided, the emergent features of the behaviour of the original system, such as multimodality or oscillations. In this regard, formal languages can be used to formalize and check such qualitative properties. In particular, we can check whether the satisfaction probability (of non rare events) is similar in real and abstract trajectories. Examples are shown in Appendix C of [6]. Furthermore, such quantification of qualitative properties can be used to measure how good the reconstruction is. As future work, we intend to use it as query strategy for an active learning approach, so that the obtained abstract model is driven in the desired direction.

5 Conclusions

In the paper we presented a technique to abstract the simulation process of stochastic trajectories for various CRNs. The WGAN-based abstraction improves considerably the computational efficiency, which is no more related to the complexity of the underlying CRN. This would be extremely helpful in all those applications in which a large number of simulations is required, i.e., applications whose solution is unfeasible via SSA simulation. It would enable the simulation of multi-scale models for very large populations, it would speed-up statistical model checking [18] and it can be used in particular cases of parameter estimation, for example when only few parameters have to be estimated multiple times. In conclusion, the c-WCGAN-based solution to model abstraction perform well in scenarios that are very complex and challenging, requiring relatively little data and very little fine-tuning.

As future work, we plan to study how our abstraction technique works on real data. In this regard, we do not aim at capturing the underlying dynamical system, but we would rather be able to reproduce the trajectories observed in real applications. A great strength of our method, compared to state of the art solutions, is that it is able to generate trajectories only for a subset of the species present in the system domain, ignoring the information that is not observable, even during the training phase. Another interesting extension is to adapt our technique to sample bridging trajectories, where both the initial and the terminal states are fixed. Typically, the simulation of such trajectories requires expensive Monte Carlo simulations, which makes clear the benefits of resorting to model abstraction.

Acknowledgements. This work has been partially supported by the Italian PRIN project "SEDUCE" n. 2017TWRCNB.

References

1. Arjovsky, M., Chintala, S., Bottou, L.: Wasserstein GAN. arXiv preprint arXiv:1701.07875 (2017)
2. Bengio, Y.: RMSProp and equilibrated adaptive learning rates for nonconvex optimization. Corr abs/1502.04390 (2015)
3. Bishop, C.M.: Pattern Recognition and Machine Learning. Information Science and Statistics. Springer, New York (2006)
4. Bortolussi, L., Cairoli, F.: Bayesian abstraction of Markov population models. In: Parker, D., Wolf, V. (eds.) QEST 2019. LNCS, vol. 11785, pp. 259–276. Springer, Cham (2019). https://doi.org/10.1007/978-3-030-30281-8_15
5. Bortolussi, L., Palmieri, L.: Deep abstractions of chemical reaction networks. In: Češka, M., Šafránek, D. (eds.) CMSB 2018. LNCS, vol. 11095, pp. 21–38. Springer, Cham (2018). https://doi.org/10.1007/978-3-319-99429-1_2
6. Cairoli, F., Carbone, G., Bortolussi, L.: Abstraction of Markov population dynamics via generative adversarial nets. CoRR abs/2106.12981 (2021). https://arxiv.org/abs/2106.12981

7. Dauphin, Y.N., De Vries, H., Bengio, Y.: RMSProp and equilibrated adaptive learning rates for non-convex optimization. arXiv preprint arXiv:1502.04390v1 (2015)
8. Gillespie, D.T.: Exact stochastic simulation of coupled chemical reactions. J. Phys. Chem. **81**(25), 2340–2361 (1977)
9. Goodfellow, I., Bengio, Y., Courville, A., Bengio, Y.: Deep Learning, vol. 1. MIT Press, Cambridge (2016)
10. Goodfellow, I., et al.: Generative adversarial nets. In: Advances in Neural Information Processing Systems, pp. 2672–2680 (2014)
11. Gulrajani, I., Ahmed, F., Arjovsky, M., Dumoulin, V., Courville, A.C.: Improved training of Wasserstein GANs. In: Advances in Neural Information Processing Systems, pp. 5767–5777 (2017)
12. Maarleveld, T.R., Olivier, B.G., Bruggeman, F.J.: StochPy: a comprehensive, user-friendly tool for simulating stochastic biological processes. PLoS ONE **8**(11), e79345 (2013)
13. Mirza, M., Osindero, S.: Conditional generative adversarial nets. arXiv preprint arXiv:1411.1784 (2014)
14. Pahle, J.: Biochemical simulations: stochastic, approximate stochastic and hybrid approaches. Brief. Bioinform. **10**(1), 53–64 (2009)
15. Paszke, A., et al.: Automatic differentiation in PyTorch. In: NIPS-W (2017)
16. Petrov, T., Repin, D.: Automated deep abstractions for stochastic chemical reaction networks. arXiv preprint arXiv:2002.01889 (2020)
17. Villani, C.: Optimal Transport: Old and New. GL, vol. 338. Springer, Heidelberg (2008). https://doi.org/10.1007/978-3-540-71050-9
18. Younes, H.L., Simmons, R.G.: Statistical probabilistic model checking with a focus on time-bounded properties. Inf. Comput. **204**(9), 1368–1409 (2006)

Greening R. Thomas' Framework with Environment Variables: A Divide and Conquer Approach

Laetitia Gibart(✉), Hélène Collavizza, and Jean-Paul Comet

University Côte d'Azur, I3S Laboratory, UMR CNRS 7271, CS 40121, 06903 Sophia Antipolis Cedex, France
{laetitia.gibart,helene.collavizza,jean-paul.comet}@univ-cotedazur.fr

When we model a complex biological system, we try to understand the causality chains that explain the different behaviours observed. However, these observations are often made under experimental conditions which are not necessarily comparable since they depend on the culture medium for example. The construction of a right modelisation therefore depends on our ability to take into account all this information in a single framework.

In this article, we show that well-known R. Thomas' modelling framework allows the simulations of successive environmental situations in a unique global network at the expense of the use of artefacts. Therefore, it becomes possible to search for parameter settings compatible with biological knowledge for all environments by just enumerating the parameter settings. Another option we recommend here, is a green extension of R. Thomas' framework with the notion of *environments*. For each environment, the regulatory network is adapted and parameter settings compatible with the associated biological knowledge are searched on a smaller search space. Then, these sets of settings are intersected to obtain those which yield the traces consistent with observations of all environments. This "divide and conquer" approach is amazingly more efficient than the global approach.

1 Introduction

Modelling a biological system aims at understanding the underlying chains of causalities which leads the system behave as observed. Biological systems are called complex because the underlying causalities are difficult to be extracted from global observation. Thus systems biology can be seen as the study of the interactions between the components of biological systems, and of the consequences of these interactions on functions and behaviours of these systems. In order to complicate the portrait of this research field, observations are often made under experimental conditions which are not necessarily comparable (constant supply of glucose, and reduced supply of oxygen for example).

Moreover, even in a given modelisation framework, several modelling choices are possible because different instantiations of dynamical parameters which pilot the behaviour of the model can lead to traces consistent with all observations.

E. Cinquemani and L. Paulevé (Eds.): CMSB 2021, LNBI 12881, pp. 36–56, 2021.
https://doi.org/10.1007/978-3-030-85633-5_3

If the modeller chooses a particular setting, when new information is known, the parameter identification step must be restarted from the beginning. The systematic approach would then consist in characterizing, at each step, all of the parameter settings consistent with current knowledge: when a new observation becomes available, the modeller just refines the previous set of consistent parameter settings by selecting only those that are also consistent with this new information.

In the 70's, qualitative models based on discrete mathematics [10,19] have proved useful to understand the main causalities that govern observed phenotypes [20,21], and the multivalued framework of R. Thomas and H. Snoussi has become a classic for biological regulatory networks. It aroused new interest when, in the early 2000s, formal methods came to complete this formalism [2], as well as that of signaling networks [5]. For example, we developed a genetically modified Hoare Logic [1] for characterizing the set of parameter settings making possible a particular trace (if known). If only global temporal properties are available, these properties are translated into a formal temporal logic and the right parameter settings are selected via a model checking decision procedure [12]. These two approaches are combined in *TotemBioNet*, a tool which enumerates all parameter settings and selects those that are consistent with the biological properties [4].

The issue addressed in this paper is how to mix up the search for all parameter settings that are consistent with the temporal properties of multiple environments. The difficulty arises from the very wide diversity of behaviours due to diverse environments. We consider in this paper that behavioural biological knowledge have already been translated into a formal temporal logic (here CTL, Computational Tree Logic) and first show how the classic R. Thomas' framework allows the design of a unique regulatory network that mimics the different environments. That requires the duplication of the states of the internal variables to allow different behaviours according to the environments. A global property encompassing all temporal properties in all environments is built and verified for each parameter setting.

But we recommend another option based on a "divide and conquer" approach: a green extension of R. Thomas' framework with the notion of *environments*. During the divide step, a specific (and thus smaller) regulatory network is built for each environment, and the sets of settings consistent with the associated temporal property are searched. During the combine step, the intersection of these sets is compute to obtained the settings which satisfy the properties for all environments.

Extending R. Thomas' framework with environments is more efficient than using a single network which takes into account all environments. When applied to a network modelling the main regulations of cellular metabolic pathways, the method based on a single network would only give the result after an unreasonable time (estimated to 49,1 years), whereas the parameter settings are computed in 44.6 min in this new framework.

Running Example. *Pseudomonas œriginosa* is an opportunistic bacteria that can secrete mucus. Mucus production is due to the presence of the protein *mucB* which is activated through a genetic element called an *operon*. Moreover *mucB* inhibits the operon and the operon activates itself through several molecules. These individual influences are summarized in the influcnce graph of *Pseudomonas œriginosa* framed in blue in Fig. 1.

Mucus produced by mucoid *Pseudomonas œriginosa* is composed by alginate and its accumulation can lead to the creation of bacteria biofilms. When these bacteria affect the lung, they cause serious infections, particularly for Cystic Fibrosis (CF) diagnosed patient [16]. Microbiologists discovered that the transition of bacteria from non-mucoid state to mucoid one is due to a very high concentration of calcium-ion in the cell environment in lung of CF patients. This led us to add in Fig. 1 the environment variable *Calcium* (in green).

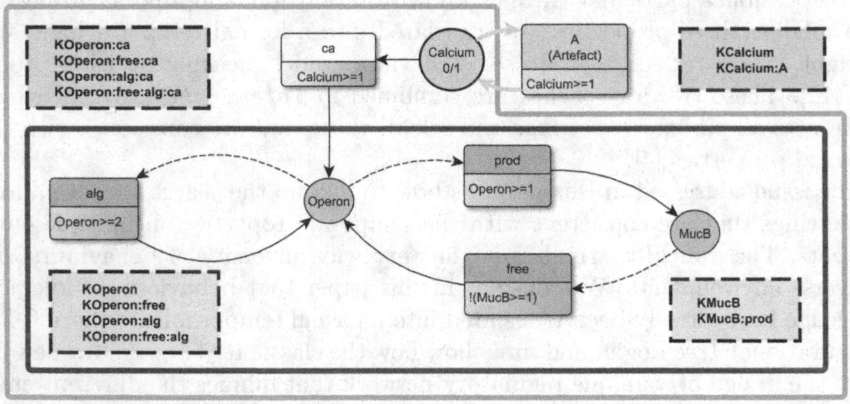

Fig. 1. Influence graph and parameters for *Pseudomonas œriginosa* mucus production system. Blue frame: initial influence graph. Grey frame: influence graph with environment variable *Calcium* (green) playing a role in the lungs of Cystic Fibrosis patients. (Color figure online)

This article is organised as follows. Section 2 sketches the framework of regulatory networks. Section 3 is dedicated to the approach based on a unique regulation network encompassing all environments. Section 4 defines the regulatory networks with environments and presents the environment by environment approach. Section 5 then compares these two approaches from a theoretical and efficiency point of view. Finally, Sect. 6 presents the case study based on an abstract model of the regulation of the cellular metabolism.

2 Adding Environment Variables to Thomas' Framework

Given a biological system, and some hypothesis on the dynamics of this system, our modelling approach is to:

- Create an *influence graph IG* that defines the *individual* influences between variables,
- Deduce from IG the set of *parameters* which express the relative strength of influences on their common target. These parameters are used to define the *global dynamics* of the system, via a parameter setting P,
- Find an appropriate translation of the biological knowledge in terms of a *temporal logic formula*,
- and at the end, use the ***TotemBioNet*** tool to find the parameter settings which make the dynamics based on IG and P consistent with this formula.

In this section, we first define the influence graph, and then introduce parameters. These two pieces of information describe the biological system and constitute a *regulatory network*. The next definitions concern the biological property: the temporal logic language (CTL) and the models of a CTL formula.

2.1 Regulatory Network with Multiplexes

Multiplexes were introduced in [11], as an extension of R. Thomas' modelling framework. They express, via a logic formula, some conditions under which an influence occurs. For example, if both variables a and b influence the variable c only when they form a protein complex, then, this coordinated influence can be represented in a multiplex which allows the effective influence only when both a and b are simultaneously present. A multiplex then combines in a unique predecessor some conditions on many variables. As the number of parameters to be identified depends directly on the number of predecessors, the introduction of multiplexes reduces the number of parameters (see Definition 2).

Definition 1 (Influence Graph with multiplexes). *An* influence graph with multiplexes $IG = (V, M, A)$ *is a directed graph such that:*

- *Vertices are variables in V or multiplexes in M ($V \cap M = \emptyset$),*
- *With each variable $v \in V$ is associated a discrete domain $D_v = [\![v_l, v_u]\!]$ where $0 \leq v_l \leq v_u$,*
- *Arcs in A go from multiplexes to variables ($A \subset M \times V$),*
- *With each* multiplex $m \in M$ *is associated a formula φ_m which expresses the condition under which m influences its target variable(s). The language of multiplex formulas is defined by:*
  - *Atoms are atomic formulas $(v \geqslant n)$ with $v \in V$ and $n \in D_v$*
  - *if φ, φ_1 and φ_2 are multiplex formulas, then $\neg\varphi$, $\varphi_1 \Box \varphi_2$ are also multiplex formulas, where $\Box$ is either $\wedge$, $\vee$ or $\Rightarrow$.*

Given an influence graph, *parameters* represent the relative strength of influences on a variable, and by allowing to build the global behaviour of the system.

Definition 2 (Parameters). *The* parameters *of a variable v are denoted $K_{v,\omega}$ where ω is a subset of the predecessors of v in IG. $K_{v,\omega}$ represents the direction of evolution of variable v when it is controlled according to the multiplexes in ω.*

- *A variable v has $2^{d^-(v)}$ parameters where $d^-(v)$ is the number of predecessors of v in IG.*
- *$\mathcal{K}(v)$ denotes the set of parameters of variable v.*
- *$\mathcal{K}$ is the set of all parameters: $\mathcal{K} = \cup_{v \in V}(\mathcal{K}(v))$.*

Definition 3 (Regulatory network). *A* Regulatory network *is a couple $\mathcal{N} = (IG, \mathcal{K})$.*

Pseudomonas æriginosa Regulatory Network. In Fig. 1, variables of IG are represented by circles and multiplexes (with their associated formulas) by round rectangles. Parameters are listed in dashed line rectangles, written in the concrete syntax of ***TotemBioNet***. For example, `K_MucB:prod` represents the case where `MucB` is influenced by `prod` while `K_Operon:alg:free` represents the case where `Operon` is influenced by both `alg` and `free`.

2.2 Formulas of Biological Properties and Their Models

We focus here on biological properties which express some global behaviour such that the existence of an attraction basin or a sustained oscillation. For such properties, we need to talk about future, and about the successive transitions of the system, in other words, to talk about the paths. Among the different temporal logics we have chosen the *Computational Tree Logic* (***CTL***). Formulas of ***CTL*** are inductively built over variables in V in the usual way, using boolean operators and using modalities on time X (neXt), F (Future), U (Until), G (Generally), and modalities on paths E (Exists), A (All), see [7,9] where the semantics of these formulas is formally defined.

Definition 4 (CTL Language). *Given an influence graph IG, the* language $\mathcal{L}_V$ of ***CTL*** formulas over V *is defined by :*

- *an* atom *is either a boolean constant ($\top$, $\bot$), or a comparison $v \Delta n$ with $v \in V$, $n \in D_V$, and Δ an operator among $<, >, \leq, \geq, =$*
- *a* ***CTL*** formula *is either an atom or :*
  - *$\neg\varphi_1$, $\varphi_1 \nabla \varphi_2$ where ∇ is a boolean operator among $\wedge, \vee, \Rightarrow, \Leftrightarrow$*
  - *$E(\varphi_1 U \varphi_2)$ or $A(\varphi_1 U \varphi_2)$*
  - *$\odot\varphi_1$ where $\odot$ is in EX, EF, EG, AX, AF, AG*

  *with φ_1 and φ_2 **CTL** formulas*

Operon Stable States. From biologists we know that a non-mucoid *Pseudomonas æriginosa* will never create mucus, and that when *Pseudomonas æriginosa* is mucoid, it cannot turn off again. This can be translated by the following ***CTL*** formula:

$$\begin{aligned}\varphi^0_{bacteria} \equiv\ & ((Operon = 0) \Rightarrow AG(\neg(Operon = 2))) \quad \wedge \\ & ((Operon = 2) \Rightarrow AG(\neg(Operon = 0)))\end{aligned}$$

It expresses that starting from a state where $Operon = 0$ (non-mucoid) then there is no path leading to a state where $Operon = 2$ (mucoid), and vice versa.

The dynamics of a regulatory network $\mathcal{N} = (IG, \mathcal{K})$ is determined by a valuation of its parameters in $\mathcal{K}$, which is called a *parameter setting.* Our aim is therefore to find the parameter settings which lead to dynamics consistent with the ***CTL*** formula expressing a temporal biological knowledge.

Definition 5 (Parameter Setting). *Given a regulatory network $\mathcal{N} = (IG, \mathcal{K})$, a* parameter setting *$P \in \mathcal{PN}$ assigns a value in D_v to each parameter $K_{v,\omega}$ associated with variable v:*

$$\begin{aligned} P : \mathcal{K} &\rightarrow [\![0, m]\!] \\ K_{v,\omega} &\mapsto k_{v,\omega} \in D_v \end{aligned}$$

where $m = max\{v_u |\ v \in V\}$ and $\mathcal{PN}$ denotes the set of parameter settings of $\mathcal{N}$.

A parameter setting $P \in \mathcal{PN}$ sets the global dynamics of the system. On each state, the possible transitions of the system depend on the *applicable parameters* of the variables. Indeed, the applicable parameter of a variable v in a particular state is determined by the combination of multiplexes targeting v whose formulas are true in that state. The value given by P to the applicable parameter of a variable defines how this variable can evolve.

Definition 6 (States). *The* set of states *of the system is the Cartesian product $S = \Pi_{v \in V} D_v$. We denote $s \in S$ a state.*

Definition 7 (Applicable parameter on a state). *The* applicable parameter *of variable v on a state $s \in S$, is the only parameter $K_{v,\omega}$ s.t.*

- *$\forall m \in \omega$, φ_m is true on state s,*
- *$\forall m \in pred(v), m \notin \omega$, φ_m is false on state s (where $pred(v)$ is the set of predecessors of v).*

The applicable parameter of v in state s is denoted $K_v[s]$. Each multiplex m, predecessor of v, such that φ_m is true on s is called a resource *of v in s.*

Applicable Parameters for Pseudomonas æriginosa. The applicable parameter of `Operon` for state *01* (where 0 is the value of *Operon* and 1 the value of *Mucus*) is `K_Operon` because neither $\varphi_{alg} = (Operon \geq 2)$ nor $\varphi_{free} = \neg(Mucus \geq 1)$ are true. The applicable parameter of `MucB` for state *21* is `K_MucB:prod` because $\varphi_{prod} = \neg(Operon \geq 1)$ is true.

Given a parameter setting P, a variable v and a state s, the value of the applicable parameter for v on s indicates if, on state s, v tends to increase, decrease or stay stable. This allows the definition of the (asynchronous) state transition graph which sums up the global behaviour of the system for this chosen parameter setting.

Definition 8 (*ASTG* for a parameter setting *P*). *Given a regulatory network $\mathcal{N} = (IG, \mathcal{K})$ whose variables are denoted $v_1, \ldots, v_n$ and $P \in \mathcal{PN}$ a parameter setting for $\mathcal{N}$, the associated* Asynchronous State Transition Graph *$ASTG_P$ is defined as follow:*

- **Vertices** *are states $s \in S$,*
- **Loops***: there is an arc from s to itself if $P(K_{v_i}[s]) = s_i, \forall i = 1 \ldots n$. This expresses that each variable v_i has reached its* focal *value toward which it tends.*
- **Arcs***: there is an arc from $s^p = (s_1^p, ..., s_n^p)$ to $s^q = (s_1^q, ..., s_n^q)$ if there exists one and only one index i s.t. $s_i^p \neq s_i^q$ with either : $s_i^q = s_i^p + 1$ and $P(K_{v_i}[s^p]) > s_i^p$ or $s_i^q = s_i^p - 1$ and $P(K_{v_i}[s^p]) < s_i^p$. This expresses that v_i has not reached its focal value and will increase (or decrease) by one level.*

Finally, the two last definitions characterize what is a model of a property on the dynamics of the system. They are based on a decision procedure (such as a model checking procedure).

Definition 9 (Decision procedure for properties on $\mathcal{N}$). *Given a regulatory network $\mathcal{N}$ and the language $\mathcal{L}_V$ of* CTL *formulas over variables of $\mathcal{N}$, $check_{\mathcal{N}}$ is a* decision procedure *for formulas in $\mathcal{L}_V$:*

$$\begin{aligned} check_{\mathcal{N}} : \mathcal{PN} \times \mathcal{L}_v &\to Bool \\ (P, \varphi) &\mapsto true/false \end{aligned}$$

such that $check_{\mathcal{N}}(P, \varphi) = true$ iff φ is true on $ASTG_P$.

Definition 10 (Model of $\varphi \in \mathcal{L}_V$). *$P \in \mathcal{PN}$ is a* model of φ *in the regulatory network $\mathcal{N}$ iff $check_{\mathcal{N}}(P, \varphi) = true$. The set of models of φ is denoted $\mathcal{M}(\varphi) = \{P \in \mathcal{PN} \mid check_{\mathcal{N}}(P, \varphi) = true\}$.*

2.3 Environmental Regulatory Networks

We are now ready to introduce *environments*, the new concept proposed in this article. In an influence graph with *environments*, some variables can be used to set the system into a specific configuration.

Definition 11 (Influence graph with environment variables). *An* influence graph with environment variables *$IG_{EV} = (V, EV, M, A)$ is an influence graph s.t.:*

- *(V, M, A) is an influence graph,*
- *$EV \subsetneq V$ is an ordered set of environment variables,*
- *The environment variables have no predecessors in IG: $\forall v \in EV, d^-(v) = 0$,*
- *Environment variables appear in atoms of multiplex formulas.*

Calcium Environment Variable. In the grey frame of Fig. 1, the `Calcium` variable is an environment variable which can be set to 0 or 1. It does not have any associated parameter, but the presence of `Calcium` adds new parameters to `Operon` variable which is targeted by `Calcium` through the multiplex `ca` (in the grey dashed-line rectangle).

Definition 12 (Environments). *Given $IG_{EV} = (V, EV, M, A)$, an* environment *assigns a value to each environment variable of EV. It is a tuple of values $(e_1, \dots, e_p)$ s.t. for all k in $1 \dots p$, $e_k \in D_{v_k}$, where $p = |EV|$ is the number of environment variables. The set of environments is denoted E.*

Definition 13 (Environmental property). *An* environmental property *is a couple (φ, e) where φ is a* CTL *formula in $\mathcal{L}_{V \setminus EV}$, the language of* CTL *formulas over systemic variables in $V \setminus EV$ and $e \in E$ is an environment. φ expresses a biological temporal property which occurs under the environmental condition represented by e.*

Environmental Properties for Pseudomonas œriginosa. Because `Calcium` variable can be set to 0 or 1, there are two environments: e^0 when $Calcium = 0$ and e^1 when $Calcium = 1$. In e^1 the bacteria will become mucoid, expressed by the CTL formula:

$$\varphi^1_{bacteria} \equiv AG(AF(mucB = 1))$$

In e^0, the bacteria does not change its phenotype: a mucoid (resp. non mucoid) bacterium will remain mucoid (resp. non mucoid); this is the formula $\varphi^0_{bacteria}$ used to illustrate Definition 4. Thus, there are two environmental properties $(\varphi^0_{bacteria}, e^0)$ and $(\varphi^1_{bacteria}, e^1)$.

Our main objective is now to study a system for which different dynamic behaviours are known under different environment contexts. So, the temporal biological knowledge that we search exhaustively for all models is a list of environmental properties $\Psi = [(\varphi^i, e^i)]_{i=1}^n$. We present in the next sections two ways to compute these models. The first one is a global approach which uses a modelling artefact to simulate the different environments in a single network. The second approach begins by defining as many regulatory networks with environment as there are environments : In each network, the number of parameters is reduced. It then computes the global models by intersection of the sets of models found for each of the environments.

3 All Environments' Coexistence in Thomas' Framework

In order to handle the biological knowledge on the behaviours of the system in different environments, the first idea is to design a unique regulatory network that takes into account the different environments. In this section, we present the formalisation of such regulatory network.

3.1 Regulatory Network

Because each environment can make the system behave in a peculiar way, the *ASTG* under construction has to handle "copies" of the useful states, one copy per environment. An environment variable is then considered as an inner variable which can take as many different qualitative levels as imposed by the environment. In the influence graph with environments presented in the previous section, environment variables have no predecessors. Then a unique parameter is associated with them, and each environment variable is attracted towards the value to its unique parameter, whatever its initial value, leading to an implicit change of value. To guarantee the stability of environment variables, we complete the influence graph by adding auto-regulations on each environment variable.

Auto-regulations for Simulating the Stability of Environment Variables. Let us suppose that the environment variable $v \in EV$ takes its value in $[\![v_l, v_u]\!]$. Then, for each value $n \in [\![v_l + 1, v_u]\!]$, one adds a multiplex $\mathtt{A}_v^n$ whose formula is simply $v \geqslant n$. Thus when $v < n$ (resp. $v \geqslant n$), the multiplex $\mathtt{A}_v^n$ is not a resource (resp. is a resource) of v. In Fig. 1, multiplex $\mathtt{A}$ allows the variable *Calcium* to stay at level 0 (resp. 1) when initialised at 0 (resp. 1).

Multiplexes $\mathtt{A}_v^n$ play a particular role inside this formalisation, because they do not represent any particular aspect of the biological system: they do allow the *ASTG* to simulate the existence of different stable values for the environment variables. In that meaning, they are artefactual.

Parameters Controlling the Stability of Each Environment. For each environment variable $v \in EV$, the set of parameters controlling its dynamics is (see Definition 2)

$$\mathcal{K}(v) = \{K_{v,\omega} \mid \omega \text{ is a subset of the predecessors of } v\}.$$

Since v is an environment variable, the only predecessors are the $n_v = v_u - v_l$ artefactual multiplexes $\mathtt{A}_v^k$, $k \in [\![v_{l+1}, v_u]\!]$. Thus, there are 2^{n_v} parameters. Fortunately some of them are structurally inoperable (not at all useful for determining dynamics) and have not to be instantiated: Indeed when $v = k$ ($k \in [\![v_l, v_u]\!]$), the only resources of v are the multiplexes $\mathtt{A}_v^i$ with $i \leqslant k$. Thus parameters $K_{v,\omega}$ such that $\mathtt{A}_v^i \in \omega$ and $\mathtt{A}_v^j \notin \omega$ with $j < i$ are inoperable. All in all, when $v \in EV$ can take its value in $[\![v_l, v_u]\!]$, there exists exactly $v_u - v_l + 1$ operable parameters: $K_v, K_{v,\{A_v^{v_l+1}\}}, K_{v,\{A_v^{v_l+1}, A_v^{v_l+2}\}}, \ldots,$ $K_{v,\{A_v^{v_l+1}, A_v^{v_l+2} \ldots A_v^{v_u}\}}$.

Moreover the stability of environment variables are strong properties that impose the values of previously defined operable parameters. Indeed, when environment variable v is set to k ($k \in [\![v_l, v_u]\!]$), it cannot change, leading to deduce that the value of the effective parameter is k. Thus $K_v = v_l$, $K_{v,\{A_v^{v_l+1}\}} = v_l + 1$, $K_{v,\{A_v^{v_l+1}, A_v^{v_l+2}\}} = v_l + 2$, $\ldots$, $K_{v,\{A_v^{v_l+1}, A_v^{v_l+2} \ldots A_v^{v_u}\}} = v_u$.

This regulatory network, including auto-regulations of environment variables and values of associated parameters that guarantee their stability, is denoted $\mathcal{N}_{global}$ in the sequel.

3.2 Formula Summing Up all Behavioural Properties

Last but not the least, the different behavioural properties have to be expressed for this regulatory network containing all environments. We consider here that the biological knowledge has been summed up in a list of environmental properties $[(\varphi^i, e^i)]_{i=1}^n$. Obviously, the characterisation of the states corresponding to environment e^i is given by the formula: $\varepsilon^i \equiv \bigwedge_{v_k \in EV} (v_k = e_k^i)$ where e_k^i represents the value of environment variable v_k in the environment e^i. Naturally, the list of environmental properties can be transcribed in CTL formula:

$$\Phi_{global} \equiv \bigwedge_{i \in [\![1,n]\!]} \left(\varepsilon^i \Rightarrow \varphi^i\right)$$

where ε^i characterises initialisation of environment variables. The values of parameters of environment variables used to build $\mathcal{N}_{global}$ guarantee the stability of theses variables. Finally one can use the decision procedure of ***TotemBioNet*** using the modified influence graph and Formula Φ_{global}: it enumerates all possible parameter settings, and selects only those that are consistent with Φ_{global}.

3.3 Application to *Pseudomonas æriginosa*

We apply now this global approach for determining all parameter settings consistent with known behavioural properties of the *Pseudomonas æriginosa* system. The auto-regulation of *Calcium* variable (multiplex and parameters) governing the *Calcium* environment variable is represented outside the grey frame in Fig. 1. Variable *operon* can take 3 different levels ($operon \in [\![0, 2]\!]$) and has 3 predecessors (itself, *mucB* and *Calcium*) and thus 2^3 parameters. Variable *mucB* can take 2 different levels ($mucB \in [\![0, 1]\!]$) and has a unique predecessor (*operon*) and thus 2^1 parameters. Finally, the number of parameter settings to consider is: $3^{2^3} \times 2^{2^1} = 26,244$.

Finally, the formula Φ_{global} is defined from $\varphi_{bacteria}^0$ and $\varphi_{bacteria}^1$ of Subsect. 2.3:

$$\Phi_{global} \equiv ((Calcium = 0) \Rightarrow \varphi_{bacteria}^0) \quad \wedge \quad ((Calcium = 1) \Rightarrow \varphi_{bacteria}^1)$$

4 Divide with Environments, Combine with Intersection

This "divide and conquer" approach works environment by environment. For each environment, a smaller regulatory network is used, and only the environmental property associated with the environment is checked. Afterwards, the models of the global system are built by abstracting and then intersecting the models found in each specific environment.

4.1 Regulatory Networks with Environments

Setting a value for an environment variable reduces the state space to the hyperplane defined by this value. This has a major impact on the size of the search space: some parameters of the targets of the environment variables become inoperable, leading to a drastic reduction of the number of parameter settings to consider.

Definition 14 (State space for an environment). *Given an influence graph with environment variables* $IG_{EV} = (V, EV, M, A)$ *and an environment* $e \in E$, *the* state space of the system for e *is* $S_e = \prod_{v \notin EV} D_v \times \coprod_{v_k \in EV} \{e_k\}$.

Definition 15 (Operable parameters for an environment). *Given* IG_{EV}, *and an environment* $e \in E$, *a parameter* $K_{v,\omega}$ *is* operable *if there exists at least a state* $s \in S_e$ *where* $K_{v,\omega}$ *is applicable.*

Pseudomonas æriginosa's Operable Parameters. For e^0 environment, the operable parameters are the original parameters in dashed rectangles in the blue frame of Fig. 1. Since *Calcium* targets variable *Operon*, the parameters associated with *Operon* change for e^1 environment, they are `K_Operon:ca`, `K_Operon:alg:ca`, `K_Operon:free:ca`, `K_Operon:alg:free:ca` listed in the grey frame.

Definition 16 (Regulatory network with environment). *A* regulation network for environment $e \in E$ *is the couple* $\mathcal{N}_e = (IG_{EV}, \mathcal{K}_e)$ *where* $\mathcal{K}_e \subset \mathcal{K}$ *is the subset of operable parameters for e. A parameter setting* P_e *assigns to each* $K_v \in \mathcal{K}_e$ *a value in* D_v*. The set of all parameter settings is denoted* $\mathcal{PN}_e$

Definition 17 (*ASTG* for a parameter setting P_e in environment *e*). *Given* $\mathcal{N}_e = (IG_{EV}, \mathcal{K}_e)$ *a regulatory network for environment e and* $P_e \in \mathcal{PN}_e$ *a parameter setting for* $\mathcal{N}_e$, *the associated* $ASTG_{P_e}$ *is defined as follow:*

- **Vertices** *are states* $s \in S_e$,
- **Loops***: there is an arc from s to itself if* $P_e(K_{v_i}[s]) = s_i, \forall v_i \in V \setminus EV$.
- **Arcs***: there is an arc from* $s^p = (s_1^p, ..., s_n^p)$ *to* $s^q = (s_1^q, ..., s_n^q)$ *if there exists one and only one index i s.t.* $v_i \in V \setminus EV$, *and* $s_i^p \neq s_i^q$ *with either :* $s_i^q = s_i^p + 1$ *and* $P_e(K_{v_i}[s^p]) > s_i^p$ *or* $s_i^q = s_i^p - 1$ *and* $P_e(K_{v_i}[s^p]) < s_i^p$.

4.2 Formulas and Abstraction of Models

In this approach, for each environmental property (φ, e), we successively search the models of the formulas φ associated with e. From a regulatory network with environment $\mathcal{N}_e$ and a particular parameter setting $P_e \in \mathcal{PN}_e$, the associated transition graph is built and the decision procedure for formula φ is launched on this reduced graph.

Definition 18 (Decision procedure for properties on $\mathcal{N}_e$). *Given a regulatory network $\mathcal{N}_e$, a parameter setting for this network $P_e \in \mathcal{PN}_e$, the language $\mathcal{L}_{V\setminus EV}$ of* CTL *formulas over systemic variables in $V \setminus EV$, $check_{\mathcal{N}_e}$ is defined by:*

$$\begin{array}{rl} check_{\mathcal{N}_e} : \mathcal{PN}_e \times \mathcal{L}_{V\setminus EV} & \to Bool \\ (P_e, \varphi) & \mapsto true/false \end{array}$$

such that $check_{\mathcal{N}_e}(P_e, \varphi) = true$ iff φ is true on $ASTG_{P_e}$.

Definition 19 (Model of an environmental property). *A* model of an environmental property (φ, e) *is the set of parameter settings which validate φ in $\mathcal{N}_e$: $\mathcal{M}_e(\varphi) = \{P_e \in \mathcal{PN}_e \mid check_{\mathcal{N}_e}(P_e, \varphi) = true\}$.*

To be able to combine the models $\mathcal{M}_{e^i}(\varphi^i)$ for several e^i, which relate on different operable parameter sets, one needs to *abstract* the set of operable parameter settings to a common superset. Since $\mathcal{K}_{e^i} \subset \mathcal{K}$ for all e^i, each parameter setting P_{e^i} are abstracted by a subset of parameter settings in $\mathcal{PN}$.

Definition 20 (Abstraction of a parameter setting). *Let $\mathcal{PN}$ be the set of parameter settings for the regulation network with environment variables $\mathcal{N} = (IG_{EV}, \mathcal{K})$. Let $\mathcal{N}_e = (IG_{EV}, \mathcal{K}_e)$ the regulation network for the particular environment $e \in E$, and $\mathcal{PN}_e$ its set of parameter settings.*

An abstraction of a parameter setting $P_e \in \mathcal{PN}_e$ *to $\mathcal{PN}$ is the subset $\mathcal{AP}_e \subset \mathcal{PN}$ such that : $\forall P \in \mathcal{AP}_e, \forall K_{v,\omega} \in \mathcal{K}_e, P(K_{v,\omega}) = P_e(K_{v,\omega})$, and $\forall K_{v,\omega} \notin \mathcal{K}_e, P(K_{v,\omega}) \in D_v$.*

In other words, P_e is the *projection* of $\mathcal{AP}_e$ on parameters of $\mathcal{K}_e$.

Definition 21 (Abstraction of a model). *The* abstraction of a model *of an environmental property (φ, e) is the union of abstractions of the parameter settings in $\mathcal{M}_e(\varphi)$: $\mathcal{AM}_e(\varphi) = \bigcup_{P_e \in \mathcal{M}_e(\varphi)} \mathcal{AP}_e$.*

Given a list of environmental properties $[(\varphi^i, e^i)]_{i=1}^n$, the parameter settings satisfying all these properties is the intersection of the abstractions of the models of each (φ^i, e^i).

Definition 22 (Model of environmental properties). *Let $\Psi = [(\varphi^i, e^i)]_{i=1}^n$ a list of environmental properties. The* model of Ψ *is the set : $\mathcal{M}(\Psi) = \bigcap_{i=1}^{n} \mathcal{AM}_{e^i}(\varphi^i)$.*

Example of abstraction. Let $\mathcal{K} = \{K_1, K_2, K_3, K_4\}$ be the parameters of $\mathcal{N}$ with $D_1 = [\![0,1]\!]$, $D_2 = [\![0,2]\!]$, $D_3 = [\![0,3]\!]$ and $D_4 = [\![0,1]\!]$ the domains of their associated variables. Assume that $\mathcal{K}_{e^1} = \{K_1, K_4\}$ and $\mathcal{K}_{e^2} = \{K_1, K_2\}$ are the operable parameters for environments e^1 and e^2.

Let $P_{e^1} \in \mathcal{PN}_{e^1}$ be a parameter setting of $\mathcal{N}_{e^1}$ which assigns 0 to K_1 and 1 to K_4 (denoted $P_{e^1} = (0, -, -, 1)$). Let $P_{e^2} \in \mathcal{PN}_{e^2}, P_{e^2} = (0, 1, -, -)$ be a parameter setting of $\mathcal{N}_{e^2}$. Then $AP = (0, 1, 3, 1) \in \mathcal{PN}$ which assigns 0 to K_1, 1 to K_2, 3 to K_3 and 1 to K_4 belongs to $\mathcal{AP}_{e^1}$ and to $\mathcal{AP}_{e^2}$. $AP' = (0, 0, 2, 1)$ also belongs to $\mathcal{AP}_{e^1}$. Furthermore, if P_{e^1} is a model of φ^1 and P_{e^2} is a model of φ^2, then AP is a model of the list of environmental properties $[(\varphi^i, e^i)]_{i=1}^2$.

4.3 Application to *Pseudomonas œriginosa*

We illustrate here the environment by environment approach on *Pseudomonas œriginosa*. According to the values the environment variable *Calcium* can take, two environmental regulatory networks have to be constructed. Formula $\varphi^0_{bacteria}$ must be checked on the first one, and $\varphi^1_{bacteria}$ on the second one.

Some parameters become *inoperable* for some environments (see Def. 15). For example here, `Operon`, has three predecessors in the global approach. But in e^0, the multiplex `ca` is not a resource anymore, and all parameters containing `ca` are in that case inoperable. This divides by two ther numberof parameter of `Operon`.

In e^0, `Operon` has 2 predecessors (so has 2^2 parameters) and can take its value in $D_{Operon} = [\![0, 2]\!]$, thus it has 3^{2^2} parameter settings. `mucB` is not directly regulated by a environment variable, so the number of parameters which is 2^{2^1}, does not change compared to a not extended Thomas' Framework. Thus, for environment e^0, the formula $\varphi^0_{bacteria}$ must be checked on $3^{2^2} \times 2^{2^1} = 324$ parameter settings. Similarly, there is the same amount of parameter settings to handle for checking $\varphi^1_{bacteria}$ in e^1. Finally, there are $324 + 324 = 648$ parameter settings to consider.

5 Comparing the Two Approaches

5.1 Theoretical Point of View

The question which naturally arises, is to know if the two approaches compute the same models. The answer relies firstly on the link between the transition graph obtained for an environment and the subgraph of the global *ASTG* induced by an environment, and secondly on the link between the global formula and the environmental formulas.

Lemma 1 (Isomorphism between $ASTG_{P_e}$ and a subgraph of $ASTG_P$). *For each environment e, each parameter setting P_e of $\mathcal{PN}_e$ and each $P \in \mathcal{AP}_e$, there exists a canonical isomorphism between $ASTG_{P_e}$ and the subgraph of $ASTG_P \in \mathcal{N}_{global}$ reduced to S_e.*

Proof of the lemma is given in Appendices. Following notations of Subsect. 2.2 and Definition 22, we denote $\mathcal{M}(\Phi_{global})$ the set of models of Φ_{global} on the regulatory network completed with auto-regulations on environment variables and associated parameters.

Theorem 1 ($\mathcal{M}(\Psi) = \mathcal{M}(\Phi_{global})$). *Given a list $\Psi \equiv [(\varphi^i, c^i)]_{i=1}^n$ of environmental properties on an influence graph with environment variables $IG_{EV} = (V, EV, M, A)$, the set $\mathcal{M}(\Psi)$ of models of Ψ (computed environment by environment) is equal to the set of models of Φ_{global} on $\mathcal{N}_{global}$.*

Proof. (1) Let us consider a parameter setting P selected by the environment by environment approach. $P \in \bigcap_{i=1}^{n} \mathcal{AM}_{e^i}(\varphi^i)$ (Def. 22). For all $i \in [\![1, n]\!]$, φ^i is satisfied in all states of $ASTG_{P_{e^i}}$, then, by Lemma 1, φ^i is satisfied in states corresponding to e^i in $ASTG_P$. The formula $(\varepsilon^i \wedge AG(\varepsilon^i)) \Rightarrow \varphi^i$ is then satisfied in all states of $ASTG_P$. We conclude that Φ_{global} is satisfied ($\mathcal{M}(\Psi) \subset \mathcal{M}(\Phi_{global})$).

(2) Conversely, let us consider now a parameter setting P selected by the classical approach: $P \in \mathcal{M}(\Phi_{global})$. Since $\Phi_{global} \equiv \bigwedge_{i=1}^{n} ((\varepsilon^i \wedge AG(\varepsilon^i)) \Rightarrow \varphi^i)$, for all $i \in$

$[\![1,n]\!]$, φ^i is satisfied in all states defining the environment e^i. Thus, by Lemma 1, $P \in \bigcap_{i=1}^{n} \mathcal{AM}_{e^i}(\varphi^i)$, in other words, $\mathcal{M}(\Phi_{global}) \subset \mathcal{M}(\Psi)$. □

Let us just remark that this proof does not suppose that all e^i are different in the list Ψ. Then, if two behaviours (φ^1 and φ^2) are known in a common environment e, one can represent this information either by a unique environmental property $(\varphi^1 \wedge \varphi^2, e)$ or by the list of environmental properties $[(\varphi^1, e), (\varphi^2, e)]$.

5.2 Practical Results

In this subsection, we illustrate the benefit of the environment by environment approach in terms of efficiency. We first give a short description of the tools used to search the models and to compute the intersection.

TotemBioNet**: A Tool to Compute the Models of a Formula.* The computation of the models of a formula is implemented in our tool ***TotemBioNet [4,14], which is dedicated to the identification of parameters in R. Thomas' Modelling framework. Its inputs are an influence graph with environment variables, an environment e, the values of known parameters, and formalised behaviours, expressed either as Hoare triples for trace properties or as ***CTL*** formulas for temporal properties. ***TotemBioNet*** enumerates the parameter settings and for each $P_e \in \mathcal{PN}_e$ builds the $ASTG_{P_e}$ associated with P_e, and calls the model-checker $NuSMV$ [6] as a decision procedure for a ***CTL*** property φ. The final ***TotemBioNet*** output is therefore $\mathcal{M}_e(\varphi)$. Successive environments are treated by calling ***TotemBioNet*** as many times as the number of environments, after having set the environment variables to their specific values. The outputs are written in .csv files which are the inputs of the intersection module.

MDDs to Compute the Intersection of Models. We choose to use *Multi Valued Decision Diagrams* (MDDs) to compute the intersection of sets of models from different environments. MDDs provide a compact representation for discrete data sets, with efficient set operations: intersection, union and set complement. Furthermore, it is easy (and efficient) to express that inside a common set of variables, certain variables can take any value. In this way, it is very convenient to abstract and combine sets of parameter settings which do not relate to the same variables because of inoperable parameters. ***TotemBioNet*** calls the *Colomoto mddlib* library[1] developed by A. Naldi to compute the intersection of the models obtained in successive environments. This library was designed for modelling biological systems, and notably to find stable states and analyse circuits [17].

Pseudomonas œriginosa Execution Time. The 26,244 parameter settings for the global approach (see Sect. 3.3) are enumerated and checked against the global formula Φ_{global} in 147.238 seconds on a personal computer.[2]

For the environment by environment approach, consistent parameter settings are computed by ***TotemBioNet*** in about 700 ms for each environment. Intersection of sets of models (using MDD to find the final result) needs 9.61 ms and thus the total time to

[1] https://github.com/colomoto/mddlib.

[2] All the given execution times are means over 20 ***TotemBioNet*** runs on an Intel® Core™ i7-7700 CPU/3.60 GHz × 8, RAM: 32 Go, under Linux. Interested readers can get the input files and results for the examples presented in this paper: https://gitlab.com/totembionet/totembionet/-/tree/master/examples/CMSB2021.

compute the models compatible for the two environments is 1.41 s, which is 105 times faster.

This example is a very small regulation network which involves only few variables. It already shows that the second approach is less time consuming. With a bigger regulatory network involving many environment variables, the relevance of the second approach is even more visible. The next section chooses a cell metabolism regulatory network as case study to show the scaling of the approach.

6 Case Study: Cell Metabolism

Cellular metabolism is a set of chemical reactions that occur in living cells. It involves intertwined biochemical reaction series, better known as *metabolic pathways*, which are thinly regulated. These processes allow cells to grow, multiply and maintain their structures [3]. Starting from classical biological knowledge of an healthy cell, we already proposed a qualitative regulatory network of the metabolism regulation [8] based on a precursor model [13]. Blue frame in Fig. 2 in Appendices represents the initial influence graph when the cell is in a healthy context. Unfortunately, when the context changes (nutrient lacks for example) some of these regulations are affected.

6.1 Metabolism Regulations According to Environments

Dependence of the metabolism on nutrient availability has been largely studied, and we decided to incorporate these dependences at a coarse grained level by adding four kinds of nutrients: sugar, amino acids, lipids and also oxygen. The level of their availability affects the regulation: For example, with oxygen supply the cell uses the oxidative phosphorylation pathway while without oxygen supply it uses fermentative pathways [15]. Grey frame in Fig. 2 incorporates environment variables representing nutrient supplies (GLC $\in [\![0,2]\!]$ for sugar, AA $\in [\![0,2]\!]$ for Amino Acids, $exO_2 \in [\![0,1]\!]$ for oxygen and FA $\in [\![0,1]\!]$ for Fatty Acids) and Table 1 (Appendices) lists all operable parameters of this regulation network. The parameters of variables which do not depend on environment variables have been all determined from biological knowledge, only parameters concerning NCD, O2 and GLYC remain to be identified (see below).

Globally, one has to consider $3 \times 3 \times 2 \times 2 = 36$ different environments, and for each environment, some properties are associated. The list $\Psi = [(\varphi^i, e^i)]_{i=1}^{36}$ of environmental properties is represented in Tables 2 and 3 (Appendices).

6.2 All Environments Coexistence in Thomas' Framework

In this approach, the time necessary for extracting all the models of Φ_{global} essentially depends on the number of parameter settings which *exponentially* depends on the number of unknown parameters. Thus, let us first count the number of unidentified parameters.

Taking into account FA does not add new multiplex (FA participates to an already present multiplex). So, it does not increase the number of parameters but the three others do. GLC acts on GLYC through two distinct multiplexes: glc1 and gl2 which respectively specify the external sugar at level 1 and 2. Thus, with the 2 previously existing predecessors, the GLYC in-degree becomes 4 , leading to a number of parameters for GLYC equal to $2^4 = 16$ (instead of $2^2 = 4$). Among them, 4 parameters are structurally inoperable because when the formula of `glc2` is true, the formula of `glc1` is also true. Consequently, a parameter $K_{GLYC,\omega}$ where $\texttt{glc2} \in \omega$ and $\texttt{glc1} \notin \omega$ is not operable. Finally, `GLYC` has 12 parameters.

The target of `AA`, called `NCD`, gains also 2 predecessors (because `AA` has three levels) and has now 3 predecessors: $|\mathcal{K}(NCD)| = 2^3 = 8$. Two parameters are also structurally inoperable for `NCD`, because `AA` cannot reach level 2 without passing level 1, decreasing the number of parameters to 6. Finally, the target of exO_2, that is O_2, has now 2 predecessors leading to $|\mathcal{K}(O_2)| = 2^2 = 4$ parameters.

The number of parameter settings is equal to the product of the numbers of values that each of the parameters can take: the 12 (resp. 6, 4) parameters associated to `GLYC` (resp. `NCD`, O_2) can take 3 (resp. 3, 2) different values ($D_{\texttt{GLYC}} = D_{\texttt{NCD}} = [\![0, 2]\!]$ and $D_{O_2} = [\![0, 1]\!]$). This gives rise to a number of parameter settings equal to: $3^{12} \times 3^6 \times 2^4 = 6,198,727,824$.

Knowing that ***TotemBioNet*** performs 4.2 decisions $check_{\mathcal{N}}(P, \Phi_{global})$ per second (for this regulatory network $\mathcal{N}$, the formula Φ_{global} and any parameter setting $P \in \mathcal{NP}$), enumeration of all parameter settings would take approximatively 49, 1 years.[3]

6.3 Divide with Environments, Combine with Intersection

The second option treats in an independent way each of the 36 environments on exponentially smaller search spaces, drastically decreasing the number of calls to the decision procedure.

Indeed, for each environment where exO_2= 0 (resp. = 1), the number of operable parameters associated with O_2 in only two: $K_{O_2,\emptyset}$, $K_{O_2,\{\texttt{PHOX}\}}$ (resp. $K_{O_2,\{\texttt{exO}_2\}}$, $K_{O_2,\{\texttt{exO}_2,\texttt{PHOX}\}}$). It is the same for the other targets of environment variables. For each environment where `AA`= 0 (resp. = 1, = 2), the number of operable parameters associated with `NCD` in only two: $K_{\texttt{NCD},\emptyset}$ and $K_{\texttt{NCD},\{\texttt{KREBS}\}}$ (resp. $K_{\texttt{NCD},\{\texttt{AA1}\}}$ and $K_{\texttt{NCD},\{\texttt{AA1},\texttt{KREBS}\}}$, $K_{\texttt{NCD},\{\texttt{AA1},\texttt{AA2}\}}$ and $K_{\texttt{NCD},\{\texttt{AA1},\texttt{AA2},\texttt{KREBS}\}}$). In a similar manner, for each environment where `GLC`= 0 (resp. = 1, = 2), the number of operable parameters associated with `GLYC` in only 4. All in all, in each environment, ***TotemBioNet*** has to consider is $3^4 \times 3^2 \times 2^2 = 2916$ parameter settings.

For each environment, the 2916 decisions are computed by ***TotemBioNet*** in approximately 74,365 s. For the 36 environments, ***TotemBioNet*** needs 2677.142 s. Adding 789 ms for the computation of the intersection between all sets of models, the total execution time for extracting the exhaustive set of models is $(2677.142 + 0.789)/60 = 44.6$ min.

This proves the usefulness of the second approach when modelling larger influence graphs for which the first approach is unable to compute the models in an acceptable time. This second option is 579,103 times faster.

7 Conclusion

Our "divide and conquer" approach allows to reduce the time necessary for searching all models of a list of environmental properties in an unthinkable way. In fact, for a given influence graph, the global execution time quasi-linearly depends on the number of parameter settings which *exponentially* depends on the number of unknown parameters. The *environment by environment* approach seeks to reduce as much as possible

[3] $49.1 = 6,198,727,824/(4.2 \times 3600 \times 24 \times 365.25)$, where 365.25 is for leap years.

the number of parameters to be identified by taking advantage of the fact that each environment (and by the way each environmental property) does not involve the whole set of parameters. This allows us to process large examples that were not yet accessible.

To process even more complex networks, it becomes manifest to parallelise the whole process: A coarse-grained parallelization is very easy because the searches of the models for each environment are completely independent.

From a modelling point of view, the environment variables are very powerful, because they constitute a good tool for exploring hypotheses. In particular, the consequences of Knock-Outs can be studied *via* such variables: If a Knock-Out leads to stopping a metabolic pathway, one can add an environment variable regulating this metabolic pathway, and impose *via* parameters of its target the decrease in the activity of its target.

In a longer term perspective, these environment variables constitute a first step towards a coupling of several sub-systems: Before embarking on the coupling, we can consider each of the studied subsystems with environment variables which control them differently in different part of the global phase space.

Appendices

Proof of the lemma. Each state of $ASTG_{P_e}$ is trivially unequivocally associated to a state of $ASTG_P$ (see Definitions 6 and 14). Let us show that transitions are the same. Let us consider a common state s. For determining the applicable parameters at s in $ASTG_{P_e}$, one has to evaluate the formulas of multiplexes controlling each non environment variable. Atoms concerning either environment or non environment variables are evaluated in the same way in $ASTG_{P_e}$ and $ASTG_P$ (the tuple representing s in $ASTG_{P_e}$ equals the one representing s in $ASTG_P$). Thus applicable parameters of non-environment variables at s are the same, leading to the same transitions that do not change the environment variables.

Moreover, because of our choice of parameter values for controling the evolution of environment variables, there does not exist any transition in $ASTG_P$ that change the values of environment variables.

Thus, when all environment variables are fixed, the subgraph of $ASTG_P$ reduced to the states corresponding to environment e, and $ASTG_{P_e}$ are isomorphic. □

Table 1. Operable Parameters associated with the metabolism regulation Influence Graph. The majority of parameters have been identified from biological knowledge, only parameters concerning NCD, O_2 and GLYC remain to be identified.

 # Parameters for ATP
$K_{\text{ATP}} = 0$
$K_{\text{ATP, LBP}} = 0$
$K_{\text{ATP, nLBP}} = 0$
$K_{\text{ATP, LBP nLBP}} = 0$
$K_{\text{ATP, PHOX}} = 1$
$K_{\text{ATP, nLBP PHOX}} = 1$
$K_{\text{ATP, LBP nLBP PHOX}} = 2$
$K_{\text{ATP, GLYC1}} = 0$
$K_{\text{ATP, GLYC1 nLBP}} = 0$
$K_{\text{ATP, GLYC1 LBP nLBP}} = 2$
$K_{\text{ATP, GLYC1 GLYC2}} = 1$
$K_{\text{ATP, GLYC1 GLYC2 nLBP}} = 1$
$K_{\text{ATP, GLYC1 GLYC2 LBP nLBP}} = 2$
$K_{\text{ATP, GLYC1 PHOX}} = 1$
$K_{\text{ATP, GLYC1 nLBP PHOX}} = 1$
$K_{\text{ATP, GLYC1 LBP nLBP PHOX}} = 2$
$K_{\text{ATP, GLYC1 GLYC2 PHOX}} = 1$
$K_{\text{ATP, GLYC1 GLYC2 nLBP PHOX}} = 1$
$K_{\text{ATP, GLYC1 GLYC2 LBP nLBP PHOX}} = 2$

 # Parameters for LBP
$K_{\text{LBP}} = 0$
$K_{\text{LBP, LS}} = 1$
$K_{\text{LBP, BOX}} = 0$
$K_{\text{LBP, LS BOX}} = 1$

 # Parameters for NCD
K_{NCD}
$K_{\text{NCD, KREBS}}$
$K_{\text{NCD, AA1}}$
$K_{\text{NCD, AA1 KREBS}}$
$K_{\text{NCD, AA1 AA2}}$
$K_{\text{NCD, AA1 AA2 KREBS}}$

 # Parameters for O2
K_{O2}
$K_{\text{O2, PHOX}}$
$K_{\text{O2, exO2}}$
$K_{\text{O2, exO2 PHOX}}$

 # Parameters for GLYC
K_{GLYC}
$K_{\text{GLYC, GR}}$
$K_{\text{GLYC, GLC1}}$
$K_{\text{GLYC, GLC1 GR}}$
$K_{\text{GLYC, GLC1 GLC2}}$
$K_{\text{GLYC, GLC1 GLC2 GR}}$
$K_{\text{GLYC, COF}}$
$K_{\text{GLYC, COF GR}}$
$K_{\text{GLYC, COF GLC1}}$
$K_{\text{GLYC, COF GLC1 GR}}$
$K_{\text{GLYC, COF GLC1 GLC2}}$
$K_{\text{GLYC, COF GLC1 GLC2 GR}}$

 # Parameters for nLBP
$K_{\text{nLBP}} = 0$
$K_{\text{nLBP, PPP}} = 1$
$K_{\text{nLBP, AAS}} = 1$
$K_{\text{nLBP, AAS PPP}} = 1$

 # Parameters for KREBS
$K_{\text{KREBS}} = 0$
$K_{\text{KREBS, AnO}} = 1$
$K_{\text{KREBS, AnO aKG}} = 2$
$K_{\text{KREBS, BOX}} = 1$

 # Parameters for PHOX
$K_{\text{PHOX}} = 0$
$K_{\text{PHOX, PC}} = 1$

 # Parameters for NADH
$K_{\text{NADH}} = 0$
$K_{\text{NADH, FERM}} = 0$
$K_{\text{NADH, PHOX}} = 0$
$K_{\text{NADH, AAS}} = 0$
$K_{\text{NADH, FERM PHOX}} = 0$
$K_{\text{NADH, FERM AAS}} = 0$
$K_{\text{NADH, AAS PHOX}} = 0$
$K_{\text{NADH, FERM AAS PHOX}} = 0$
$K_{\text{NADH, FERM KREBS AAS}} = 0$
$K_{\text{NADH, GLYC AAS PHOX}} = 0$
$K_{\text{NADH, GLYC FERM AAS PHOX}} = 1$
$K_{\text{NADH, KREBS FERM AAS PHOX}} = 1$
$K_{\text{NADH, GLYC KREBS FERM AAS PHOX}} = 1$
$K_{\text{NADH, GLYC KREBS AAS}} = 0$
$K_{\text{NADH, FERM GLYC AAS}} = 0$
$K_{\text{NADH, GLYC KREBS}} = 0$
$K_{\text{NADH, FERM KREBS}} = 0$
$K_{\text{NADH, KREBS AAS}} = 0$
$K_{\text{NADH, GLYC AAS}} = 0$
$K_{\text{NADH, GLYC PHOX}} = 0$
$K_{\text{NADH, FERM GLYC}} = 0$
$K_{\text{NADH, KREBS}} = 0$
$K_{\text{NADH, GLYC}} = 0$
$K_{\text{NADH, FERM GLYC KREBS AAS}} = 1$
$K_{\text{NADH, KREBS PHOX}} = 1$
$K_{\text{NADH, KREBS FERM PHOX}} = 1$
$K_{\text{NADH, KREBS AAS PHOX}} = 1$
$K_{\text{NADH, GLYC KREBS PHOX}} = 1$
$K_{\text{NADH, FERM GLYC KREBS PHOX}} = 1$
$K_{\text{NADH, GLYC KREBS AAS PHOX}} = 1$

 # Parameters for FERM
$K_{\text{FERM}} = 0$
$K_{\text{FERM, EP}} = 1$

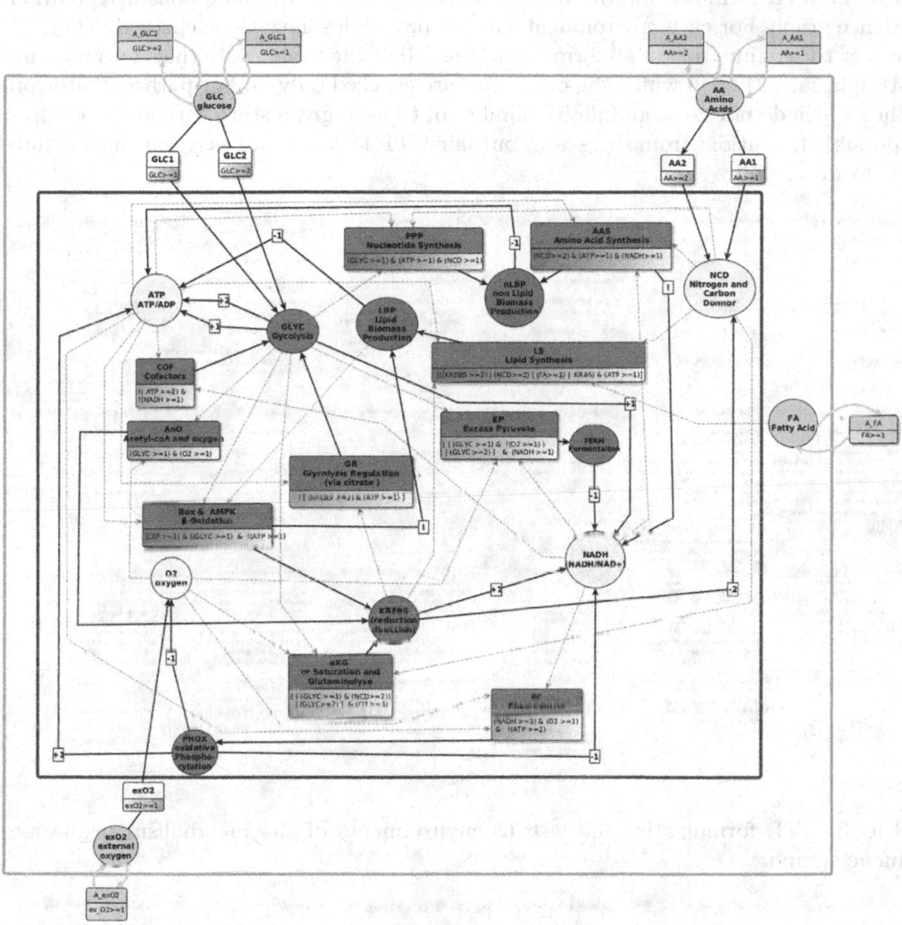

Fig. 2. Influence Graph with environment variables. Blue frame: initial influence graph. Grey frame: interaction graph with environment variables. Outside: artefactual regulations guaranteeing the stability of environment variables. Little squares with a number s inside are compact descriptions of multiplexes: a positive (resp. negative) number represents an activation (resp. inhibition) at level s. (Color figure online)

Table 2. CTL formulas for the first 18 environments of the metabolism regulation influence graph. For each environment, several properties have to be checked. Thus, φ^i is in fact the conjunction of all formulas in the cell of the table. Note that, formulas are written in fair-CTL (in which the properties are checked only on fair pathes, that is on pathes which do not cross an infinite number of times a given state without firetaking all possible transitions from this state) but fair-CTL formulas are easy to translate into CTL formulas [18].

#1 (FA=0, exO2=0, GLC=0, AA=0)	#2 (FA=0, exO2=0, GLC=0, AA=1)	#3 (FA=0, exO2=0, GLC=0, AA=2)	#4 (FA=0, exO2=0, GLC=1, AA=0)	#5 (FA=0, exO2=0, GLC=1, AA=1)	#6 (FA=0, exO2=0, GLC=1, AA=2)
f111=(AF(AG(ATP=0))) f121=(AF(AG(O2=0))) f131=(AF(AG(GLYC=0))) f141=(AF(AG(nLBP=0))) f151=(AF(AG(LBP=0))) f161=(AF(AG(FERM=0))) f171=(AF(AG(PHOX=0))) f181=(AF(AG(NADH=0)))	f211=(AF(AG(ATP=0))) f221=(AF(AG(O2=0))) f231=(AF(AG(GLYC=0))) f241=(AF(AG(nLBP=0) f251=(AF(AG(LBP=0))) f261=(AF(AG(FERM=0))) f271=(AF(AG(PHOX=0)))	f311=(AF(AG(ATP=0))) f321=(AF(AG(O2=0))) f331=(AF(AG(GLYC=0))) f341=(AF(AG(nLBP=0))) f351=(AF(AG(LBP=0))) f361=(AF(AG(FERM=0))) f371=(AF(AG(PHOX=0)))	f411=(!(AF(AG(ATP=0)))) f421=(AF(AG(O2=0))) f431=(AG(AF(GLYC=0)&AF(GLYC=1))) f461=(!(AF(AG(FERM=0)))) f471=(AF(AG(PHOX=0))) f481=(AG(AF(NADH>=0)&AF(NADH>=1))| AG(AF(NADH>=0)&AF(NADH>=2))| AG(AF(NADH>=2)&AF(NADH>=1)))	f511=(AG(AF(ATP>=0)&AF(ATP>=1))| AG(AF(ATP>=0)&AF(ATP>=2))| AG(AF(ATP>=2)&AF(ATP>=1))) f521=(AF(AG(O2=0))) f531=(AG(AF(GLYC=0)&AF(GLYC=1))) f541=(!(AF(AG(nLBP=0)))) f561=(!(AF(AG(FERM=0)))) f571=(AF(AG(PHOX=0))) f581=(AG(AF(NADH>=0)&AF(NADH>=1))| AG(AF(NADH>=0)&AF(NADH>=2))| AG(AF(NADH>=2)&AF(NADH>=1)))	f611=(AG(AF(ATP>=0)&AF(ATP>=1))| AG(AF(ATP>=0)&AF(ATP>=2))| AG(AF(ATP>=2)&AF(ATP>=1))) f621=(AF(AG(O2=0))) f631=(AG(AF(GLYC>=0)&AF(GLYC>=1))| AG(AF(GLYC>=0)&AF(GLYC>=2))| AG(AF(GLYC>=2)&AF(GLYC>=1))) f641=(!(AF(AG(nLBP=0)))) f651=(!(AF(AG(LBP=0)))) f661=(!(AF(AG(FERM=0)))) f671=(AF(AG(PHOX=0))) f681=(AG(AF(NADH>=0)&AF(NADH>=1))| AG(AF(NADH>=0)&AF(NADH>=2))| AG(AF(NADH>=2)&AF(NADH>=1)))
#7 (FA=0, exO2=0, GLC=2, AA=0)	**#8 (FA=0, exO2=0, GLC=2, AA=1)**	**#9 (FA=0, exO2=0, GLC=2, AA=2)**	**#10 (FA=0, exO2=1, GLC=0, AA=0)**	**#11 (FA=0, exO2=1, GLC=0, AA=1)**	**#12 (FA=0, exO2=1, GLC=0, AA=2)**
f711=(!(AF(AG(ATP=0)))) f721=(AF(AG(O2=0))) f731=(AG(AF(GLYC>=0)&AF(GLYC>=1))| AG(AF(GLYC>=0)&AF(GLYC>=2))| AG(AF(GLYC>=2)&AF(GLYC>=1))) f761=(!(AF(AG(FERM=0)))) f771=(AF(AG(PHOX=0))) f781=(AG(AF(NADH>=0)&AF(NADH>=1))| AG(AF(NADH>=0)&AF(NADH>=2))| AG(AF(NADH>=2)&AF(NADH>=1)))	f811=(AG(AF(ATP>=0)&AF(ATP>=1))| AG(AF(ATP>=0)&AF(ATP>=2))| AG(AF(ATP>=2)&AF(ATP>=1))) f821=(AF(AG(O2=0))) f831=(AG(AF(GLYC>=0)&AF(GLYC>=1))| AG(AF(GLYC>=0)&AF(GLYC>=2))| AG(AF(GLYC>=2)&AF(GLYC>=1))) f841=(!(AF(AG(nLBP=0)))) f861=(!(AF(AG(FERM=0)))) f871=(AF(AG(PHOX=0))) F981=(AG(AF(NADH=1)&AF(NADH=2))) f881=(AG(AF(NADH>=0)&AF(NADH>=1))| AG(AF(NADH>=0)&AF(NADH>=2))| AG(AF(NADH>=2)&AF(NADH>=1)))	f911=(AG(AF(ATP>=0)&AF(ATP>=1))| AG(AF(ATP>=0)&AF(ATP>=2))| AG(AF(ATP>=2)&AF(ATP>=1))) f921=(AF(AG(O2=0))) f931=(AG(AF(GLYC>=0)&AF(GLYC>=1))| AG(AF(GLYC>=0)&AF(GLYC>=2))| AG(AF(GLYC>=2)&AF(GLYC>=1))) f941=(!(AF(AG(nLBP=0)))) f951=(!(AF(AG(LBP=0)))) f961=(!(AF(AG(FERM=0)))) f971=(AF(AG(PHOX=0)))	f1011=(AF(AG(ATP=0))) f1021=(!(AF(AG(O2=0)))) f1031=(AF(AG(GLYC=0))) f1041=(AF(AG(nLBP=0))) f1051=(AF(AG(LBP=0))) f1071=(AF(AG(PHOX=0))) f1081=(AF(AG(NADH=0)))	f1111=(AF(AG(ATP=0))) f1121=(!(AF(AG(O2=0)))) f1141=(AF(AG(nLBP=0))) f1151=(AF(AG(LBP=0)))	f1211=(AF(AG(ATP=0))) f1221=(!(AF(AG(O2=0)))) f1241=(AF(AG(nLBP=0))) f1251=(AF(AG(LBP=0)))
#13 (FA=0, exO2=1, GLC=1, AA=0)	**#14 (FA=0, exO2=1, GLC=1, AA=1)**	**#15 (FA=0, exO2=1, GLC=1, AA=2)**	**#16 (FA=0, exO2=1, GLC=2, AA=0)**	**#17 (FA=0, exO2=1, GLC=2, AA=1)**	**#18 (FA=0, exO2=1, GLC=2, AA=2)**
f1311=(AG(AF(ATP>=0)&AF(ATP>=1))| AG(AF(ATP>=0)&AF(ATP>=2))| AG(AF(ATP>=2)&AF(ATP>=1))) f1321=(AG(AF(O2>=0)&AF(O2>=1))) f1331=(AG(AF(GLYC>=0)&AF(GLYC>=1))| AG(AF(GLYC>=0)&AF(GLYC>=2))| AG(AF(GLYC>=2)&AF(GLYC>=1))) f1361=(AF(AG(FERM=0))) f1371=(!(AF(AG(PHOX=0)))) f1381=(AG(AF(NADH>=0)&AF(NADH>=1))| AG(AF(NADH>=0)&AF(NADH>=2))| AG(AF(NADH>=2)&AF(NADH>=1)))	f1411=(AG(AF(ATP>=0)&AF(ATP>=1))| AG(AF(ATP>=0)&AF(ATP>=2))| AG(AF(ATP>=2)&AF(ATP>=1))) f1421=(AG(AF(O2>=0)&AF(O2>=1))) f1431=(AG(AF(GLYC>=0)&AF(GLYC>=1))| AG(AF(GLYC>=0)&AF(GLYC>=2))| AG(AF(GLYC>=2)&AF(GLYC>=1))) f1441=(!(AF(AG(nLBP=0)))) f1461=(AF(AG(FERM=0))) f1471=(AG(AF(PHOX>=0)&AF(PHOX>=1))) f1481=(AG(AF(NADH>=0)&AF(NADH>=1))| AG(AF(NADH>=0)&AF(NADH>=2))| AG(AF(NADH>=2)&AF(NADH>=1)))	f1511=(AG(AF(ATP>=0)&AF(ATP>=1))| AG(AF(ATP>=0)&AF(ATP>=2))| AG(AF(ATP>=2)&AF(ATP>=1))) f1521=(AG(AF(O2>=0)&AF(O2>=1))) f1531=(AG(AF(GLYC>=0)&AF(GLYC>=1))| AG(AF(GLYC>=0)&AF(GLYC>=2))| AG(AF(GLYC>=2)&AF(GLYC>=1))) f1541=(!(AF(AG(nLBP=0)))) f1551=(!(AF(AG(LBP=0)))) f1561=(AF(AG(FERM=0))) f1571=(AG(AF(PHOX>=0)&AF(PHOX>=1))) f1581=(AG(AF(NADH>=0)&AF(NADH>=1))| AG(AF(NADH>=0)&AF(NADH>=2))| AG(AF(NADH>=2)&AF(NADH>=1)))	f1611=(AG(AF(ATP>=0)&AF(ATP>=1))| AG(AF(ATP>=0)&AF(ATP>=2))| AG(AF(ATP>=2)&AF(ATP>=1))) f1621=(!(AF(AG(O2=0)))) f1631=(AG(AF(GLYC>=0)&AF(GLYC>=1))| AG(AF(GLYC>=0)&AF(GLYC>=2))| AG(AF(GLYC>=2)&AF(GLYC>=1))) f1641=(AF(AG(nLBP=0))) f1651=(!(AF(AG(LBP=0)))) f1661=(!(AF(AG(FERM=0)))) f1671=(!(AF(AG(PHOX=1)))) f1681=(AG(AF(NADH>=0)&AF(NADH>=1))| AG(AF(NADH>=0)&AF(NADH>=2))| AG(AF(NADH>=2)&AF(NADH>=1)))	f1711=(AG(AF(ATP>=0)&AF(ATP>=1))| AG(AF(ATP>=0)&AF(ATP>=2))| AG(AF(ATP>=2)&AF(ATP>=1))) f1721=(!(AF(AG(O2=0)))) f1731=(AG(AF(GLYC>=0)&AF(GLYC>=1))| AG(AF(GLYC>=0)&AF(GLYC>=2))| AG(AF(GLYC>=2)&AF(GLYC>=1))) f1741=(!(AF(AG(nLBP=0)))) f1751=(!(AF(AG(LBP=0)))) f1761=(!(AF(AG(FERM=0)))) f1771=(!(AF(AG(PHOX=1)))) f1781=(AG(AF(NADH>=0)&AF(NADH>=1))| AG(AF(NADH>=0)&AF(NADH>=2))| AG(AF(NADH>=2)&AF(NADH>=1)))	f1811=(AG(AF(ATP>=0)&AF(ATP>=1))| AG(AF(ATP>=0)&AF(ATP>=2))| AG(AF(ATP>=2)&AF(ATP>=1))) f1821=(!(AF(AG(O2=0)))) f1831=(AG(AF(GLYC>=0)&AF(GLYC>=1))| AG(AF(GLYC>=0)&AF(GLYC>=2))| AG(AF(GLYC>=2)&AF(GLYC>=1))) f1841=(!(AF(AG(nLBP=0)))) f1851=(!(AF(AG(LBP=0)))) f1861=(!(AF(AG(FERM=0)))) f1871=(!(AF(AG(PHOX=1)))) f1881=(AG(AF(NADH=1)&AF(NADH=2)))

Table 3. CTL formulas for the last 18 environments of the metabolism regulation influence graph.

#19 (FA=1, exO2=0, GLC=0, AA=0)	#20 (FA=1, exO2=0, GLC=0, AA=1)	#21 (FA=1, exO2=0, GLC=0, AA=2)	#22 (FA=1, exO2=0, GLC=1, AA=0)	#23 (FA=1, exO2=0, GLC=1, AA=1)	#24 (FA=1, exO2=0, GLC=1, AA=2)
f1911=(AF(AG(ATP=0))) f1921=(AF(AG(O2=0))) f1931=(AF(AG(GLYC=0))) f1941=(AF(AG(nLBP=0))) f1951=(AF(AG(LBP=0))) f1961=(AF(AG(FERM=0))) f1971=(AF(AG(PHOX=0))) f1981=(AF(AG(NADH=0)))	f2011=(AF(AG(ATP=0))) f2021=(AF(AG(O2=0))) f2031=(AF(AG(GLYC=0))) f2041=(AF(AG(nLBP=0))) f2051=(AF(AG(LBP=0))) f2061=(AF(AG(FERM=0))) f2071=(AF(AG(PHOX=0)))	f2111=(AF(AG(ATP=0))) f2121=(AF(AG(O2=0))) f2131=(AF(AG(GLYC=0))) f2141=(AF(AG(nLBP=0))) f2151=(AF(AG(LBP=0))) f2161=(AF(AG(FERM=0))) f2171=(AF(AG(PHOX=0)))	f2211=(!(AF(AG(ATP=0)))) f2221=(AF(AG(O2=0))) f2231=(AG(AF(GLYC=0)&AF(GLYC=1))) f2261=(!(AF(AG(FERM=0)))) f2271=(AF(AG(PHOX=0))) f2281=(AG(AF(NADH>=0)&AF(NADH>=1))| AG(AF(NADH>=0)&AF(NADH>=2))| AG(AF(NADH>=2)&AF(NADH>=1)))	f2311=(AG(AF(ATP>=0)&AF(ATP>=1))| AG(AF(ATP>=0)&AF(ATP>=2))| AG(AF(ATP>=2)&AF(ATP>=1))) f2321=(AF(AG(O2=0))) f2331=(AG(AF(GLYC=0)&AF(GLYC=1))) f2341=(!(AF(AG(nLBP=0)))) f2351=(!(AF(AG(LBP=0)))) f2361=(!(AF(AG(FERM=0)))) f2371=(AF(AG(PHOX=0))) f2381=(AG(AF(NADH>=0)&AF(NADH>=1))| AG(AF(NADH>=0)&AF(NADH>=2))| AG(AF(NADH>=2)&AF(NADH>=1)))	f2411=(AG(AF(ATP>=0)&AF(ATP>=1))| AG(AF(ATP>=0)&AF(ATP>=2))| AG(AF(ATP>=2)&AF(ATP>=1))) f2421=(AF(AG(O2=0))) f2431=(AG(AF(GLYC>=0)&AF(GLYC>=1))| AG(AF(GLYC>=0)&AF(GLYC>=2))| AG(AF(GLYC>=2)&AF(GLYC>=1))) f2441=(!(AF(AG(nLBP=0)))) f2451=(!(AF(AG(LBP=0)))) f2461=(!(AF(AG(FERM=0)))) f2471=(AF(AG(PHOX=0))) f2481=(AG(AF(NADH>=0)&AF(NADH>=1))| AG(AF(NADH>=0)&AF(NADH>=2))| AG(AF(NADH>=2)&AF(NADH>=1)))
#25 (FA=1, exO2=0, GLC=2, AA=0)	**#26 (FA=1, exO2=0, GLC=2, AA=1)**	**#27 (FA=1, exO2=0, GLC=2, AA=2)**	**#28 (FA=1, exO2=1, GLC=0, AA=0)**	**#29 (FA=1, exO2=1, GLC=0, AA=1)**	**#30 (FA=1, exO2=1, GLC=0, AA=2)**
f2511=(!(AF(AG(ATP=0)))) f2521=(AF(AG(O2=0))) f2531=(AG(AF(GLYC>=0)&AF(GLYC>=1))| AG(AF(GLYC>=0)&AF(GLYC>=2))| AG(AF(GLYC>=2)&AF(GLYC>=1))) f2561=(!(AF(AG(FERM=0)))) f2571=(AF(AG(PHOX=0))) f2581=(AG(AF(NADH>=0)&AF(NADH>=1))| AG(AF(NADH>=0)&AF(NADH>=2))| AG(AF(NADH>=2)&AF(NADH>=1)))	f2611=(AG(AF(ATP>=0)&AF(ATP>=1))| AG(AF(ATP>=0)&AF(ATP>=2))| AG(AF(ATP>=2)&AF(ATP>=1))) f2621=(AF(AG(O2=0))) f2631=(AG(AF(GLYC>=0)&AF(GLYC>=1))| AG(AF(GLYC>=0)&AF(GLYC>=2))| AG(AF(GLYC>=2)&AF(GLYC>=1))) f2641=(!(AF(AG(nLBP=0)))) f2651=(!(AF(AG(LBP=0)))) f2661=(!(AF(AG(FERM=0)))) f2671=(AF(AG(PHOX=0))) f2681=(AG(AF(NADH>=0)&AF(NADH>=1))| AG(AF(NADH>=0)&AF(NADH>=2))| AG(AF(NADH>=2)&AF(NADH>=1)))	f2711=(AG(AF(ATP>=0)&AF(ATP>=1))| AG(AF(ATP>=0)&AF(ATP>=2))| AG(AF(ATP>=2)&AF(ATP>=1))) f2721=(AF(AG(O2=0))) f2731=(AG(AF(GLYC>=0)&AF(GLYC>=1))| AG(AF(GLYC>=0)&AF(GLYC>=2))| AG(AF(GLYC>=2)&AF(GLYC>=1))) f2741=(!(AF(AG(nLBP=0)))) f2751=(!(AF(AG(LBP=0)))) f2761=(!(AF(AG(FERM=0)))) f2771=(AF(AG(PHOX=0))) f2781=(AG(AF(NADH=1)&AF(NADH=2)))	f2811=(AF(AG(ATP=0))) f2821=(!(AF(AG(O2=0)))) f2831=(AF(AG(GLYC=0))) f2841=(AF(AG(nLBP=0))) f2851=(AF(AG(LBP=0))) f2871=(AF(AG(PHOX=0))) f2881=(AF(AG(NADH=0)))	f2911=(AF(AG(ATP=0))) f2921=(!(AF(AG(O2=0)))) f2941=(AF(AG(nLBP=0))) f2951=(AF(AG(LBP=0)))	f3011=(AF(AG(ATP=0))) f3021=(!(AF(AG(O2=0)))) f3041=(AF(AG(nLBP=0))) f3051=(AF(AG(LBP=0)))
#31 (FA=1, exO2=1, GLC=1, AA=0)	**#32 (FA=1, exO2=1, GLC=1, AA=1)**	**#33 (FA=1, exO2=1, GLC=1, AA=2)**	**#34 (FA=1, exO2=1, GLC=2, AA=0)**	**#35 (FA=1, exO2=1, GLC=2, AA=1)**	**#36 (FA=1, exO2=1, GLC=2, AA=2)**
f3111=(AG(AF(ATP>=0)&AF(ATP>=1))| AG(AF(ATP>=0)&AF(ATP>=2))| AG(AF(ATP>=2)&AF(ATP>=1))) f3121=(AG(AF(O2>=0)&AF(O2>=1))) f3131=(AG(AF(GLYC>=0)&AF(GLYC>=1))| AG(AF(GLYC>=0)&AF(GLYC>=2))| AG(AF(GLYC>=2)&AF(GLYC>=1))) f3161=(AF(AG(FERM=0))) f3171=(!(AF(AG(PHOX=0)))) f3181=(AG(AF(NADH>=0)&AF(NADH>=1))| AG(AF(NADH>=0)&AF(NADH>=2))| AG(AF(NADH>=2)&AF(NADH>=1)))	f3211=(AG(AF(ATP>=0)&AF(ATP>=1))| AG(AF(ATP>=0)&AF(ATP>=2))| AG(AF(ATP>=2)&AF(ATP>=1))) f3221=(AG(AF(O2>=0)&AF(O2>=1))) f3231=(AG(AF(GLYC>=0)&AF(GLYC>=1))| AG(AF(GLYC>=0)&AF(GLYC>=2))| AG(AF(GLYC>=2)&AF(GLYC>=1))) f3241=(!(AF(AG(nLBP=0)))) f3251=(!(AF(AG(LBP=0)))) f3261=(AF(AG(FERM=0))) f3271=(AG(AF(PHOX>=0)&AF(PHOX>=1))) f3281=(AG(AF(NADH>=0)&AF(NADH>=1))| AG(AF(NADH>=0)&AF(NADH>=2))| AG(AF(NADH>=2)&AF(NADH>=1)))	f3311=(AG(AF(ATP>=0)&AF(ATP>=1))| AG(AF(ATP>=0)&AF(ATP>=2))| AG(AF(ATP>=2)&AF(ATP>=1))) f3321=(AG(AF(O2>=0)&AF(O2>=1))) f3331=(AG(AF(GLYC>=0)&AF(GLYC>=1))| AG(AF(GLYC>=0)&AF(GLYC>=2))| AG(AF(GLYC>=2)&AF(GLYC>=1))) f3341=(!(AF(AG(nLBP=0)))) f3351=(!(AF(AG(LBP=0)))) f3361=(AF(AG(FERM=0))) f3371=(AG(AF(PHOX>=0)&AF(PHOX>=1))) f3381=(AG(AF(NADH>=0)&AF(NADH>=1))| AG(AF(NADH>=0)&AF(NADH>=2))| AG(AF(NADH>=2)&AF(NADH>=1)))	f3411=(AG(AF(ATP>=0)&AF(ATP>=1))| AG(AF(ATP>=0)&AF(ATP>=2))| AG(AF(ATP>=2)&AF(ATP>=1))) f3421=(!(AF(AG(O2=0)))) f3431=(AG(AF(GLYC>=0)&AF(GLYC>=1))| AG(AF(GLYC>=0)&AF(GLYC>=2))| AG(AF(GLYC>=2)&AF(GLYC>=1))) f3441=(AF(AG(nLBP=0))) f3451=(!(AF(AG(LBP=0)))) f3461=(AG(AF(FERM>=0)&AF(FERM>=1))) f3471=(!(AF(AG(PHOX=1)))) f3481=(AG(AF(NADH>=0)&AF(NADH>=1))| AG(AF(NADH>=0)&AF(NADH>=2))| AG(AF(NADH>=2)&AF(NADH>=1)))	f3511=(AG(AF(ATP>=0)&AF(ATP>=1))| AG(AF(ATP>=0)&AF(ATP>=2))| AG(AF(ATP>=2)&AF(ATP>=1))) f3521=(!(AF(AG(O2=0)))) f3531=(AG(AF(GLYC>=0)&AF(GLYC>=1))| AG(AF(GLYC>=0)&AF(GLYC>=2))| AG(AF(GLYC>=2)&AF(GLYC>=1))) f3541=(!(AF(AG(nLBP=0)))) f3551=(!(AF(AG(LBP=0)))) f3561=(AG(AF(FERM>=0)&AF(FERM>=1))) f3571=(!(AF(AG(PHOX=1)))) f3581=(AG(AF(NADH>=0)&AF(NADH>=1))| AG(AF(NADH>=0)&AF(NADH>=2))| AG(AF(NADH>=2)&AF(NADH>=1)))	f3611=(AG(AF(ATP>=0)&AF(ATP>=1))| AG(AF(ATP>=0)&AF(ATP>=2))| AG(AF(ATP>=2)&AF(ATP>=1))) f3621=(!(AF(AG(O2=0)))) f3631=(AG(AF(GLYC>=0)&AF(GLYC>=1))| AG(AF(GLYC>=0)&AF(GLYC>=2))| AG(AF(GLYC>=2)&AF(GLYC>=1))) f3641=(!(AF(AG(nLBP=0)))) f3651=(!(AF(AG(LBP=0)))) f3661=(AG(AF(FERM>=0)&AF(FERM>=1))) f3671=(!(AF(AG(PHOX=1)))) f3681=(AG(AF(NADH=1)&AF(NADH=2)))

References

1. Bernot, G., Comet, J.P., Khalis, Z., Richard, A., Roux, O.F.: A genetically modified Hoare logic. Theoret. Comput. Sci. **765**, 145–157 (2019)
2. Bernot, G., Comet, J.P., Richard, A., Guespin, J.: Application of formal methods to biological regulatory networks: extending Thomas' asynchronous logical approach with temporal logic. J. Theor. Biol **229**(3), 339–347 (2004)
3. Blanco, A., Blanco, G.: Chapter 13 - metabolism. In: Blanco, A., Blanco, G. (eds.) Medical Biochemistry, pp. 275–281. Academic Press, January 2017. https://doi.org/10.1016/B978-0-12-803550-4.00013-6
4. Boyenval, D., Bernot, G., Collavizza, H., Comet, J.P.: What is a cell cycle checkpoint? the TotemBioNet answer. In: CMSB, pp. 362–372 (2020)
5. Chabrier, N., Fages, F.: Symbolic model checking of biochemical networks. In: Priami, C. (ed.) CMSB 2003. LNCS, vol. 2602, pp. 149–162. Springer, Heidelberg (2003). https://doi.org/10.1007/3-540-36481-1_13
6. Cimatti, A., et al.: NuSMV 2: an OpenSource tool for symbolic model checking. In: Brinksma, E., Larsen, K.G. (eds.) CAV 2002. LNCS, vol. 2404, pp. 359–364. Springer, Heidelberg (2002). https://doi.org/10.1007/3-540-45657-0_29
7. Clarke, E.M., Emerson, E.A.: Design and synthesis of synchronization skeletons using branching time temporal logic. In: Kozen, D. (ed.) Logic of Programs 1981. LNCS, vol. 131, pp. 52–71. Springer, Heidelberg (1982). https://doi.org/10.1007/BFb0025774
8. Gibart, L., Khoodeeram, R., Bernot, G., Comet, J.P., Trosset, J.Y.: Regulation of eukaryote metabolism: an abstract model. submitted (2021)
9. Huth, M., Ryan, M.: Logic in Computer Science: Modelling and reasoning about systems. Cambridge University Press (2000)
10. Kauffman, S.A.: Metabolic stability and epigenesis in randomly constructed genetic nets. J. Theor. Biol. **22**(3), 437–467 (1969)
11. Khalis, Z., Bernot, G., Comet, J.P.: Gene Regulatory Networks: Introduction of multiplexes into R. Thomas' modelling. In: Proceedings of the Nice Spring school on Modelling complex biological systems in the context of genomics, pp. 139–151. EDP Science, ISBN: 978-2-7598-0437-5 (2009)
12. Khalis, Z., Comet, J.P., Richard, A., Bernot, G.: The SMBioNet method for discovering models of gene regulatory networks. Genes, Genomes Genomics **3**(special issue 1), 15–22 (2009)
13. Khoodeeram, R., Bernot, G., Trosset, J.Y.: An Ockham Razor model of energy metabolism. In: Amar, P., Képès, F., Norris, V. (eds.) Proceedings of the Thematic Research School on Advances in Systems and Synthetic Biology, pp. 81–101. EDP Science (2017), ISBN: 978-2-7598-2116-7
14. Laetitia, G., Bernot, G., Collavizza, H., Comet, J.P.: TotemBioNet enrichment methodology: Application to the qualitative regulatory network of the cell metabolism. In: BIOINFORMATICS 2021 (2021)
15. Liberti, M.V., Locasale, J.W.: The warburg effect: how does it benefit cancer cells? Trends Biochem. Sci. **41**(3), 211–218 (2016). https://doi.org/10.1016/j.tibs.2015.12.001
16. Malhotra, S., Hayes, D., Wozniak, D.J.: Cystic fibrosis and pseudomonas aeruginosa: the host-microbe interface. Clin. Microbiol. Rev. **32**(3), June 2019. https://doi.org/10.1128/CMR.00138-18

17. Naldi, A., Thieffry, D., Chaouiya, C.: Decision diagrams for the representation and analysis of logical models of genetic networks. In: Calder, M., Gilmore, S. (eds.) CMSB 2007. LNCS, vol. 4695, pp. 233–247. Springer, Heidelberg (2007). https://doi.org/10.1007/978-3-540-75140-3_16
18. Richard, A.: Fair paths in CTL (2008), personnal communication. https://gitlab.com/totembionet/totembionet
19. Thomas, R.: Boolean formalization of genetic control circuits. J. Theor. Biol. **42**(3), 563–585 (1973)
20. Thomas, R.: Logical analysis of systems comprising feedback loops. J. Theor. Biol. **73**(4), 631–56 (1978)
21. Thomas, R., Gathoye, A., Lambert, L.: A complex control circuit. Regulation of immunity in temperate bacteriophages. Eur. J. Biochem. **71**(1), 211–227 (1976)

Automated Inference of Production Rules for Glycans

Ansuman Biswas[2], Ashutosh Gupta[1], Meghana Missula[1](✉), and Mukund Thattai[2]

[1] IITB, Mumbai, India
[2] NCBS, Bengaluru, India

Abstract. Glycans are tree-like polymers made up of sugar monomer building blocks. They are found on the surface of all living cells, and distinct glycan trees act as identity markers for distinct cell types. Proteins called GTase enzymes assemble glycans via the successive addition of monomer building blocks. The rules by which the enzymes operate are not fully understood. In this paper, we present the first SMT-solver-based iterative method that infers the assembly process of the glycans by analyzing the set of glycans from a cell. We have built a tool based on the method and applied it to infer rules based on published glycan data.

1 Introduction

The ability to control the assembly of small building blocks into large structures is of fundamental importance in biology and engineering. Familiar examples of this process from biology include the synthesis of linear DNA from nucleotide building blocks and the synthesis of linear proteins from amino-acid building blocks. In both these examples, the synthesis is templated: the new DNA or protein molecule is essentially copied from an existing molecule. However, most biological assembly proceeds without a template. For example, when an adult animal is grown from a fertilized egg, the genome within the egg contains a dynamical recipe encoding the animal rather than a template. The genome restricts and controls the set of events that can take place subsequent to fertilization.

While the process of animal development is too complex to study comprehensively, the same themes arise in the synthesis of complex tree-like sugar polymers known as glycans [1] that are covalently attached to proteins. Unlike linear proteins and DNA, glycans are tree-like structures: their nodes are sugar monomers, and their edges are covalent carbon-carbon bonds. The tree-like structure of a glycan is a direct consequence of the fact that a sugar monomer can directly bond to at least three neighboring sugar monomers (in contrast to nucleotides or amino acids, which can only bind to two neighbors and are constrained to make a chain).

A given cell produces a specific set of glycan molecules that are present in the cell. Since different cells produce different sets of molecules, the assembly process must be programmable: the assembly process includes a set of production

E. Cinquemani and L. Paulevé (Eds.): CMSB 2021, LNBI 12881, pp. 57–73, 2021.
https://doi.org/10.1007/978-3-030-85633-5_4

rules. The reactions that underlie glycan production are carried out by enzymes known as GTases [1]. There are hundreds of enzymes present in a given cell: each enzyme is a protein encoded by a distinct gene, which carries out a distinct biochemical reaction. The enzymes thus execute the production rules. A glycan tree is assembled piece by piece in successive steps. At each step, a production rule adds a small piece of a tree at the leaves or internal nodes of the current tree. Not all the rules are applicable at all the leaves. The monomer at a leaf and current surroundings of the leaf controls the applicability of a rule on the leaf. Identifying the exact set of the production rules by extensive biochemical experiments are costly and often needs an initial hypothesis for the rules to test. Biologists must identify the production rules and their control conditions by manually analyzing the set of observed glycans in a cell and using prior knowledge of biochemistry. We estimate that there may be more then 10^{70} possible rule sets if we consider all biological variations.[1] This is an error-prone process since the production rules must generate exactly the set of molecules in the cells and nothing else; and moreover, the data sets about which glycans are present in which cells are often incomplete. Manually comprehending all possible tree generation rules is difficult and ad-hoc. It would be useful to automate the process of learning which rules are operating in a given cell, based on incomplete data.

In this paper, we are presenting the *first* automated synthesis method for the production rules. Our method takes the observed glycan molecules in a cell as input and synthesizes the possible production rules that may explain the observation. To our knowledge, our work is the *first* to consider the computational problem. Our method of synthesis is similar to counterexample guided inductive synthesis(CEGIS) [2]. Several methods for solving problems of searching in a complex combinatorial space use templates to define and limit their search space, such as learning invariants of programs [3], and synthesizing missing components in programs [4,5]. We also use templates to model the production rules.

Our method is iterative. We first construct constraints encoding that a set of unknown rules defined by templates can assemble the input set of molecules. The generated constraints involve Boolean variables, finite range variables, and integer ordering constraints. We solve the constraints using an off-the-shelf SMT solver. We call the query to the solver *synthesis query*. If the constraints are unsatisfiable, there are no production rules with the search space defined by the templates. Otherwise, a solution of the constraints gives a set of rules.

However, there is also an additional requirement that a molecule that is not in the input set must not be producible by the synthesized rules. Therefore, the method looks for the producible molecules that are not in the input set. Again the search of the molecule is assisted by a template, which bounds the height of searched molecules. We generate another set of constraints using the templates for the unknown molecule. We again solve the constraints using an SMT solver. We call the query to the solver *counterexample query*. If there is no such molecule,

---

[1] For a problem having 10 monomers, 10 rules, 3 as rule size, 3 compartments and fast-slow reactions, the search space is $\approx 10^{74}$ rules $(2^{10} * \binom{(10+3-1)}{(3-1)} * 10^{(2^3-1)*10})$.

our method reports the synthesized rules. Otherwise, we have found a producible molecule that is not in the input set.

We append our synthesis constraints with additional constraints stating that no matter how we apply the synthesized rules, they will not produce the extra molecule. Since there is a requirement that *all* possible applications of rules must satisfy a condition, we have quantifiers in the constraints. We use a solver that handles quantifiers over finite range variable in the synthesis query. We go to the next iteration of the method. The method always terminates because the search space of rules is finite. The set of rules synthesized need not be minimal or unique. The solver reports the first set which satisfies the constraints. However, our method is adoptable. We can add various optimization criteria to find optimal rules for the given objectives, e.g., smallest rule sizes, number of rules, etc.

Our encoding to constraints depends on the model of execution of the rules. The current biological information is not sufficient to make a precise and definite model, and do the synthesis. We have also explored the variations of the models. For example, all rules may apply simultaneously. They apply in batches because they stay in different compartments. The molecule under assembly may flow through the compartments. The distribution of the stay of the molecules in a compartment also affects the execution model. Furthermore, we may have variations in the type and quality of data available to us. For example, we may have missed a produced molecule in experiments. We support the variations, which is presented in the extended version of this paper [6].

We have implemented the method in our tool GlySynth. We have applied the tool on data sets from published sources (available in the database UniCarbKB [7]). The output rules are within the expectations of biological intuition.

We organize the paper as follows. In Sect. 2, we introduce the biological background. In Sect. 3, we present a motivating example to illustrate our method. In Sect. 4, we present the formal model of the glycans and their production rules. In Sect. 5, we present our method for the synthesis problem. In Sect. 6 and 7, we present our experiments and conclude the paper.

2 Production of Glycans

DNA and RNA are made by copying template DNA, in a process called transcription carried out by an enzyme called RNA polymerase. Proteins are made by copying a template messenger RNA, in a process called translation carried out by a molecular machine called a ribosome [8]. In contrast, glycans are grown without a template, in a process called glycosylation. Glycosylation is carried out, not by a single enzyme, but by a large collection of so-called GTase enzymes that assemble one sugar monomer at a time into a final glycan tree. This process involves an ordered series of reactions, in which an enzyme first recruits the correct monomer, the enzyme-monomer complex binds to the target glycan at the appropriate motif, and finally a chemical reaction occurs which serves to bind the new monomer at the correct place on the glycan. The enzyme's binding motif can corresponding to a single monomer, or a large sub-structure of the entire glycan several nodes deep [9]. This process is reminiscent of a factory assembly line to make a car [10]. However, the assembly process operates without a blueprint: the final glycan structure is determined by the behavior of the enzymes themselves.

The process of glycosylation is stochastic, governed by the Poisson statistics of single-step chemical reactions. One result of this stochasticity is that the enzymes can operate in different time orders [11]. It is as if factory workers could operate in many different orders while building the car, first adding doors and later windows. Moreover, the enzymes are promiscuous: they can add new monomers to many different places on the growing tree. This is as if the factory workers could add headlights at many different points on the car. Since there is no template, the existing tree determines where new monomers are added. Given the stochastic and promiscuous nature of the GTase enzymes, it is not surprising that the final product is highly variable [12]. The same set of enzymes can build many different glycan trees.

This variability is evident in the glycans observed to be produced by living cells. In a typical experiment, a protein is purified from a cell and the glycans attached to it are separated and their structure is characterized. Such an experiment produces a spectrum of glycan trees termed the protein's glycan profile [12]. A single glycan profile typically contains ten to twenty trees in measurable abundance, each tree being a tree of depth two to ten bonds.

In [13], the authors had reported a method to infer the production rules when a single glycan is produced. However, the biologically interesting case is when the data set contains many glycan trees. This raises the following question: given a set of glycan trees produced by a cell, can we infer the set of enzymes that produce the glycans? This is the problem we tackle here.

In Fig. 1, we present details of glycan production. A glycan is a tree-like sugar tree (nodes linked by edges) attached to a substrate protein at the root (labeled 'R'). Distinct edge orientations correspond to covalent bonds of distinct carbons on the sugar monomer. Curved boxes represent reaction compartments within cells, which are the site of glycan production. Each step of glycan growth (black arrows) represents the addition of a single new monomer to a specific attachment point on the tree. Each such step is catalyzed by an enzyme, labeled E_i. At any stage of growth, the tree can exit the reaction compartment as an output. Alternatively, it can be passed to a subsequent reaction compartment for further growth driven by different enzymes. Note that the enzymatic rule is sensitive to the two monomers being linked by a bond, as well as any branches. For example, enzyme E_2 will add a Galactose to a GalNAc only if the GlcNAc branch is present; otherwise, the reaction will not proceed ('X'). The structures, reactions, and enzymes shown here are illustrative, and they do not correspond to any measured data set; see the following section for a real example. In biological experiments, the combined outputs of every compartment are measured; the underlying reactions must be inferred.

3 Motivating Example

In this section, we first present a motivating example to illustrate our method. In Fig. 2, we consider the glycan oligomers associated with human chorionic gonadotropin [14]. The data set has four glycan oligomers (shown in boxes and numbered). We assume that all these oligomers are built by starting from a root GalNAc (yellow square) by adding one monomer at a time (lines between glycans represent monomer addition reactions). At the top of the figure, we illustrate if all enzymes (rules) operate in a single compartment, a large number of glycans can potentially be made in addition to the measured ones. In the lower part of the figure, we illustrate if the enzymes are split into three compartments (separated by dotted lines), then certain reactions are prevented from occurring. Thus, reducing the set of structures. In this case, we assume that only the terminal (bottom-most) structures will be produced as outputs. Here we have assumed certain rules of operation that are most consistent with the observed glycan data set. The goal of this paper is to infer the rules.

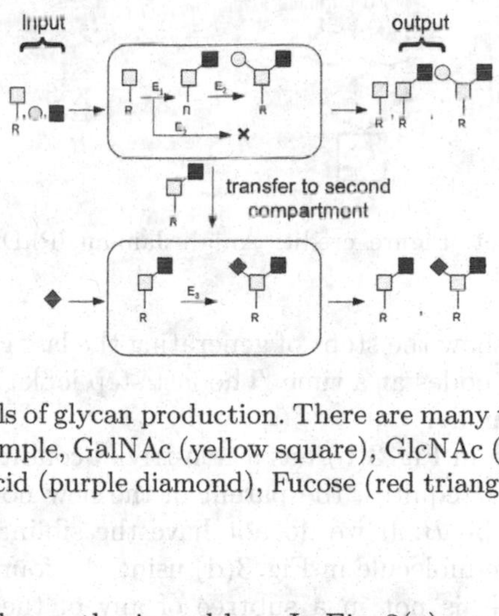

Fig. 1. Biological details of glycan production. There are many types of sugar monomer building blocks; for example, GalNAc (yellow square), GlcNAc (blue square), Galactose (yellow circle), Sialic Acid (purple diamond), Fucose (red triangle) and so on [1]. (Color figure online)

Now we consider the synthesis problem. In Fig. 3(a), we present a set of glycan molecules present in a cell consists of three molecules, which are structurally similar to the three glycan molecules in Fig. 2. To keep illustration simple, we have dropped the third glycan molecule from the earlier set. The molecules contain four types of monomers. As we are considering the abstract case, we have named them A, B, C, D. Each monomer is associated with an arity, i.e., the maximum number of potential children. The arities of the monomers are 2, 1, 1, and 0, respectively.

Let us first consider six rules in Fig. 3(b) that produce the molecules. All the rules are in the same compartment, i.e., they can be applied in arbitrary order. The rules have two kinds of nodes. If the circular nodes are present around a node, the rule is enabled and may append the molecule at the node with the square

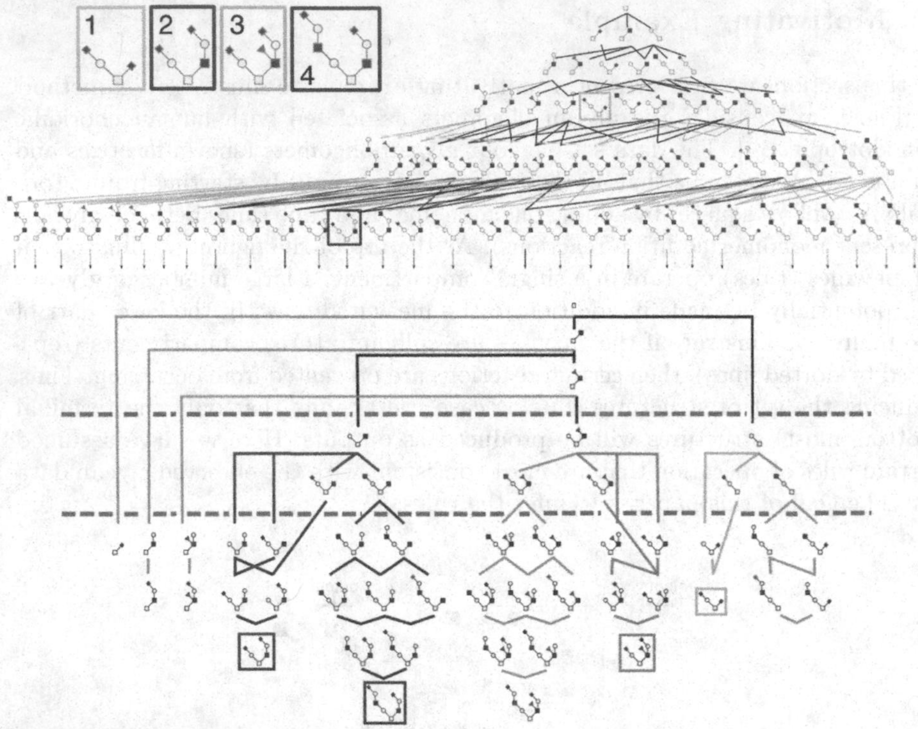

Fig. 2. A glycan data set. Figure credit: Anjali Jaiman, PhD thesis. (Color figure online)

nodes. In Fig. 3(c), we show the steps of generating the last glycan molecule. The first two steps add two nodes at a time. The last step looks at the two ancestors before adding a single node.

The second last rule in Fig. 3(b) has a non-trivial condition on the sibling of the anchor leaf node. It requires, the parent of the new node should be A and the right sibling must be B. If we do not have the sibling condition, we may be able to construct the molecule in Fig. 3(d) using the fourth and the modified fifth rule. The molecule is not in a subtree of any of the three input glycan molecules. Therefore, there are scenarios where rules must look into the context before applying themselves.

Our method for synthesis takes the three glycan molecules as input. It also needs the budget of resources to search for the production rules. If we allow an arbitrary number of rules, and the rules to look at their context up to an arbitrary depth, then there is a trivial solution. Therefore, our method limits the number and size of rules. For this illustration, we searched for the seven production rules with three as the limit on the rule heights. All rules are in a single compartment.

GLYSYNTH, the tool that implements the method, reported the synthesized rules from Fig. 3(b) in 0.85 s. In our tool, we first construct a synthesis query using the templates to encode that a set of rules produces the input molecules. We call a solver to solve the synthesis query. After the first query, we obtain the rules presented in Fig. 3(e). The rule set can produce molecules that are not in input. We need to iterate further. After 8 iterations, our tool synthesizes a set of rules that satisfies the requirements.

4 Modelling of the Synthesis Problem

In this section, we present the formal model for the synthesis problem. We model glycan molecules and production rules as labeled trees. The glycan molecules are assembled by applying the production rules repeatedly. Our synthesis problem reduces into finding the pieces of trees that represent the production rules.

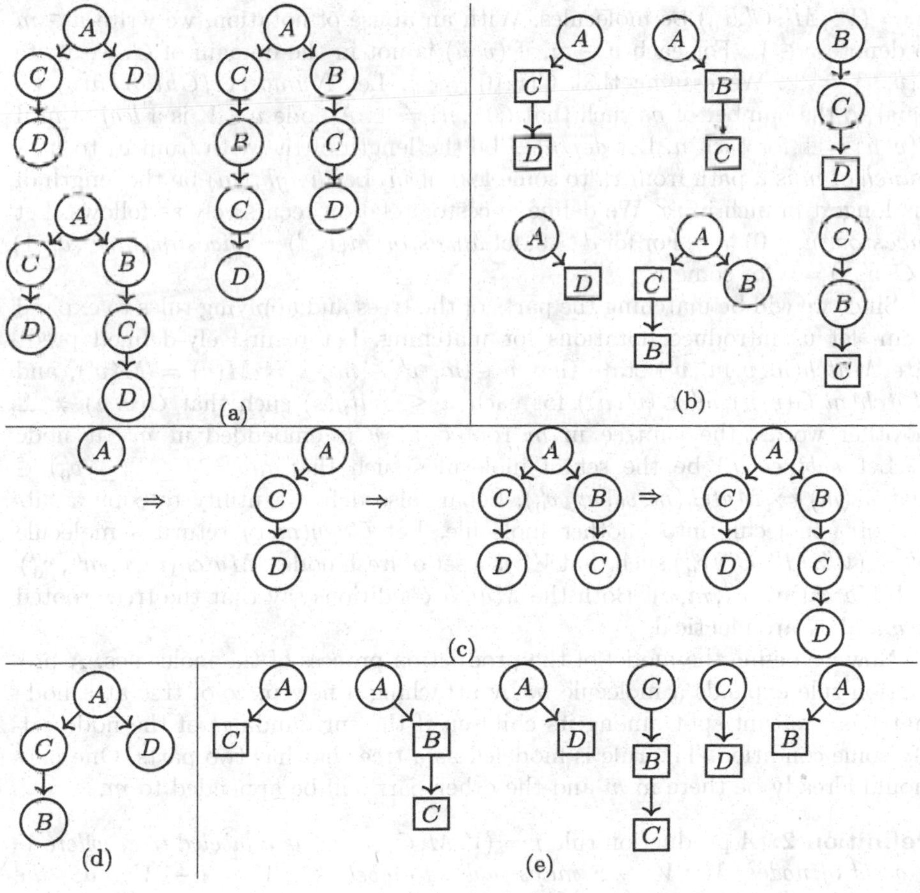

Fig. 3. (a) A schematic example of a data set that includes three glycan oligomers. (b) A set of production rules for the glycan molecules (c) The steps of producing the middle glycan molecule (d) An undesired molecule. (e) The synthesized rules at the first iteration.

Let S be the set of sugar monomers that builds glycans, the oligomer molecules. Each $s \in S$ is associated with arity m (written $arity(s) = m$). The children of the monomers are indexed. We refer to the kth child of s for some $k \leq arity(s)$. They correspond to bonds at specific positions in the monomers where children are connected. Now we define the glycan molecules as labeled trees. Now onward we refer to the glycans simply as molecules.

Definition 1. *A* molecule $m = (V, M, C, v_0)$ *is a labeled tree, where* V *is a set of nodes in the tree,* $M : V \rightarrow S$ *maps nodes to their label,* $C : V \times \mathbb{N} \hookrightarrow V$ *maps the indexed children of nodes, and* $v_0 \in V$ *is the root of* m. *A molecule must respect the arity of monomers, i.e., if* $M(v) = s$ *and* $C(v, n) = v'$ *then* $n \leq arity(s)$.

Let us define notations related to the tree structure. Let $m = (V, M, C, v_0)$ and $m' = (V', M', C', v'_0)$ be molecules. With an abuse of notation, we write $v \in m$ to denote $v \in V$. For each $v \in m$, if (v, n) is not in the domain of C, we write $C(v, n) = \perp$. We assume that $C(v, 0) = \perp$. Let $NumberOfChildren(v)$ be equal to the number of ns such that $C(v, n) \neq \perp$. A node $v \in V$ is a *leaf* of m if $C(v, n) = \perp$ for each n. Let $depth(v)$ be the length of the path from v_0 to v. A *branch* of m is a path from v_0 to some leaf of m. Let $height(m)$ be the length of the longest branch in m. We define ancestor relation recursively as follows. Let $ancestor(m, v, 0) = v$. For for $d > 0$, let $ancestor(m, v, d) = ancestor(m, v', d-1)$ if $C(v', i) = v$ for some i.

Since we will be matching the parts of the trees and applying rules to expand them, let us introduce notations for matching. Let recursively-defined predicate $Match(m, v, m', v')$ state that $v \in m$, $v' \in m'$, $s = M(v) = M(v')$, and $Match(m, C(v, n), m', C(v', n))$ for each $n \leq arity(s)$ such that $C(v, n) \neq \perp$. In other words, the subtree in m rooted at v is embedded in m' at node v'. Let $subtree(m)$ be the set of molecules such that $m' = (\_, \_, \_, v'_0) \in subtree(m) \Leftrightarrow Match(m, v_0, m', v'_0)$. Let us also define a utility to copy a subtree of a molecule into another molecule. Let $Copy(m, v)$ return a molecule $m'' = (V'', M'', C'', v''_0)$ such that V'' is a set of fresh nodes, $Match(m, v, m'', v''_0)$, and $Match(m'', v''_0, m, v)$. Both the $Match$ conditions say that the trees rooted at v and v''_0 are identical.

Now we define the model of the production process of the molecules. A production rule expands a molecule m by attaching a new piece of tree at a node that has a vacant spot among its children if the surroundings of the node satisfy some condition. The rule is modeled as a tree that has two parts. One part should already be there in m and the other part will be appended to m.

Definition 2. *A* production rule $r = (V, M, C, v_0, v_e)$ *is a labeled tree, where* V *is a set of nodes,* $M : V \rightarrow S$ *maps nodes to labels,* $C : V \times \mathbb{N} \hookrightarrow V$ *maps the indexed children of nodes,* $v_0 \in V$ *is the root, and* $v_e \in V$ *is the root of expanding part of the rule.*

If we *apply*, a rule r on a molecule m, then it is extended at some node $v \in m$. A copy of the descendants of v_e will be attached to v in m, and the rest of the nodes in the rule have to match v and above. We call the descendants of v_e as *expanding nodes* and all the other nodes as *matching nodes*.

Example: In Fig. 4(a), we present a rule. It has two kinds of nodes. The rule adds the square node (v_e). The circular nodes are the pattern, which must be present in the molecule to apply the rule. In Fig. 4(b), we present an application of the rule. The solid tree with three nodes is the initial molecule. The middle node A and its right child B form a pattern, where the rule is applicable. The rule adds a left child with the label A to the middle node. The rule is not applicable at the root A due to pattern mismatch.

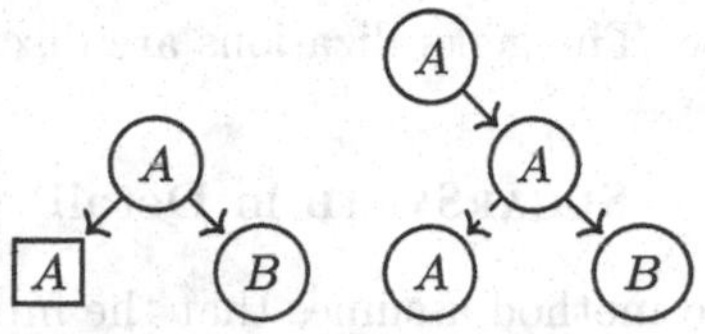

(a) (b)

Fig. 4. (a) A rule. (b) An application of the rule.

We naturally extend the definitions related to molecules, including $Match$ and *Copy*, to the production rules. Let us formally define the molecule productions using the rules. Let $m = (V, M, C, v_0)$ be a molecule and $r = (V_r, M_r, C_r, v_{0r}, v_e)$ be a production rule. Let d be such that $v_{0r} = ancestor(r, v_e, d)$, i.e., v_e is at the depth d in r. Let i be such that $C_r(v', i) = v_{0r}$ for some $v' \in r$. We apply r on m at node $v \in m$ such that $C(v, i) = \bot$. We obtain an expanded molecule as follows. Let $(V', M', C', v_0') = Copy(r, v_e)$. The expanded molecule is $m' = (V \uplus V', M \uplus M', C \uplus C' \uplus \{(v, i) \mapsto v_0'\}, v_0)$ if $Match(r, v_{0r}, m', ancestor(m', v_0', d))$ where $\uplus$ is the disjoint union. The match condition states that after attaching the new nodes V' the rule tree must be embedded in m' at the dth ancestor of v_0'. We write $m' = Apply(m, v, r)$ to indicate the application of r on molecule m at node v that results in m'. If r is not applicable at v, we write $Apply(m, v, r) = \bot$. We write $m' = Apply(m, r)$ if there is a $v \in m$ such that $m' = Apply(m, v, r)$.

Let R be a set of rules. A molecule m is *producible* by R from a set of molecules Q if there is sequence of molecules $m_0, ..., m_k$ such that $m_0 \in Q$, $m_k = m$, and for each $0 < i \leq k$, $m_i = Apply(m_{i-1}, r)$ for some $r \in R$. Let $P(Q, R)$ denote the set of molecules that are producible from rules R from a set of molecules Q. We have discussed in Sect. 2 that all the production rules are not applied at the same time. The rules may live in compartments and the rule sets of the compartments are applied one after another. To model compartments for the rules, let us suppose we have a sequence $R_1, ..., R_k$ of set of rules. Let $P(Q, R_1, .., R_k) = P(..P(P(Q, R_1), R_2), .., R_k)$ denoting the trees obtained after applying the rule sets one after another.

In nature, we observe a set of glycan molecules μ present in a cell. However, we may not know the production rules to produce the molecules. We will be developing a method to find the rules. The *synthesis problem* is to find a set R of production rules such that $\mu = P(S, R)$, where S is the set of monomers.

5 Method for the Synthesis Problem

In this section, we present a method to solve the synthesis problem of finding a set of production rules that produce a given set μ of molecules. Our method SUGARSYNTH is Algorithm 1. Here, we are considering only the single compartment case. The generalizations are discussed in the extended version [6].

5.1 SUGARSYNTH in Detail

The method assumes that the input set μ is finite. This is a reasonable assumption because even if a set of rules can produce an unbounded number of molecules, no biology will exhibit an infinite set in a cell. Our method bounds the search space of production rules. The method also takes two numbers as input: d is the maximum height of the learned rules and n is the maximum number of them. If the method fails to find production rules, the user may call the method with larger parameters. First, the method initializes S with the set of monomers occurring in μ and sets w to be the maximum arity of any monomer in S.

We use templates to model the search space of rules. A template is a tree that has a depth and the internal nodes of the tree have the same number of children. Two variables label each node. One variable is for choosing the sugar at the node

Algorithm 1. SUGARSYNTH(μ, d, n)

Input : μ : molecules, d : maximum rule depth, n : number of rules
Output: R: synthesized rules

S := the set of monomers appear in μ, w := the maximum arity of a monomer in S
T := MAKETEMPLATESRULE(S, d, w, n)
tCons := RULETEMPLATECORRECTNESS(T)
Let h is maximum of the heights of molecules in μ
$\hat{m}$:= MAKETEMPLATEMOL(S, h, w)
$mCons$:= MOLTEMPLATECORRECTNESS($\hat{m}$, μ)
$pCons$:= $\bigwedge_{m \in \mu}$ ENCODEPRODUCE(m,T)
$nCons$:= **tt**
while tt do ▷ while True
  if a = getModel($tCons \wedge pCons \wedge nCons$) **then**
    R := READRULES(a)
  else
    throw Failed to synthesize the rules!
  end if
  $rCons$:= ENCODEPRODUCE($\hat{m}$,Rs)
  if a = getModel($mCons \wedge rCons$) **then**
    m' := GETNEGMOL($\hat{m}$, a)
    $nCons$:∧= $\forall \tau, cuts. \neg$ENCODEPRODUCE(m', T)
  else
    return synthesized rules R
  end if
end while

and the other is for describing the 'situation' of the node. The domain of the first variables is $S \cup \{\bot\}$. Let $SVars$ be the unbounded set of variables with the domain. We will use the pool of $SVars$ to add variables to the templates.

A node in a production rule can be in four situations. In Fig. 5, we illustrate the situations. The first situation is when a node is in the expanding part, which are shown in dark gray. The second situation is when a node is not in the rule. The dashed area are the absent nodes. Among the matching nodes, we have two cases.

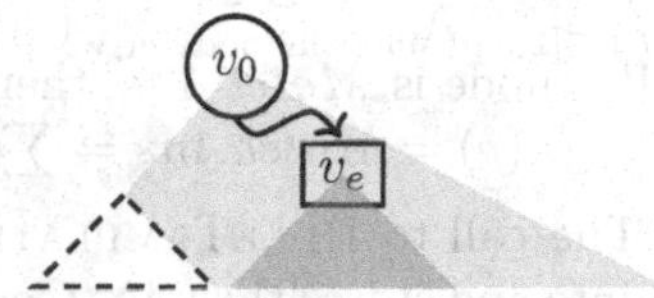

Fig. 5. Parts of production rules in the rule templates.

The third situation is when a node is in the matching part and has expanding descendants. The nodes on the solid path from v_0 to the root v_e of the expanding part are in the third situation. Finally, the fourth situation is the rest of the nodes in the matching part, which is in light gray. A variable is mapped to a node to encode the four situations. Let $K = \{Expand, Absent, MatchAns, Match\}$ be the set of symbols to indicate the situations. Let $KVars$ be an unbounded set of variables with domain K. Our templates are sufficiently expressive to cover all aspects of biology, which are defined as follows.

Definition 3. *For given integers d and w, a* rule template $t = (V, \nu, \kappa, C, v_{0r})$ *is a labeled full tree with depth d and each internal node has w children, where V is a set of nodes, $\nu : V \to SVars$ maps nodes to distinct sugar choice variables, $\kappa : V \to KVars$ maps nodes to distinct situation variables, $C : V \times \mathbb{N} \hookrightarrow V$ maps the indexed children of nodes, and $v_{0r} \in V$ is the root of the tree.*

For a node v in a template if we assign $\kappa(v) = Absent$, we call the node *absent*. Otherwise, we call the node *present*. We will also be searching for the molecules that may be produced by the learned rules. Therefore, we need to define the search space for the molecules. We use templates for defining the search space. We limit the template size using a parameter, namely the height of the template.

Definition 4. *For given integers h and w, a molecule template $\hat{m} = (V, \nu, C, v_{0m})$ is a labeled full tree with height h and each internal node has w children, where V is a set of nodes, $\nu : V \to SVars$ maps nodes to sugar choice variables, $C : V \times \mathbb{N} \hookrightarrow V$ maps the indexed children of nodes, and $v_{0m} \in V$ is the root.*

In the Algorithm at line 2, we call MakeTemplatesRule(S, d, w, n) to create n templates for height d and children width w. Since w is the maximum arity of any sugar, we can map any node to any sugar. Next at line 3, we will construct constraints that encode the set of valid rules. A valid assignment to the variables in a template $t = (V, \nu, \kappa, C, r)$ must satisfy the following six conditions.

1. If a node is present, then it is labeled with a sugar.
$\bigwedge_{s \in S} \bigwedge_{v \in V} (\kappa(v) \neq Absent \Rightarrow \nu(v) \neq \bot)$

2. The children that are at greater arity than that of the label are absent.
   $\bigwedge_{s\in S}\bigwedge_{i\in(arity(s),w]}\bigwedge_{v\in V}(\nu(v)=s\Rightarrow\kappa(C(v,i))=Absent)$
3. If a node is present, then the parent of the node is also present.
   $\bigwedge_{\text{internal node } v\in V}\bigwedge_{i\in[1,w]}(\kappa(C(v,i))\neq Absent\Rightarrow\kappa(v)\neq Absent)$
4. If a node is *Expand*, then its children are also *Expand* if present.
   $\bigwedge_{i\in[1,w]}\bigwedge_{\text{internal node } v\in V}(\kappa(v)=Expand\Rightarrow\kappa(C(v,i))\in\{Expand, Absent\})$
5. If a node is *Match*, then its children are also *Match* if present.
   $\bigwedge_{i\in[1,w]}\bigwedge_{\text{internal node } v\in V}(\kappa(v)=Match\Rightarrow\kappa(C(v,i))\in\{Match, Absent\})$
6. If a node is *MatchAns*, then exactly one child is *MatchAns* or *Expand*.
   $\bigwedge_{v\in V}(\kappa(v)=MatchAns\Rightarrow\sum_{i\in[1,w]}(\kappa(C(v,i))\in\{MatchAns, Expand\})=1$

The call to RULETEMPLATECORRECTNESS at line 3 creates the above constraints and stores them in *tCons*. At line 5, we use the call to MAKETEMPLATEMOL(S, h, w) to create a molecule template of height h and children width w. Our method searches for unwanted producible molecules up to the height of h, which we set to the maximum of the heights of molecules in μ. The choice of h is arbitrary. Similar to the rule templates, not all assignments to molecule template variables are valid. We add the following conditions for valid assignments for molecule template $\hat{m}=(V,\nu,C,\hat{v_0})$.

1. If a node is present, then the parent of the nodes is also present.
   $\bigwedge_{i\in[1,w]}\bigwedge_{\text{internal node } v\in V}(\nu(C(v,i))\neq\bot\Rightarrow\nu(v)\neq\bot)$
2. The children count of a node matches with the arity of the labeled sugar.
   $\bigwedge_{s\in S}\bigwedge_{i\in(arity(s),w]}\bigwedge_{v\in V}(\nu(v)=s\Rightarrow\kappa(C(v,i))=Absent)$
3. We find a molecule that is not in μ. We encode $\bigwedge_{(\_,\_,\_,v_0)\in\mu}Neq(\hat{v_0},v_0)$, where predicate Neq be recursively defined as follows.
   $Neq(v,v') := \nu(v)\neq M(v')\vee\bigvee_{i\in[1,w]}Neq(C(v,i),C(v',i))$
   $Neq(v,\bot) := \nu(v)\neq\bot$

In our method, the call to MOLTEMPLATECORRECTNESS at line 6 generates the above constraints and stores them in *mCons*.

We need to encode that the rules do generate the molecules in μ and do not generate any other. Using procedure ENCODEPRODUCE, we generate the corresponding constraints. We will discuss the procedure in-depth shortly. Let us continue with SUGARSYNTH. At line 7, we call ENCODEPRODUCE for each molecule in μ and generate constraints *pCons* stating that the solutions of the templates will produce the molecules in μ.

After producing the constraints *tCons*, *mCons*, and *pCons*, the method enters in the loop. It first checks the satisfiability of conjunction $tCons\wedge pCons\wedge nCons$, where *nCons* is **tt** in the first iteration and will encode constraints related to counterexample molecules. If the conjunction is unsatisfiable, there are no rules of the input number and height, and the method returns failure of synthesis. If it is satisfiable, it constructs the rules at line 11 from and stores them in R.

At line 15, we construct constraints *rCons* again using ENOCDEPRODUCE that says that template molecule $\hat{m}$ is generated by rules R. We check the satisfiability of $rCons\wedge mCons$. If it is not satisfiable, we have found the rules that generate exactly the molecules in μ and the loop terminates. Otherwise, we use the satisfying assignment to create a counterexample molecule m'. At line 18, we add constraints to *nCons* stating that all possible applications of template rules in T must not produce m'. We use shorthand $F :\wedge= G$ for $F := F \wedge G$. As we will see that EncodeProduce introduces fresh variable maps τ and *cuts* in the encoding. Since we negate the returned formula by EncodeProduce and then we check the satisfiability, we need to introduce universal quantifiers over the fresh variables. Afterwards, the loop goes to the next iteration.

5.2 EncodeProduce in Detail

Now let us look at the encoding generated by EncodeProduce. The process of production of molecules adds pieces of trees one after another. In order to show that a molecule is producible by a set of production rules, we need to find the nodes where the production rules are applied, the rules that are applied on the nodes, and the order of the application of the rules. To model the production due to the application of the rules, we attach three maps to the molecule nodes.

- Let *cuts* map each node to a Boolean variable indicating the node is the point where a rule is applied to expand the molecule. We say points of the applications of the production rules as *cuts* of the tree.
- Let *rmatch* map each node to a rule indicating the rule that is applied to expand at the node. We need to match a rule to a node if it is a cut point.
- Let τ map each node to an integer variable indicating the time point when the node was added to the molecule. Since already added nodes in a molecule decide what can be added later, we need to record the order of the addition.

Algorithm 2 presents function EncodeProduce that returns the encoding. It takes a molecule m and a set of rules T. Both the inputs can be template or concrete. Our presentation assumes that the molecule is concrete and the rules are templates. This will cover the case when the EncodeProduce is called at line 7 in Algorithm 1. However, at line 15 the molecule is a template and the rules are concrete, which we will discuss later. EncodeProduce uses the help of three other supporting functions, EncodeP, MatchTree, and MatchCut.

EncodeProduce returns constraints stating for each node v and rule template t, if v is at a cut and t is applied at v, then t must match at the node v. We require $rmatch(v)$ to be equal to some rule. Since we are matching with a template rule, we do not know the position of the root of expanding nodes. We enumerate to all possible depths from 1 to $d-1$ for finding the root. For each $\ell \in [1, d)$, we call EncodeP(v, t, ℓ) to construct the constraints encoding that t is applied at v and the depth of the root of the expanding nodes is at depth ℓ.

In EncodeP, we can traverse up from v for ℓ steps to find the node to match the root of t. It returns ff if there is no ℓth ancestor of v. The variable *mark* is the timestamp for v. The variable c collects the constraints as they are generated. The variable v_r is initially equal to the root v_{0r} of t and traverses the nodes in t. The while loop at line 3 starts with the ℓth ancestor from v, matches all the ancestors up to the parent of v. As the loop traverses down, it also traverses t using variable v_r. In each iteration, v_r is updated to the jth child due to lines 8 and 13 if the ancestors of v also traverse along the jth child.

Algorithm 2. ENCODEPRODUCE(m : molecule (template), T : rule (template))

```
1: return ⋀_{v∈m} ⋀_{t∈T} [rmatch(v) = t ∧ cuts(v) ⇒ ⋁_{ℓ∈[1,d)} ENCODEP(v,t,ℓ)]
```

ENCODEP(v, $t = (\_, \nu, \kappa, v_{0r})$, ℓ)

```
1: if ancestor(v,ℓ) = ⊥ then return ff end if
2: mark := τ(v),  c := tt,  v_r := v_0r,  i = ℓ
3: while i > 0 do
4:     v' := ancestor(v,i)
5:     c :∧= κ(v_r) = MatchAns ∧ ν(v_r) = M(v') ∧ τ(v') < mark
6:     for j ∈ [1, NumberChildren(v')] do
7:         if C(v',j) = ancestor(v,i − 1) then
8:             v'_r := C(v_r,j)
9:         else
10:            c :∧= MATCHTREE(C(v',j), C(v_r,j), mark, ff)
11:        end if
12:    end for
13:    v_r := v'_r,  i := i − 1
14: end while
15: c :∧= κ(v_r) = Expand ∧ MATCHTREE(v, v_r, mark, tt) ∧ MATCHCUT(v, v_r, ff)
16: return c
```

MATCHTREE(v, v_r, $mark$, $isExpand$)

```
1: if v_r = ⊥ then return tt end if
2: if v = ⊥ then return κ(v_r) = Absent end if
3: tCons := isExpand ? (mark ≤ τ(v)) : (τ(v) < mark)
4: c := κ(v_r) ≠ Absent ⇒ tCons ∧ ν(v_r) = M(v)
5: c :∧= ⋀_{i∈[1,NumberOfChidren(v)]} MATCHTREE(C(v,i), C(v_r,i), mark, isExpand)
6: return c
```

MATCHCUT(v, v_r, $ruleParentIsNotAbsent$)

```
1: if v = ⊥ then return tt end if
2: if v_r = ⊥ then return parentIsNotAbsent ⇒ cuts(v) end if
3: c := ruleParentIsNotAbsent ⇒ (κ(v_r) = Absent) = cuts(v)
4: c :∧= ⋀_{i∈[1,NumberOfChidren(v)]} MATCHCUT(C(v,i), C(v_r,i), κ(v_r) ≠ Absent)
5: return cons
```

Let v' be the ith ancestor at some iteration of the loop. At line 5, the loop adds constraints stating that the corresponding node v_r in t is $MatchAns$, sugar matches in v_r and v', and node v' was added before $mark$. The loop at line 6, iterates over children of v'. If there is a child of v' that is not along the path to v, it is matched at line 10 with the corresponding child of v_r by calling MATCHTREE, which will be discussed shortly. If the jth child of v' is along the path to v, we get node v'_r for updating v_r for the next iteration of the while loop. After the while loop in EncodeP at line 15, we declare v_r is *Expand*, match the node v with the corresponding node v_r in the template rule by calling MatchTree, and also call to match the cut pattern at the subtree of v with the template rule.

MatchTree is a recursive procedure and matches sugar assignments between the molecule and the rule template. If the rule template node v_r does not exist, then there is nothing to match and it returns **tt** at line 1. At line 2, we encode if the molecule node v does not exist, then the rule node must also be flagged absent. Otherwise, we add constraints that if v_r is not absent in the template, then the labels of v and v_r must match at line 4. At the same line, MatchTree also adds constraints *tCons* that nodes are added in the molecule in the correct order. The last two inputs of the function are timestamp *mark* and a bit *isExpand*, which tells us that the matching nodes should be added before or after *mark*. The calls to MatchTree from EncodeP use the *isExpand* appropriately. Afterwords at line 5, the procedure calls itself for the children of v and v_r.

Once a rule is applied to a molecule, all the nodes inside the expanding part are added together. Therefore, there should be no cuts within the set of added nodes. Furthermore, if there are children nodes below the added nodes, they must be added due to the application of some other rule. Therefore, the children are at the cut points. We call the above requirement *cut pattern*. MatchCuts is a recursive function over the trees, and matches cut patterns between the molecule node v and the rule template node v_r. The cuts must occur whenever nodes of the rule template transition from present to absent. For helping to detect the transition, the third parameter is a constraint that encodes that the parent of v_r is absent or not. The call to MatchCuts in EncodeP passes **ff** as the third input. Even if the parent of v_r is present, v_r is a cut point adding cut pattern constraints for the node will create inconsistency. Therefore, we are passing **ff**.

In the case when we call EncodeProduce with template molecule and concrete rules. We swap the roles of ν and M at their occurrences at line 5 in EncodeP and line 4 in MatchTree. Furthermore, $\kappa(v_r)$ is a concrete value. In the functions, the variable is to be replaced by the evaluated value of $\kappa(v_r)$.

Theorem 1. *If* SugarSynth*(μ, d, n) returns rules R, then if R produces a molecule that has depth less than $h+1$, the it is in μ (soundness). If it fails to find rules, there are no rule sets within the budget of d and n (completeness).*

The soundness is true by construction. At each iteration, at least one solution, i.e., a set of rules is discarded. In fact, we discard many because in each iteration we reject all rule sets that may produce a counterexample molecule. Since the set of all possible rules is finite, we terminate. Therefore, the completeness holds.

Table 1. Results of applying GlySynth on data sets.

	#molecules	#Rules	Rule depth	#Compartments	Success?	Time (in secs.)		#molecules	#Rules	Rule depth	#Compartments	Success?	Time (in secs.)
D1	6	7	3	1	Yes	3.02	D5	3	6	2	1	No	0.64
		7	4	2	Yes	**1.60**			7	2	1	Yes	**0.72**
		6	3	3	Yes	9.36			8	4	1	Yes	2.39
D2	3	7	3	2	Yes	14.37	D6	2	5	3	2	Yes	0.86
		5	3	2	Yes	**7.97**			5	3	1	Yes	**0.73**
		5	3	3	Yes	13.42			4	2	1	No	0.69
D3	6	6	4	2	Yes	1.02	D7	3	5	3	2	Yes	0.72
		5	2	1	Yes	**0.57**			5	3	1	Yes	**0.65**
		5	4	1	Yes	0.71			6	2	1	No	0.69
D4	3	8	4	1	Yes	4.35	D8	3	4	3	2	No	0.79
		6	3	1	Yes	**0.85**			5	3	2	Yes	**0.84**
		6	2	2	No	1.17			8	4	3	Yes	1.53

6 Experiments

Implementation: We have implemented SugarSynth in a tool GlySynth [15]. The tool, written in C++, uses Z3[16] as the SMT solver to solve the satisfiability queries. We ran the experiments on a laptop with 8 GB RAM and 2.4 GHz CPU.

Benchmark: We have applied our tool to three sets of real data (D1, D2, D3, D4) and two sets of synthetic data (D5, D6). The molecules have been obtained from respiratory mucins of a cystic fibrosis patient (D1), horse chorionic gonadotropin (D2), SARS-CoV-2 spike protein T323/S325 (D3), and human chorionic gonadotropin from a cancer cell line (D4) [10,17]. The availability of clean data, where we are clear about the source, limits our choices (Table 1).

Results: We have applied GlySynth on the data set. For each data set, we choose several parameter combinations to illustrate the relative performance of the tool. If we did not budget large enough parameters such as the size of unknown molecules, number of rules etc., then the tool fails to synthesize the rules. We present the rules learned after giving minimum resources. However, the set of rules reported need not be either unique or minimal.

For D1, we synthesize the rules in 1.60 s. Even by reducing the first two parameters, we were able to synthesize the rules but it took longer time. Giving an extra compartment in the third row did not impact the performance. We also observe the trade-off between the number of compartments and the depth of the rules at the second and third row of D1 and how it impacts performance. We synthesize the rules for D2, involving runaway reactions, in 7.97 s. For D2, we also learned large rules, i.e., they are adding many nodes at a single time. We synthesize the rules for D3 in 0.57 s. We synthesize the rules for D4, which is also our motivating example, in 0.85 s. However, if we use too many resources or too little, the tool runs for a long time as the search in combinatorial space is highly sensitive to the parameters. The synthesis for D5 takes 0.72 s. We can observe that by reducing the number of rules to learn, the tool fails to learn rules as it required minimum 7 rules. Our synthesized production is in line with the reported rules in the literature [10].

7 Conclusion and Future Work

We have presented a novel method for synthesizing production rules for glycans. We have applied our method to real-world data sets. We are planning to work in biological labs to check the viability of the solutions found by our method. We are the *first* to apply formal methods for the synthesis problem.

References

1. Varki, A. (ed.): Essentials of Glycobiology, 3rd edn. Cold Spring Harbor Laboratory Press, New York (2017)
2. Solar-Lezama, A., Jones, C.G., Bodík, R.: Sketching concurrent data structures. In: Gupta, R., Amarasinghe, S.P. (eds.) Proceedings of the ACM SIGPLAN 2008 Conference on Programming Language Design and Implementation, Tucson, AZ, USA, 7–13 June 2008, pp. 136–148. ACM (2008)
3. Gupta, A., Majumdar, R., Rybalchenko, A.: From tests to proofs. In: TACAS (2009)
4. Alur, R., et al.: Syntax-guided synthesis. In: Formal Methods in Computer-Aided Design, FMCAD 2013, Portland, OR, USA, 20–23 October 2013, pp. 1–8 (2013)
5. Solar Lezama, A., Rabbah, R., Bodík, R., Ebcioğlu, K.: Programming by sketching for bit-streaming programs. In: Proceedings of the 2005 ACM SIGPLAN Conference on Programming Language Design and Implementation, PLDI '05, pp. 281–294. ACM, New York (2005)
6. Biswas, A., Gupta, A., Missula, M., Thattai, M.: Automated inference of production rules for glycans (extended version). http://arxiv.org/abs/2107.02203 (2021)
7. Campbell, M.P., et al.: UniCarbKB: building a knowledge platform for glycoproteomics. Nucleic Acids Res. **42**(D1), D215–D221 (2013)
8. Alberts, B., et al.: Essential cell biology. Garland Sci. (2013)
9. Biswas, A., Thattai, M.: Promiscuity and specificity of eukaryotic glycosyltransferases. Biochem. Soc. Trans. **48**(3), 891–900 (2020)
10. Jaiman, A., Thattai, M.: Algorithmic biosynthesis of eukaryotic glycans (2018)
11. Spahn, P.N., et al.: A markov chain model for n-linked protein glycosylation - towards a low-parameter tool for model-driven glycoengineering. Metab. Eng. **33**, 52–66 (2016)
12. Spahn, P.N., Lewis, N.E.: Systems glycobiology for glycoengineering. Curr. Opin. Biotechnol. **30**, 218–224 (2014)
13. Jaiman, A., Thattai, M.: Algorithmic biosynthesis of eukaryotic glycans. bioRxiv (2018)
14. Harrd, K., et al.: The carbohydrate chains of the beta subunit of human chorionic gonadotropin produced by the choriocarcinoma cell line BeWo. Novel o-linked and novel bisecting-GlcNAc-containing n-linked carbohydrates. Eur. J. Biochem. **205**(2), 785–798 (1992)
15. Gupta, A., Meghana, M.: GlySynth (2021). https://github.com/ashutosh0gupta/sugar-synth
16. de Moura, L., Bjørner, N.: Z3: an efficient SMT solver. In: Ramakrishnan, C.R., Rehof, J. (eds.) TACAS 2008. LNCS, vol. 4963, pp. 337–340. Springer, Heidelberg (2008). https://doi.org/10.1007/978-3-540-78800-3_24
17. Shajahan, A., Supekar, N.T., Gleinich, A.S., Azadi, P.: Deducing the N- and O-glycosylation profile of the spike protein of novel coronavirus SARS-CoV-2. Glycobiology (2020). cwaa042

Compiling Elementary Mathematical Functions into Finite Chemical Reaction Networks via a Polynomialization Algorithm for ODEs

Mathieu Hemery, François Fages(✉), and Sylvain Soliman

Inria Saclay Île-de-France, EPI Lifeware, Palaiseau, France
{mathieu.hemery,Francois.Fages,sylvain.soliman}@inria.fr

Abstract. The Turing completeness result for continuous chemical reaction networks (CRN) shows that any computable function over the real numbers can be computed by a CRN over a finite set of formal molecular species using at most bimolecular reactions with mass action law kinetics. The proof uses a previous result of Turing completeness for functions defined by polynomial ordinary differential equations (PODE), the dual-rail encoding of real variables by the difference of concentration between two molecular species, and a back-end quadratization transformation to restrict to elementary reactions with at most two reactants. In this paper, we present a polynomialization algorithm of quadratic time complexity to transform a system of elementary differential equations in PODE. This algorithm is used as a front-end transformation to compile any elementary mathematical function, either of time or of some input species, into a finite CRN. We illustrate the performance of our compiler on a benchmark of elementary functions relevant to CRN design problems in synthetic biology specified by mathematical functions. In particular, the abstract CRN obtained by compilation of the Hill function of order 5 is compared to the natural CRN structure of MAPK signalling networks.

1 Introduction

Chemical reaction networks (CRN) provide a standard formalism in chemistry and biology to describe, analyze, and also design complex molecular interaction networks. In the perspective of systems biology, they are a central tool to analyze the high-level functions of the cell in terms of their low-level molecular interactions. In the perspective of synthetic biology, they constitute a target programming language to implement in chemistry new functions in either living cells or artificial devices.

A CRN can be interpreted in a hierarchy of Boolean, discrete, stochastic and differential semantics [7,13] which is at the basis of a rich theory for the analysis of their dynamical properties [1,9,14], and more recently, of their computational power [6,7,11]. In particular, their interpretation by Ordinary Differential Equations (ODE) allows us to give a precise mathematical meaning to the notion of

E. Cinquemani and L. Paulevé (Eds.): CMSB 2021, LNBI 12881, pp. 74–90, 2021.
https://doi.org/10.1007/978-3-030-85633-5_5

analog computation and high-level functions computed by cells [10,23,25], with the following definitions:

Definition 1. *[11,16,26] A function* $f : \mathbb{R}_+ \to \mathbb{R}_+$ *is* generated by a CRN *on some species* y *with given initial concentrations for all species, if the ODE associated to the CRN has a unique solution verifying* $\forall t \geq 0\ y(t) = f(t)$.

That first definition states that a positive real function of one positive argument is *generated* by a CRN for some given initial concentration values, if the graph of that function is given by the temporal evolution of the concentration of one molecular species in that CRN.

Definition 2. *[11,16][1] A function* $f : \mathbb{R}_+ \to \mathbb{R}_+$ *is* computed by a CRN *for some input species* x*, output species* y*, and initial concentrations given for all species apart from* x*, if for any input concentration value* $x(0)$ *for* x*, the ODE initial value problem associated to the CRN has a unique solution satisfying* $\lim_{t\to\infty} y(t) = f(x(0))$.

The second definition states that the same function is *computed* by a CRN if for any input $x \geq 0$, and initialization of the CRN input species to value x, the CRN output species converges to the result $f(x)$. That definition for input/output functions computed by a CRN is used in [11] to show the Turing completeness of continuous CRNs in the sense that any computable function over the real numbers can be computed by a CRN over a finite set of formal molecular species using at most bimolecular reactions with mass action law kinetics. The proof uses a previous result of Turing completeness for functions defined by polynomial ordinary differential equation initial value problems (PIVP) [2], the dual-rail encoding of real variables by the difference of concentration between two molecular species [18,21], and a back-end quadratization transformation to restrict to elementary reactions with at most two reactants [3,5,19]. This proof gives rise to a pipeline, implemented in BIOCHAM-4[2], to compile any computable real function presented by a PIVP into a finite CRN.

However in practice, it is not immediate to define a PIVP that generates or computes a desired function. In this article, we solve this problem for elementary functions over the reals, by adding to our compilation pipeline a front-end module to transform any elementary function in a PIVP which either generates or computes that function, as schematized in Fig. 1.

More precisely, we present a polynomialization algorithm to transform any Elementary ODE system (EODE), i.e. ODE system in explicit form made of elementary differential functions, into a polynomial one (PODE). This algorithm proceeds by introducing variables for computing the non-polynomial terms of the input, eliminating such terms from the ODE by rewriting, and obtaining the

[1] For the sake of simplicity of the definition given here, we omit the error control mechanism that requires one extra CRN species z verifying: $\forall t > 1\ |y(t) - f(x(0))| \leq z(t)$, $\forall t' > t\ z(t') < z(t)$ and $\lim_{t\to\infty} z(t) = 0$.

[2] http://lifeware.inria.fr/biocham/. All experiments described in this paper are available at https://lifeware.inria.fr/wiki/Main/Software#CMSB21.

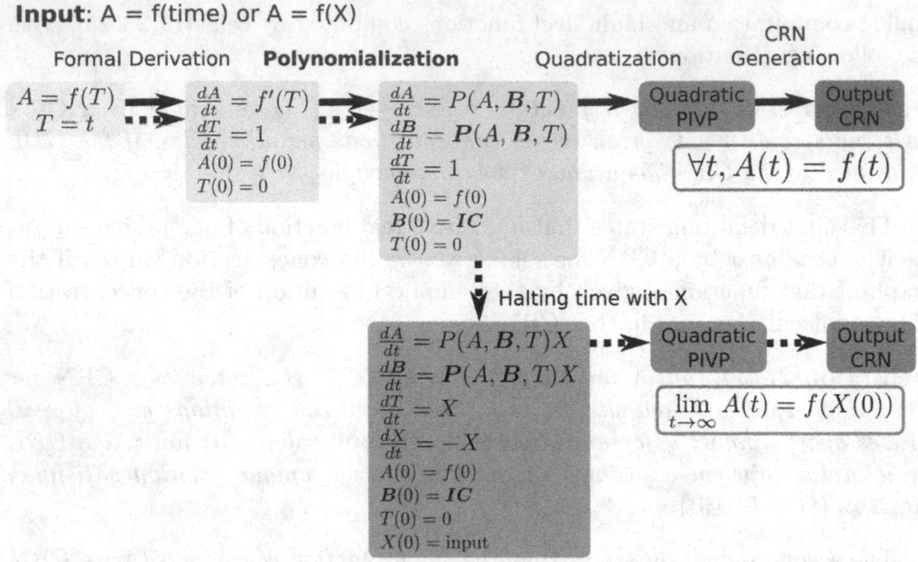

Fig. 1. Polynomialization step in the complete pipeline for compiling a formally differentiable function f (termination is proved here for elementary functions) into a finite CRN, either a function of time (plain arrows) or an input/output function (dashed arrows): P and $\boldsymbol{P}$ are polynomials, and $\boldsymbol{B}$ denotes the set of species introduced by polynomialization given with initial conditions $\boldsymbol{IC}$.

ODE for the new variables by formal derivation. The derivation steps may bring new non-polynomial terms requiring new variables. We show the termination of that algorithm, with quadratic time complexity, and that only a linear number of new variables, in terms of the size of the input expression, are actually needed.

Related work includes one method described in [24] to compute some polynomial abstractions of a non-polynomial system. That method progresses top-down from some set of possible functions given a priori, and progressively sieves them down to a polynomial system. This permits to choose the degree of the polynomial abstraction, but that method may fail if either the starting set of functions or the chosen degree are too small. On the other hand, our algorithm proceeds bottom-up, by introducing the functions when needed, with no choice a priori. A bottom-up approach closer to ours is also mentioned in [17] but for a very restricted grammar of functions while we develop here a general algorithm and prove its termination on the whole class of elementary mathematical functions. One can also cite the function `dpolyform` of Maple PDETools package which returns a PODE in implicit form for which the expression given as input is guaranteed to be a solution, but that does not provide a polynomial expression for the derivative of each variable, i.e. a PODE in explicit form, as required for our compilation pipeline.

The rest of the paper is organized as follows. In the next section, we describe the language of elementary function that are accepted as input of our compilation pipeline into CRN. In Sect. 3, we present a general polynomialization algorithm for elementary ODEs, prove its termination, and show its quadratic time complexity. In Sect. 4 we describe the use of this algorithm as a front-end transformation in our compilation pipeline to compile any elementary mathematical function into a finite CRN. In Sect. 5 we evaluate this approach on a benchmark of elementary functions relevant to CRN design problems in synthetic biology used in [19]. In particular, we compare the CRN synthesized for the Hill function of order 5 to the structure of the natural MAPK signalling CRNs that have been shown to implement a similar input/output function [20]. Finally, we conclude on the results achieved and several perspectives for future work.

2 Input Language of Elementary Functions

2.1 Example

Let us consider the problem of synthesizing a CRN to *generate* the function of time: $A(t) = \log(1 + t^2)$ in the sense of Definition 1. The compilation method described in [11] takes as input a PIVP which admits that function as solution on variable A. Here, we want to automate the construction of such a PIVP.

The first step of our front-end transformation schematized in Fig. 1, is to determine an ODE which admits that function of time as solution. For this, one can simply take the derivative of the equation with respect to time and set as initial condition the value for the desired function at 0, giving:

$$\frac{dA}{dt} = \frac{2t}{1+t^2} \qquad A(0) = 0$$

Then we need to transform this ODE into a PODE. Our polynomialization algorithm will introduce a new variable $B = \frac{1}{1+t^2}$, and similarly compute its derivative and its initial value, as follows:

$$\frac{dB}{dt} = \frac{-2t}{(1+t^2)^2} = -2tB^2 \qquad B(0) = 1$$

We also need to introduce a variable T for time with $\frac{dT}{dt} = 1$ and $T(0) = 0$. giving the following PIVP:

$$\frac{dA}{dt} = 2.T.B \qquad \frac{dB}{dt} = -2.T.B^2 \qquad \frac{dT}{dt} = 1$$
$$A(0) = 0 \qquad B(0) = 1 \qquad T(0) = 0$$

Note that the termination of those transformations is not obvious in general. It is proved in the next section. That PIVP of degree 3 can now be used as input of our previous compilation pipeline [11]. It is first transformed in quadratic form [19], in this case by introducing one variable, $BT = B.T$, and removing the time T, giving the following quadratic PIVP:

$$\frac{dA}{dt} = 2.BT \qquad \frac{dB}{dt} = -2.BT.B \qquad \frac{d(BT)}{dt} = \frac{dB}{dt}.T + B\frac{dT}{t} = -2.BT^2 + B$$
$$A(0) = 0 \qquad B(0) = 1 \qquad BT(0) = 0$$

One reaction with mass action law kinetics is then generated for each monomial of the ODE. Since in this example the reactions are well-formed and strict the system is positive (lemma 1 in [12]). There is thus no need to introduce dual-rail variables for negative values, and the generated elementary CRN (with rate constants written above the arrow) is:

$$B + BT \xrightarrow{2} BT \qquad B \xrightarrow{1} B + BT$$
$$2.BT \xrightarrow{2} BT \qquad BT \xrightarrow{2} A + BT \qquad B(0) = 1$$

Now, it is worth noting that if we want to synthesize a CRN that *computes* (instead of generating) the function $\log(1 + x^2)$ of some input x in the sense of Definition 2, the general method described in [11,22] consists in introducing a variable X, initialized to value x, multiplying the terms of the PIVP for generating the function, and following a decreasing exponential to halt the PIVP on the prescribed input. The previous PIVP of degree 3 thus becomes a PIVP of degree 4 by this transformation:

$$\frac{dX}{dt} = -X \qquad \frac{dA}{dt} = 2.T.B.X \qquad \frac{dB}{dt} = -2.T.B^2.X \qquad \frac{dT}{dt} = X$$
$$X(0) = x \qquad A(0) = 0 \qquad B(0) = 1 \qquad T(0) = 0$$

The quadratization algorithm [19] now introduces new variables BX and TBX for the corresponding monomials, and removes variables T and B. This generates the following quadratic PIVP:

$$\frac{dX}{dt} = -X \qquad \frac{dBX}{dt} = \frac{dB}{dt}X + B\frac{dX}{dt} = -2.BX.TBX - BX$$
$$\frac{dA}{dt} = 2.TBX \qquad \frac{dTBX}{dt} = BX.X - 2.TBX^2 - TBX$$
$$X(0) = x \qquad BX(0) = x \quad A(0) = 0 \quad TBX(0) = 0$$

and finally the following elementary CRN which computes the $\log(1 + x^2)$ function:

$$BX + TBX \xrightarrow{2} TBX \qquad X \xrightarrow{1} \emptyset \qquad BX \xrightarrow{1} \emptyset$$
$$BX + X \xrightarrow{1} BX + TBX + X \qquad TBX \xrightarrow{1} \emptyset$$
$$2.TBX \xrightarrow{2} TBX \qquad TBX \xrightarrow{2} A + TBX$$
$$X(0) = x \qquad BX(0) = x \qquad A(0) = 0$$

2.2 Elementary Functions as Compilation Pipeline Input Language

In mathematics, elementary functions refer to unary functions (over the reals in our case) that are defined as a sum, product or composition of finitely many polynomial, rational, trigonometric, hyperbolic, exponential functions and their inverses. Most of these functions are defined on the real axis but a few exceptions are worth mentioning: the inverse of some function is restricted to the image of $\mathbb{R}$ by the function (e.g., arccos is only defined on $[-1, 1]$) and the exponentiation may be non-analytic in 0 and is thus considered elementary only on an open interval that does not include 0.

The set of elementary functions of x is formally defined as the least set of functions containing:

- Constants: 2, π, e, etc.
- Polynomials of $x : x + 1,\ x^2,\ x^3 - 42.x$, etc.
- Powers of $x : \sqrt{x},\ \sqrt[3]{x}, x^{-4}$, etc.
- Exponential and logarithm functions: $e^x, \ln x$
- Trigonometric functions: $\sin x,\ \cos x,\ \tan x$, etc.
- Inverse trigonometric functions: $\arcsin x,\ \arccos x$, etc.
- Hyperbolic functions: $\sinh x,\ \cosh x$, etc.
- Inverse hyperbolic functions: $\operatorname{arsinh} x,\ \operatorname{arcosh} x$, etc.

and closed by arithmetic operations (addition, subtraction, multiplication, division) and composition. Elementary functions are also closed by differentiation but not necessarily by integration. On the other hand hyper-geometric functions, Bessel functions, gamma, zeta functions, are examples of (computable) non-elementary functions.

3 Polynomialization Algorithm for Elementary ODEs

3.1 Polynomialization Algorithm

The core of Algorithm 1 for polynomializing an EODE system is the detection of the elements of the derivatives that are not polynomial and their introduction as new variables. Then symbolic derivation and syntactic substitution allow us to compute the derivatives of the new variables and to modify the system of equations accordingly.

It is worth noting that the list of substitutions has to be memorized along the way, therefore handling an algebraic-differential system during the execution of the algorithm, since they may reappear during the derivation step. This typically occurs when the derivation graph harbors a cycle like: cos → sin → cos (Fig. 2).

Nevertheless, a particular treatment has to be applied to the case of non-integer or negative power as they form an infinite set and may thus produce infinite chains of derivations. This can be seen if we try to apply naively Algorithm 1 on this simple example:

$$\frac{dA}{dt} = A^{0.4}$$

Algorithm 1. Polynomialization of an EODE system

Input: A set of ODEs of the form $\{x' = f_x(x, y, \ldots), y' = f_y(x, y \ldots), \ldots\}$.
Output: A set of PODEs where the initial variables $x, y, \ldots$ are still present and accept the same solutions.
Initialize $Transformations \leftarrow \emptyset$ and $PolyODE \leftarrow \emptyset$
while ODE is not empty **do**
  take and remove $Var' = Derivative$ from ODE;
  $NewDerivative \leftarrow$ apply $Transformations$ to $Derivative$;
  $Terms \leftarrow$ set of maximal non-polynomial subterms of $NewDerivative$;
  for all $Term$ in $Terms$ **do**
    add ($Term \mapsto NewVar$) to $Transformations$;
    $TermDerivative \leftarrow$ the symbolic derivative of $Term$;
    add ($NewVar' = TermDerivative$) to ODE;
  $PolyDerivative \leftarrow$ apply $Transformations$ to $Derivatives$;
  add ($Var' = PolyDerivative$) to $PolyODE$;
return $PolyODE$

for which we introduce the new variable $B = A^{0.4}$ with

$$\frac{dB}{dt} = 0.4\,A^{-0.6}\,\frac{dA}{dt} = 0.4\,A^{-0.2}$$

At this point, it is tempting to introduce $C = A^{-0.2}$ but that would lead to an infinite loop, introducing more and more powers of A.

This can be easily avoided by introducing $C = \frac{1}{A}$ instead, then:

$$\frac{dB}{dt} = 0.4B^2\,C$$

$$\frac{dC}{dt} = -\frac{1}{A^2} \qquad \frac{dA}{dt} = -C^2 B$$

There is therefore a specific treatment to do when introducing the new variable of an exponentiation in order to force the algorithm to use the inverse variable instead of an infinite sequence of variables. For this, when adding the new variable $N = X^p$ to the system, we explicitly replace the expression X^{p-1} by N/X in the derivatives, thus making the use of $1/X$ a natural consequence. Of course, this makes the final PODE non analytic in $X = 0$. This is linked to the fact that exponentiation apart from the polynomial case is actually not analytic in 0, and it is thus not surprising that computation fails if we reach a time where $X = 0$.

3.2 Interval of Definition

Elementary functions are analytic upon open interval of their domain, but may suffer from non-analyticity on the boundary. For example exponentiation with a non integer coefficient may be extended by continuity in 0 but is not analytic here. During our polynomialization, this kind of behaviour may lead to the appearance of species that diverge on these points. This is important as it can be shown that only analytic functions can be generated by a PIVP.

In particular, the absolute value function over the reals is elementary as it can be expressed as the composition of a power and root of $x : |x| = \sqrt{x^2}$. But is not analytic on 0. Hence, if we consider the EODE $\frac{dx}{dt} = |x|$, our polynomialization will introduce the variables $y = |x|$ and $z = \frac{1}{x}$, to obtain the PODE:

$$\frac{dx}{dt} = y \qquad \frac{dy}{dt} = y^2.z \qquad \frac{dz}{dt} = -(z^2.y)$$

And when x approaches 0, the variable z, its inverse, will diverge.

The unicity of the solution of a PIVP is constrained by the initial conditions, but this unicity is not ensured when passing a non-analyticity. A consequence of this remark is that our compiled solution is defined from its initial condition ($t = 0$) up to the first non-analyticity of the compiled function.

3.3 Termination

Before proving the termination of our algorithm for the input set of elementary functions over the reals, we can show a general lemma for any set F of formally differentiable, possibly multivariate, real functions. Let $\overline{F}$ denote the closure of $F \cup \mathbb{R}$ by addition and multiplication, that is the algebra of F over $\mathbb{R}$.

Lemma 1. *For any finite set F of formally differentiable functions over the reals such that $\forall f \in F, f' \in \overline{F}$, Algorithm 1 terminates.*

Proof. In the **while** loop, since we detect all the non-polynomial parts of the derivatives in one pass, we are sure that the derivative at hand in each **while** step becomes polynomial. New variables may however be introduced in such a step, thus the only possibility of non-termination is to introduce an infinity of new variables.

The proof of termination proceeds by cases on the structure of the derivatives. Suppose we have an ODE on the set of variables $\{X_i\}$ with $i \in [1, n]$ and let us denote by $d(X_i)$ the derivative of X_i. First, it is obvious that if $d(X_i)$ is a single variable or a constant (numeric or parameter) then we have nothing to do. Similarly, for every expression composed by addition or multiplication of such terms, they are already polynomial.

If the new variable is a function of several variables (here 2): $Y = f(X_j, X_k)$, we have:

$$\frac{dY}{dt} = \frac{\partial f(X_j, X_k)}{\partial X_j} d(X_j) + \frac{\partial f(X_j, X_k)}{\partial X_k} d(X_k)$$

and as the addition is allowed in polynomial, we can consider separately the two derivatives. Thus, we can restrict ourselves to the case of functions of a single variable.

For a function of a single variable with no composition we have:

$$\frac{dy}{dt} = \frac{df(X_i)}{dt} = f'(X_i)d(X_i)$$

we have seen that $d(X_i)$ is already polynomial but nothing ensures that f' is. However by definition of our set F, $f'(X_i)$ may be expressed with other function of X in the set F, all applied to X_i, since composition is not allowed in the construction of $\overline{F}$. As the set F is finite, we are sure to terminate after introducing at most $|F|$ variables. It is thus important to not include the closure by composition in the definition of our set $\overline{F}$. Indeed a function such that its derivative would be of the form: $f'(x) = f(f(x))$ may lead to an infinite loop for our algorithm.

Finally, when facing a composition, e.g. $f(g(X))$, we replace it by a new variable, say y, with the standard derivation rule:

$$\frac{dy}{dt} = \frac{df(g(X_i))}{dt} = f'(g(X_i))g'(X_i)d(X_i)$$

We thus have two different chains of variables to introduce: at first $f(g(x))$, $f'(g(x))$, $f''(g(x))$, etc. and in a second time: $g'(x)$, $g''(x)$. The important point to remark is that there is no mixing: all derivatives of f are applied to $g(x)$ and neither to the derivatives of g. By the same argument as for the case without composition, the polynomialization of both $f'(g(X_i))$ and $g'(X_i)$ terminates.

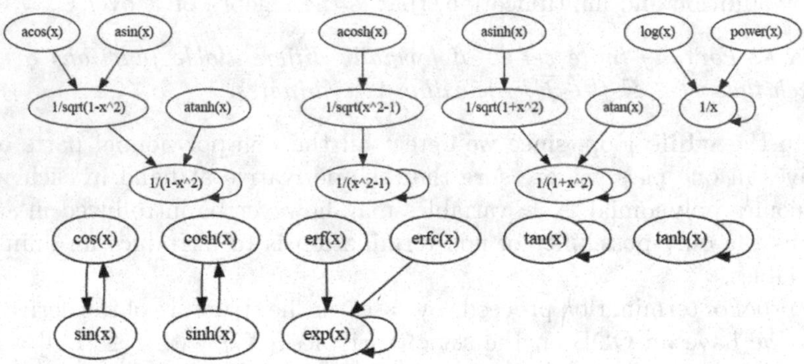

Fig. 2. Dependency graph of the derivatives for the elementary functions. Each function is given up to a polynomial composition, hence the derivative of arccos x which is $\frac{1}{\sqrt{1-x^2}}$ points to the $\frac{1}{\sqrt{x}}$ node. Note that the out-degree of each node is 1 as the derivative of each of these functions is always a single function or the composition of a polynomial and a single function.

Corollary 1. *Algorithm* 1 *terminates on elementary functions.*

Proof. Let F be the set of elementary functions. Figure 2 displays the dependency graph of the set F and its derivatives. Apart from the exponentiation, we can see that this set obeys to the condition of Lemma 1.

The only difficulty thus comes from the exponentiation as it actually defines an infinite set of functions of the form $f(x) = x^\alpha$ for any real constant α.

Nevertheless, as explained above, the algorithm terminates also in this case since we express the derivative in the form:

$$f'(x) = \alpha x^{\alpha-1} = \alpha f(x)\frac{1}{x} \tag{1}$$

and the inverse function is in F, henceforth the case of exponentation also terminates.

3.4 Complexity

To estimate the computational time complexity of Algorithm 1, let us first determine a bound on the number of new variables introduced in the system.

Proposition 1. *Algorithm* 1 *introduces at most a linear number of variables in the size of the input.*

Proof. For each elementary function f, represented in Fig. 2, we have to introduce recursively all the functions on which rely its derivative, in terms of the (directed) dependency graph, this is the reachable set starting from f. In this case the largest reachable set is of cardinal 3. More generally, let us call ℓ the cardinal of the largest set of reachable nodes starting form a single node. Then, each elementary function that is not already polynomial introduces at most ℓ variables.

Hence, we have to introduce at most ℓF variables where F is the number of functions used in the ODE. Of course, F is bounded by the size of the input.

Proposition 2. *Algorithm* 1 *has a quadratic time complexity.*

Proof. To introduce a new variable, we have to first compute its derivative, and then substitute its expression in the reminder of the ODE. Both operations are linear in the size of the current system size, and we just have seen that this one will grow at most as a linear function of the input size, giving only a linear dependency. Thus, our algorithm is quadratic in the size of the input.

3.5 Remark on the Compilation of the Exponentiation

In the compilation of the exponentiation described in Sect. 3.1 we introduce two variables, but this raises an interesting question concerning the conditions under which a differential equation of the form: $\frac{dA}{dt} = A^{\alpha}$ can be set in polynomial form by introducing only one variable. Indeed, if we introduce $B = A^{\beta}$ we have:

$$\frac{dB}{dt} = \beta A^{\beta-1} \qquad \frac{dA}{dt} = \beta A^{\alpha+\beta-1}$$

and to be polynomial we need to find four positive integers i, j, k, l such that:

$$\alpha = i + j\beta$$
$$\alpha + \beta - 1 = k + l\beta$$

We thus need that α be of the form: $\alpha = \dfrac{j(k+1)+i(1-l)}{j+1-l}$

Clearly, only a fractional power can be reduced in one step. Now suppose that $\alpha = \frac{p}{q}$ then we have by identifying numerator and denominator and setting $l = 2$:

$$j = q - 1 + l = q + 1 \quad \text{and} \quad k = \frac{p+i}{1+q} - 1$$

an equation that always admits solutions with i and k positive. For example $\alpha = \frac{1}{3}$ may be solved with $\beta = \frac{-2}{3}$ and $i = 3, j = 4, k = 0, l = 2$. But this uses a polynomial of order 7 which may impact negatively the final quadratization phase [19]. For these reasons, we chose to treat exponentiation by systematically introducing two variables.

4 CRN Compilation Pipeline for Elementary Functions

4.1 Detailed Example

To illustrate the behavior of our complete pipeline schematized in Fig. 1, let us consider the compilation of the Hill function $H = \frac{x}{1+x}$ into a CRN that computes $H(x)$ in the sense of Definition 2, i.e., such that the final concentration of species H gives the result $\lim_{t\to\infty} H(t) = \frac{x}{1+x}$.

The first step of the front-end transformation is to introduce a pseudo-time variable: $T = t$ and replace the input X by T in the expression $H = \frac{T}{1+T}$, and then to compute the formal derivative of H to obtain the ODE:

$$\frac{dH}{dt} = \frac{1}{(1+T)^2} \qquad \frac{dT}{dt} = 1$$

The second step is to polynomialize that ODE with Algorithm 1. This is done by examining the right hand side of the derivative of H and introduce as new variable $A = \frac{1}{(1+T)}$ which allows us to rewrite $\frac{dH}{dt} = A^2$. We also need to compute the derivative of this new variable:

$$\frac{dA}{dt} = -A^2, \quad A(0) = 1$$

This PIVP generates the time function $H(t)$ in the sense of Definition 1.

Now, we enter the compilation pipeline from PIVP described in [11]. To compute the function $H(x)$ of input x, a new variable X that obeys a decreasing exponential is introduced and used to halt all other derivatives at time x:

$$\begin{aligned} \frac{dH}{dt} &= A^2.X & \qquad \frac{dX}{dt} &= -X \\ \frac{dA}{dt} &= -A^2.X & \qquad \frac{dT}{dt} &= X \\ H(0) &= 0 & \qquad X(0) &= x \\ A(0) &= 1 & \qquad T(0) &= 0 \end{aligned}$$

Then the quadratization step [19], introduces the new variable $B = A.X$. Interestingly, intermediary variables X and T are no longer used and removed. We finally get:

$$\frac{dH}{dt} = A.B \qquad \frac{dB}{dt} = -B - B^2 \qquad \frac{dA}{dt} = -A.B$$
$$H(0) = 0 \qquad B(0) = x \qquad A(0) = 1$$

And the compiled CRN is:

$$B \xrightarrow{1} \emptyset \qquad A + B \xrightarrow{1} B + H \qquad 2.B \xrightarrow{1} B$$
$$A(0) = 1 \qquad B(0) = \text{input}$$

4.2 Implementation

Algorithm 1 is implemented by rewriting formal expressions using a simple algebraic normal form and standard derivation rules.

The most computationally expensive step of our complete compilation pipeline in Fig. 1 is the quadratization of the intermediate PIVP to a PIVP of order at most 2. This step is necessary to restrict ourselves to elementary reactions with at most two reactants which are more amenable to real implementations with real enzymes. While the existence of a quadratic form for any PODE can be simply shown by introducing an exponential number of variables [5], the problem of minimizing the dimension of that quadratization is NP-hard [19]. In the implementation of BIOCHAM used in the next section, we use both the MAXSAT algorithm described in [19] (option `sat_species` below) and a heuristic algorithm (option `fastnSAT` below) to first obtain a subset of variables guaranteed to contain a quadratic solution, and then call the MAXSAT solver (RC2[3]) to minimize the dimension in that quadratization.

5 Evaluation

In this section, we consider the benchmark of functions already considered in [19] for quadratization problems. Table 1 gives some performance figures about the complete compilation pipeline in terms of total computation time and size of the synthesized CRNs, with the two options discussed above for quadratization. It is worth noting that those synthetic CRNs are not unique and that other CRNs could be synthesized for the same function, by making different choices in both our polynomialization and quadratization algorithms. Even when imposing optimality in the number of introduced variables, there may exist several optimal CRNs. For example, the two CRNs obtained for Hill4 compiled with the two options for quadratization are different but both have the same number of species and reactions (Table 1).

[3] https://pysathq.github.io/docs/html/api/examples/rc2.html.

Table 1. Performance results on the benchmark of CRN design problems of [19] in terms of total compilation time, and size of the synthesized CRN with two options for quadratization.

Function	`fastnSAT`			`sat_species`		
	Time (ms)	Number of species	Number of reactions	Time (ms)	Number of species	Number of reactions
Hill1	80	4	5	85	3	3
Hill2	90	6	10	82	5	8
Hill3	100	6	10	115	6	12
Hill4	100	7	13	162	7	13
Hill5	110	8	16	550	7	11
Hill10	160	13	31	Timeout		
Hill20	380	23	61	Timeout		
Logistic	80	3	5	85	3	5
Double exp.	80	3	4	85	3	4
Gaussian	85	3	4	85	3	4
Logit	95	4	7	100	4	6

Let us examine the influence graphs of those synthetic CRNs since they provide a more compact abstract representation of the reaction graph [13]. Figure 3 depicts the influence graphs between molecular species of the synthesized CRNs for the hill functions of order 3 (Fig. 3A) and 5 (Fig. 3B), the logistic function (Fig. 3C) and the square of the cosine function (Fig. 3). One can remark on those examples that the outputs of the synthesized CRN do not participate in any feedback reaction. This is however not necessarily the case of the CRNs synthesized by our pipeline, as shown for instance by the cosine function [11].

More precisely, the Hill3 CRN in Table 1 synthesized with the `sat_species` option is the following:

$$
\begin{array}{rlrl}
x & \xrightarrow{1} \emptyset & Ax & \xrightarrow{1} \emptyset \\
TAx & \xrightarrow{1} \emptyset & T2Ax & \xrightarrow{1} \emptyset \\
Ax + T2Ax & \xrightarrow{3} T2Ax & Ax + x & \xrightarrow{1} Ax + TAx + x \\
T2Ax + TAx & \xrightarrow{3} T2Ax & TAx & \xrightarrow{2} T2A + TAx \\
T2A + T2Ax & \xrightarrow{3} T2Ax & TAx + x & \xrightarrow{2} T2Ax + TAx + x \\
2.T2Ax & \xrightarrow{3} T2Ax & Ax + T2A & \xrightarrow{2} Ax + T2A + hill3 \\
x(0) & = \text{input} & Ax(0) & = \text{input}
\end{array}
$$

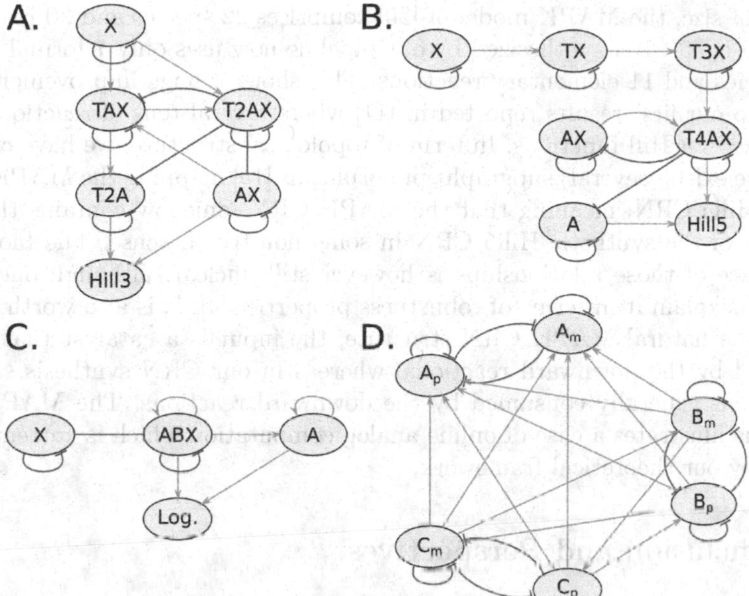

Fig. 3. Influence graphs of four of the synthetic CRNs of Table 1. **A.** and **B.** respectively implement the Hill function of order 3 and 5. **C.** corresponds to the logistic function and **D.** computes the square of the cosine of the time. In this last example, the output is only read on A_p but the presence of its negative part A_m is actually a crucial part of the computation despite having an essentially null concentration.

The Hill5 CRN in the table synthesized with the same option is:

$$
\begin{array}{ll}
x \xrightarrow{1} \emptyset & Ax \xrightarrow{1} \emptyset \\
Tx \xrightarrow{1} \emptyset & T3x \xrightarrow{1} \emptyset \\
T4Ax \xrightarrow{1} \emptyset & A + T4Ax \xrightarrow{5} T4Ax + hill5 \\
Ax + T4Ax \xrightarrow{5} T4Ax & 2.x \xrightarrow{1} Tx + 2.x \\
2.Tx \xrightarrow{3} T3x + 2.Tx & Ax + T3x \xrightarrow{4} Ax + T3x + T4Ax \\
2.T4Ax \xrightarrow{5} T4Ax & x(0) = \text{input} \\
A(0) = 1 & Ax(0) = \text{input}
\end{array}
$$

From our computational point of view, the Hill5 CRN above is one synthetic analog of the natural MAPK CRN among others. Indeed, the natural MAPK signalling network has been shown in [20] to compute an ultrasensitive input/output function which is well approximated by a Hill function of order 4.9. Both the natural MAPK CRN and the synthetic Hill5 CRN thus compute a similar input/ouput function and it makes sense to try to compare their structure.

In term of size, the MAPK model of [20] comprises 22 species and 30 reactions, while the Hill5 CRN synthesized by our pipeline now uses only 7 formal molecular species and 11 elementary reactions. This shows a huge improvement with respect to our first results reported in [11] where several tens of reactions were synthesized for Hill functions. In term of topological structure, we have checked that there exists (several) subgraph epimorphisms [15] mapping the MAPK CRN to that Hill5 CRN, meaning that the MAPK CRN somehow contains the core structure of the synthetic Hill5 CRN in some non-trivial sense. The biological significance of those relationships is however still unclear, although one could expect to explain it in terms of robustness properties [4]. It is also worth noting that in the natural MAPK CRN structure, the input is a catalyst that is not consumed by the downward reactions, whereas in our CRN synthesis scheme, the input is generally consumed by the downward reactions. The MAPK network thus illustrates a case of online analog computation which is currently not treated by our theoretical framework.

6 Conclusion and Perspectives

We have presented an algorithm to transform any system of elementary ordinary differential equations in a polynomial ordinary differential equation system preserving the solutions of the original variables. This algorithm of quadratic time complexity introduces at most a linear number of new variables. This algorithm allows us to automatically compile any elementary mathematical function into a finite CRN, using a pipeline of transformations starting from the formal derivation of the elementary function to generate or compute, the polynomialization of the elementary ODE, and continuing with the previous pipeline of [11] for the dual-rail encoding of negative values of the PODE [18,21], the quadratization of the PODE [19], and the synthesis of elementary reactions for the quadratic ODEs [12].

The implementation in BIOCHAM-4 of this complete pipeline has been used to illustrate the CRNs synthesized for a variety of elementary mathematical functions used as specification. In particular, the CRN synthesized for the Hill function of order 5 provides a synthetic analog of the MAPK signalling network which has been shown to compute a similar ultrasensitive input/output function [20]. In this compilation process, the quadratization part is the most complex one since minimizing the dimension of the result is a NP-hard problem [19]. On the benchmark presented here, our maxSAT implementation is sufficient but we also use by default a heuristic algorithm that trades optimality for better performance. It should also be noted that a new algorithm has been recently proposed for the global optimization problem in [3].

This work may be improved in several directions. We might extend this approach to multivariate functions. This is not trivial as the trick of halting the time at the input value needs be generalized to several inputs.

Another important point is to investigate is the variety of different CRNs that can be synthesized by our pipeline. As pointed out earlier, there may be several optimal solutions to the quadratization problem and there may similarly be several polynomializations of a given ODE introducing the same number of variables. Our pipeline make choices to deterministically propose one solution, but being able to explore the set of solutions and compare their properties with respect to metrics like robustness to initial conditions or reaction rates, or imposing some similarity requirement with a given biological solution would be interesting.

Furthermore, the comparison to the MAPK network also points to the interesting class of online computation which does not consume the inputs, whereas in our approach the input species are consumed, and the synthesized CRN would need to be reinitialized for another computation. This limitation is not a problem for one-shot CRN programs such as those designed for medical diagnosis applications [8], but synthesizing a CRN for computing an input/output function online appears to be a harder problem worthy of further theoretical investigation.

Acknowledgment. We acknowledge fruitful discussions with Olivier Bournez, François Lemaire, Gleb Pogudin and Amaury Pouly. This work was supported by ANR-DFG SYMBIONT "Symbolic Methods for Biological Networks" project grant ANR-17-CE40-0036, and ANR DIFFERENCE "Complexity theory with discrete ODEs" project grant ANR-20-CE48-0002.

References

1. Baudier, A., Fages, F., Soliman, S.: Graphical requirements for multistationarity in reaction networks and their verification in biomodels. J. Theor. Biol. **459**, 79–89 (2018)
2. Bournez, O., Campagnolo, M.L., Graça, D.S., Hainry, E.: Polynomial differential equations compute all real computable functions on computable compact intervals. J. Complexity **23**(3), 317–335 (2007)
3. Bychkov, A., Pogudin, G.: Optimal monomial quadratization for ode systems. In: Proceedings of the IWOCA 2021–32nd International Workshop on Combinatorial Algorithms, July 2021
4. Cardelli, L.: Morphisms of reaction networks that couple structure to function. BMC Syst. Biol. **8**(84) (2014). https://doi.org/10.1186/1752-0509-8-84
5. Carothers, D.C., Edgar Parker, G., Sochacki, J.S., Warne, P.G.: Some properties of solutions to polynomial systems of differential equations. Electron. J. Differ. Equ. **2005**(40), 1–17 (2005)
6. Chen, H.-L., Doty, D., Soloveichik, D.: Deterministic function computation with chemical reaction networks. Natural Comput. **7433**, 25–42 (2012)
7. Cook, M., Soloveichik, D., Winfree, E., Bruck, J.: Programmability of chemical reaction networks. In: Condon, A., Harel, D., Kok, J.N., Salomaa, A., Winfree, E (eds.) Algorithmic Bioprocesses, pp. 543–584. Springer, Heidelberg (2009). https://doi.org/10.1007/978-3-540-88869-7_27
8. Courbet, A., Amar, P., Fages, F., Renard, E., Molina, F.: Computer-aided biochemical programming of synthetic microreactors as diagnostic devices. Molecular Syst. Biol. **14**(4), e7845 (2018)
9. Craciun, G., Feinberg, M.: Multiple equilibria in complex chemical reaction networks: II. The species-reaction graph. SIAM J. Appl. Math. **66**(4), 1321–1338 (2006)

10. Daniel, R., Rubens, J.R., Sarpeshkar, R., Lu, T.K.: Synthetic analog computation in living cells. Nature **497**(7451), 619–623, 05 (2013)
11. Fages, F., Le Guludec, G., Bournez, O., Pouly, A.: Strong turing completeness of continuous chemical reaction networks and compilation of mixed analog-digital programs. In: Feret, J., Koeppl, H. (eds.) CMSB 2017. LNCS, vol. 10545, pp. 108–127. Springer, Cham (2017). https://doi.org/10.1007/978-3-319-67471-1_7
12. Fages, F., Gay, S., Soliman, S.: Inferring reaction systems from ordinary differential equations. Theoretical Comput. Sci. **599**, 64–78 (2015)
13. Fages, F., Soliman, S.: Abstract interpretation and types for systems biology. Theoretical Comput. Sci. **403**(1), 52–70 (2008)
14. Feinberg, M.: Mathematical aspects of mass action kinetics. In: Lapidus, L., Amundson, N.R. (eds.) Chemical Reactor Theory: A Review, chapter 1, pp. 1–78. Prentice-Hall (1977)
15. Gay, S., Soliman, S., Fages, F.: A graphical method for reducing and relating models in systems biology. Bioinformatics **26**(18), i575–i581 (2010). special issue ECCB'10
16. Graça, D.S., Costa, J.F.: Analog computers and recursive functions over the reals. J. Complexity **19**(5), 644–664 (2003)
17. Chenjie, G.: QLMOR: a projection-based nonlinear model order reduction approach using quadratic-linear representation of nonlinear systems. IEEE Trans. Comput. Aided Des. Integr. Circuits Syst. **30**(9), 1307–1320 (2011)
18. Hárs, V., Tóth, J.: On the inverse problem of reaction kinetics. In: Farkas, M. (ed.) Colloquia Mathematica Societatis János Bolyai, volume 30 of Qualitative Theory of Differential Equations, pp. 363–379 (1979)
19. Hemery, M., Fages, F., Soliman, S.: On the complexity of quadratization for polynomial differential equations. In: Abate, A., Petrov, T., Wolf, V. (eds.) CMSB 2020. LNCS, vol. 12314, pp. 120–140. Springer, Cham (2020). https://doi.org/10.1007/978-3-030-60327-4_7
20. Huang, C.-Y., Ferrell, J.E.: Ultrasensitivity in the mitogen-activated protein kinase cascade. PNAS **93**(19), 10078–10083 (1996)
21. Oishi, K., Klavins, E.: Biomolecular implementation of linear I/O systems. IET Syst. Biol. **5**(4), 252–260 (2011)
22. Pouly, A.: Continuous models of computation: from computability to complexity. PhD thesis, Ecole Polytechnique, July 2015
23. Rizik, L., Ram, Y., Danial, R.: Noise tolerance analysis for reliable analog and digital computation in living cells. J. Bioeng. Biomed. Sci. **6**, 186 (2016)
24. Sankaranarayanan, S.: Change-of-bases abstractions for non-linear systems. arXiv preprint arXiv:1204.4347 (2012)
25. Sauro, H.M., Kim, K.: Synthetic biology: it's an analog world. Nature **497**(7451), 572–573 (2013)
26. Shannon, C.E.: Mathematical theory of the differential analyser. J. Math. Phys. **20**, 337–354 (1941)

Interpretable Exact Linear Reductions via Positivity

Gleb Pogudin[1(✉)] and Xingjian Zhang[2]

[1] LIX, CNRS, École Polytechnique, Institute Polytechnique de Paris, Palaiseau, France
gleb.pogudin@polytechnique.edu

[2] École Polytechnique, Institute Polytechnique de Paris, Palaiseau, France
xingjian.zhang@polytechnique.edu

Abstract. Kinetic models of biochemical systems used in the modern literature often contain hundreds or even thousands of variables. While these models are convenient for detailed simulations, their size is often an obstacle to deriving mechanistic insights. One way to address this issue is to perform an exact model reduction by finding a self-consistent lower-dimensional projection of the corresponding dynamical system.

Recently, a new algorithm CLUE [16] has been designed and implemented, which allows one to construct an exact linear reduction of the smallest possible dimension such that the fixed variables of interest are preserved. It turned out that allowing arbitrary linear combinations (as opposed to zero-one combinations used in the prior approaches) may yield a much smaller reduction. However, there was a drawback: some of the new variables did not have clear physical meaning, thus making the reduced model harder to interpret.

We design and implement an algorithm that, given an exact linear reduction, re-parametrizes it by performing an invertible transformation of the new coordinates to improve the interpretability of the new variables. We apply our algorithm to three case studies and show that "uninterpretable" variables disappear entirely in all the case studies.

The implementation of the algorithm and the files for the case studies are available at https://github.com/xjzhaang/LumpingPostiviser.

Keywords: Exact reduction (lumping) · ODE model · Interpretability

1 Introduction

Dynamical models described by systems of polynomial ordinary differential equations (PODEs) are frequently used in systems biology and life sciences in general. One of the major classes of such models is the dynamical models of chemical reaction networks (CRN) under the mass-action kinetics in which each indeterminate corresponds to the concentration of one of the chemical species. Models

Supported by the Paris Ile-de-France region. GP was partially supported by NSF grants DMS-1853482, DMS-1760448, DMS-1853650, CCF-1564132, and CCF-1563942.

E. Cinquemani and L. Paulevé (Eds.): CMSB 2021, LNBI 12881, pp. 91–107, 2021.
https://doi.org/10.1007/978-3-030-85633-5_6

appearing in the literature often consist of hundreds or thousands of variables. While the models of this size can incorporate a substantial amount of information about the phenomena of interest, it is often hard to use them to derive mechanistic insights.

One way to address these challenges is to use *model reduction* algorithms that replace a model with a simpler one while preserving, at least approximately, some of the features of the original model. A wide range of methods has been developed for approximate model reduction, including methods based on singular value decomposition [1] and time-scale separation [15].

A complementary approach is to perform *exact model reduction*, that is, lower the dimension of the model without introducing approximation errors. For example, exact linear lumping aims at writing a self-consistent system of differential equations for a set of *macro-variables* in which each macro-variable is a linear combination of the original variables. For important classes of biochemical models, specialized lumping criteria have been developed (see, e.g., [3,6,9]), allowing the construction of macro-variables as sums of some of the original variables (that is, allowing only coefficients zero and one in the linear combinations). A general lumping algorithm has been proposed in [4,5] which is applicable to any system of PODEs (not necessarily arising from a CRN). This algorithm partitions the original variables so that the macro-variables can be the sums of the variables within the blocks in the partition. Note that the macro-variables are zero-one linear combinations of the original variables in all these cases.

In [16], an algorithm has been designed (and the corresponding software called *CLUE* presented) that, for a given set of linear forms in the state variables (the *observables*), constructs a linear lumping of the smallest possible dimension such that the observables can be written as combinations of the macro-variables (i.e., the observables are preserved). Unlike the earlier approaches, the macro-variables produced by CLUE may involve any coefficients, and this allowed to produce reductions of lower dimensions than it was possible before, see [16, Table 1]. However, there was a price to pay for this flexibility: the authors state that some of the produced macro-variables "escape physical intelligibility" (see [16, Section 4.2]). Indeed, the resulting reduction of the smallest dimension is uniquely defined up to a linear change of the coordinates, so the coordinates in the reduced state space chosen by CLUE could be not optimal in the sense of interpretability.

In this paper, we propose a post-processing step that takes an exact linear lumping (not necessarily produced by CLUE) and attempts to improve its interpretability by performing a change of variables. It has been observed in [16] that one of the sources of difficulties for interpretation is the negative coefficients in the macro-variables. We design and implement an algorithm that finds (if possible) a linear change of variables in the reduced model so that

1. the coefficients of the representations of the new macro-variables in terms of the original state variables are nonnegative
2. and the total number of nonzero coefficients in these representations is as small as possible.

Note that interpretability is not a formal mathematical property, and the conditions above is one possible formalization of the notion of a "more interpretable reduction". We do not claim that it is universal (e.g., a difference of two state variables may represent a potential), but we claim that it is useful. To support this claim, we demonstrate the efficiency of our approach on three case studies from the literature. Two of these cases are exactly the case studies from [16] in which issues with interpretability occur. We show that our method provides interpretable re-parametrizations of the optimal lumpings computed by CLUE in all three case studies. Our algorithm uses tools from convex discrete geometry and matroid theory.

2 Methods

2.1 Preliminaries on Lumping

Definition 1 (Lumping). *Consider a system of ODEs of the form*

$$\mathbf{x}' = \mathbf{f}(\mathbf{x}), \tag{1}$$

where $\mathbf{x} = (x_1, \ldots, x_n)^T$, $\mathbf{f} = (f_1, \ldots, f_n)^T$, *and* $f_1, \ldots, f_n \in \mathbb{R}[\mathbf{x}]$. *A linear transformation* $\mathbf{y} = L\mathbf{x}$ *with* $\mathbf{y} = (y_1, \ldots, y_m)^T$, $L \subset \mathbb{R}^{m \times n}$, *and* rank $L = m$ *is called* a lumping of (1) *if there exist polynomials* $g_1, \ldots, g_m \in \mathbb{R}[\mathbf{y}]$ *such that*

$$\mathbf{y}' = \mathbf{g}(\mathbf{y}), \quad \textit{where} \quad \mathbf{g} = (g_1, \ldots, g_m)^T$$

for every solution $\mathbf{x}$ *of* (1). *We say that* m *is* the dimension of the lumping. *The variables* $\mathbf{y}$ *in the reduced system are called* macro-variables. *We will call a macro-variable* nontrivial *if it is not proportional to one of the original variables.*

Remark 1. An ODE system may have many lumpings, some of them may be less useful than others. For example, if $m = n$, then the lumping is just an invertible change of variables, so no reduction happens. Another special case is when the rows of L contain the coefficients of linear first integrals of the system. In this case, the reduced ODE will be of the form $\mathbf{y}' = 0$.

Constrained linear lumping introduced in Definition 2 requires to preserve the dynamics of the variables of interest, and this is one of the ways to say that reduction is not "too coarse".

The following example is a substantially simplified version of the case study from Sect. 3.1 (see also [12]).

Example 1. We will consider a chemical reaction network consisting of

- A chemical species X.
- Species A_{UU}, A_{UX}, A_{XU}, and A_{XX}. Each of them is one of the states of a molecule A with two identical binding sites, which can be either unbound (U in the subscript) or bound (X in the subscript) to X.

For simplicity, we will assume that all the reaction rates are equal to one. The dynamics of the network is defined by the following reactions ($*$ denotes any of X and U):

$$X + A_{U*} \rightleftharpoons A_{X*}, \qquad X + A_{*U} \rightleftharpoons A_{*X}. \tag{2}$$

Under the laws of the mass-action kinetics, the reactions (2) yield the following ODE system (where $[S]$ denotes the concentration of the species S):

$$\begin{cases} [X]' = [A_{XU}] + [A_{UX}] + 2[A_{XX}] - [X]([A_{XU}] + [A_{UX}] + 2[A_{UU}]), \\ [A_{UU}]' = [A_{XU}] + [A_{UX}] - 2[X][A_{UU}], \\ [A_{XU}]' = [A_{XX}] + [X][A_{UU}] - [X][A_{XU}] - [A_{XU}], \\ [A_{UX}]' = [A_{XX}] + [X][A_{UU}] - [X][A_{UX}] - [A_{UX}], \\ [A_{XX}]' = [X][A_{XU}] + [X][A_{UX}] - 2[A_{XX}]. \end{cases} \tag{3}$$

We will show that the following matrix L and the macro-variables y_1, y_2, y_3

$$L = \begin{pmatrix} 1 & 0 & 0 & 0 & 0 \\ 0 & 0 & 1 & 1 & 2 \\ 0 & 2 & 1 & 1 & 0 \end{pmatrix} \implies \begin{cases} y_1 = [X], \\ y_2 = [A_{XU}] + [A_{UX}] + 2[A_{XX}], \\ y_3 = 2[A_{UU}] + [A_{XU}] + [A_{UX}]. \end{cases} \tag{4}$$

yield a lumping of the system (2). Indeed, a direct calculation shows that

$$\begin{cases} y_1' = [X]' = [A_{XU}] + [A_{UX}] + 2[A_{XX}] - [X]([A_{XU}] + [A_{UX}] + 2[A_{UU}]) = y_2 - y_1y_3, \\ y_2' = [A_{XU}]' + [A_{UX}]' + 2[A_{XX}]' = -y_2 + y_1y_3, \\ y_3' = 2[A_{UU}]' + [A_{XU}]' + [A_{UX}]' = y_3 - y_1y_2. \end{cases} \tag{5}$$

Since each reaction involves only one binding site, this lumping can be interpreted as follows: y_2 is the total "concentration" of the bound sites, and y_3 is the total "concentration" of the unbound sites (see also Sect. 3.1).

The lumping matrix L in the example above turns out to exactly preserve the concentration $[X]$. In general, one may fix a vector $\mathbf{x}_{\text{obs}}$ of combinations of the original variables that are to be recovered in the reduced system.

Definition 2 (Constrained linear lumping) . *Let $\mathbf{x}_{\text{obs}}$ be a vector of linearly independent forms in $\mathbf{x}$ such that $\mathbf{x}_{\text{obs}} = A\mathbf{x}$. Then we say that a lumping $\mathbf{y} = L\mathbf{x}$ is a* constrained linear lumping *with observables $\mathbf{x}_{\text{obs}}$ if each entry of $\mathbf{x}_{\text{obs}}$ is a linear combination of the entries of $\mathbf{y}$.*

2.2 The Nonuniqueness/Interpretability Issue

A recent software CLUE [16] allows to find, for a given system (1) and a vector $\mathbf{x}_{\text{obs}}$, a constrained linear lumping of the smallest possible dimension. However, such an optimal lumping is not unique in the following sense: if $\mathbf{y}_1 = L\mathbf{x}$ is a constrained linear lumping of the smallest possible dimension, then, for every invertible matrix T of the appropriate dimension, $\mathbf{y}_2 = TL\mathbf{x}$ is also such a lumping. Two such lumpings will be called *equivalent*, and one can show that all constrained linear lumpings of the smallest possible dimension are equivalent.

Because of this nonuniqueness, the lumping produced by CLUE will be optimal in terms of the dimension but not necessarily optimal in terms of the *interpretability* of the resulting macro-variables. For example, the macro-variables constructed by CLUE for the system (3) are:

$$y_1 = [X], \quad y_2 = [A_{XU}] + [A_{UX}] + 2[A_{XX}], \quad y_3 = [A_{UU}] - [A_{XX}].$$

The last macro-variable is different from the one in (4) and does not allow for the "concentration-of-sites" interpretation. Moreover, the reduced ODE system is more complicated than (5). This issue becomes more serious in more realistic (and larger) models: for the case studies in [16, Section 4.2] it has been observed that some of the resulting macro-variables "escaped physical intelligibility".

2.3 Our Approach via Nonnegativity

It has been already observed in [16, Section 4.2] that the macro-variables involving negative coefficients (such as $[A_{UU}] - [A_{XX}]$) may be an obstacle for interpretability. This is partially because such quantities cannot be naturally viewed as concentrations of some sort since they may take on negative values.

Thus, in order to improve the interpretability of a lumping, we construct an equivalent lumping with all the coefficients being nonnegative and the number of nonzero coefficients (that is, the ℓ_0-norm $\|\cdot\|_0$) being the smallest possible under the nonnegativity constraint. Mathematically, for a given lumping $\mathbf{y}_1 = L\mathbf{x}$, we find (if possible) an equivalent lumping $\mathbf{y}_2 = TL\mathbf{x}$ with invertible T satisfying:

1. the entries of TL are nonnegative and
2. $\|TL\|_0$ is as small as possible.

As we have mentioned, for fixed observables, all the constrained linear lumpings of the smallest dimension are equivalent, so the value $\|TL\|_0$ does not depend on the choice of L in the case of the optimal constrained linear lumping as in [16].

We hypothesize that the new lumping $\mathbf{y}_2 = TL\mathbf{x}$ will be typically more interpretable than the original one. We support this hypothesis by three case studies: multisite protein phosphorylation [18], Fcϵ-RI signaling pathways [8], and Jak-family protein tyrosine kinase activation [2]. The first two are exactly the case studies from [16] for which some of the macro-variables could not be properly interpreted by the authors.

2.4 Algorithmic Details

In this section, we provide and justify Algorithm 1, an algorithm for computing a new lumping described in Sect. 2.3. We will use some basic terminology from convex geometry. We refer the reader to [17, Chapters 7–8] for details. Throughout the rest of the section, for A being a vector or a matrix, $\|A\|_0$ denotes the ℓ_0-norm of A, that is, the number of nonzero entries in A.

Algorithm 1: Algorithm for constructing new lumping

Input a $m \times n$ matrix L with entries in $\mathbb{R}$ and linearly independent rows;
Output an invertible $m \times m$ matrix T such that
- the entries of TL are nonnegative
- and the number of the nonzero entries is as small as possible.

Returns NO if such matrix T does not exist.

(Step 1) Consider the row space of L and the nonegative orthant in $(\mathbb{R}_{\geqslant 0})^n$ as polyhedral cones C_1 and C_2 in $\mathbb{R}^n$.
(Step 2) Compute a polyhedral cone $C = C_1 \cap C_2$. This can be done, for example, using the Fourier-Motzkin algorithm [20, Section 1.2].
(Step 3) If $\dim C < m$, return NO
(Step 4) Let E be a set of representatives of the extreme rays of C.
(Step 5) Initialize a $0 \times n$ matrix L_1
(Step 6) While $E \neq \varnothing$
(a) choose $e \in E$ such that $\|e\|_0 = \min_{v \in E} \|v\|_0$;
(b) if e is not in the row space of L_1, append e to L_1 as a new row;
(c) remove e from E.
(Step 7) Construct an $m \times m$ matrix T such that the i-th column contains the coordinates of the i-th row of L_1 with respect to the rows of L.

Remark 2 (Implementation). Our implementation of Algorithm 1 in Julia can be found at https://github.com/xjzhaang/LumpingPostiviser. We used polymake [11] for operations with cones (at **(Step 2)** and **(Step 4)**) and Nemo [10] for symbolic linear algebra (at **(Step 6)**). Table 1 below summarizes the performance of the code on the case studies we discuss in this paper. We also provide timing for obtaining the starting reduction using CLUE. Therefore, the sum of the last two columns is the total time to obtain the final reduction for the original system. The runtimes are measured on a laptop with a 2.20 GHz CPU and 16 GB RAM using `@btime` macro in Julia. One can see that the models with hundreds of equations can be tackled in less than a minute on a commodity hardware.

Table 1. Running times of our implementation.

Model	# original variables (n)	# macro-variables (m)	Runtime (sec.)	
			CLUE	Algorithm 1
Sect. 3.1, $m = 2$	18	6	<0.01	<0.01
Sect. 3.1, $m = 3$	66	6	<0.01	<0.01
Sect. 3.1, $m = 4$	258	6	0.34	3.4
Sect. 3.2	354	69	3.3	4.7
Sect. 3.3	470	322	72	49

Remark 3 (Choice at ***(Step 6)****a).* At the **(Step 6)**a, if there are several $e \in E$ with $\|e\|_0$ being minimal possible, we choose the one with the index of the leftmost nonzero entry being the smallest one. In our experience, this makes the results slightly easier to analyze.

Remark 4 (Returning NO). Although Algorithm 1 may, in principle, return NO, we did not encounter such a situation with models from the literature. We give an artificial example with this property in Appendix.

Theorem 1 (Correctness of Algorithm *1***).** *For every matrix L over $\mathbb{R}$ with linearly independent rows, Algorithm 1 produces an invertible square matrix T such that*

- *TL has nonnegative entries*
- *and the number of nonzero entries in TL is the smallest possible under the nonnegativity constraint*

if such T exists and returns NO if there is no such T.

Proof. First, we will show that the algorithm returns NO if and only if there is no such matrix. Assume that there is such a matrix T. Then both C_1 and C_2 contain the rows of the matrix TL. Therefore, C contains m linearly independent vectors, so its dimension is at least m. In the other direction, if $\dim C \geqslant m$, then there exist m linearly independent vectors in $C = C_1 \cap C_2$. Let T be the matrix with the columns being their coordinates with respect to the rows of L. Then the rows of TL will belong to C_2 so that they will be nonnegative.

Now assume that the algorithm does not return NO. We observe that the entries of L_1 are nonnegative because all its rows belong to C_2. The rows of L_1 belong to C_1, so they are linear combinations of the rows of L. Since, by the construction on **(Step 6)**, the rows of L_1 are linearly independent, and there are $\dim C$ of them, we conclude that the row spaces of L_1 and L coincide. Therefore, the coordinates in **(Step 7)** are well-defined, so the algorithm will produce a matrix T such that $L_1 = TL$ has only nonnegative entries.

It remains to prove that the ℓ_0-norm $\|L_1\|_0$ of $L_1 = TL$ is the smallest possible. Consider any set S of m linearly independent elements of the set E of representatives of the extreme rays of C. Since **(Step 7)** is a greedy algorithm on the linear matroid defined by E, [7, (18)] implies that

$$\|L_1\|_0 \leqslant \sum_{e\in S} \|e\|_0. \tag{6}$$

Consider any invertible matrix $\widetilde{T}$ such that the entries of $\widetilde{L}_1 := \widetilde{T}L$ are nonnegative. Since the rows $r_1, \ldots, r_m$ of $\widetilde{L}_1$ belong to C, each of them can be represented as a nonnegative combination of the elements of E [17, §8.8]. For each $i = 1, \ldots, m$, we fix such a representation for r_i and denote $E_i \subseteq E$ the set of elements of E appearing in the representation with positive coefficients. We apply the generalized Hall's theorem [19, Theorem 1] to the family $\mathcal{A} = \{E_1, \ldots, E_m\}$ of subsets of E and the function μ such that $\mu(S)$ is defined to be the dimension of the linear span of the elements of S for every $S \subseteq E$. This yields linearly independent elements $e_1, \ldots, e_m \in E$ such that $e_i \in E_i$ for every $i = 1, \ldots, m$. For every $i = 1, \ldots, m$, r_i is a positive combination of e_i and maybe some other elements of E, hence $\|e_i\|_0 \leqslant \|r_i\|_0$. Using (6), we have

$$\|L_1\|_0 \leqslant \sum_{i=1}^{m} \|e_i\|_0 \leqslant \sum_{i=1}^{m} \|r_i\|_0 = \|\widetilde{L}_1\|_0,$$

and this proves the minimality of the number of the nonzero entries in $L_1 = TL$ for T constructed by the algorithm.

3 Case Studies

In this section, we demonstrate the improvements in physical intelligibility (while preserving the dimension) of reductions of biochemical models by our Algorithm 1. We analyse the results of the algorithm using models taken from the literature. We also compare the resulting reduction to the ones obtained by ERODE [4] which are always defined by zero-one linear combinations.

3.1 Multisite Protein Phosphorylation

Setup. We consider a model of multisite phosphorylation [18]. It describes a protein with m identical and independent binding sites that simultaneously undergo phosphorylation and dephosphorylation. Each binding site can be in one of the four different states (see Fig. 1):

1. unphosphorylated and unbound,
2. unphosphorylated and bound to a kinase,
3. phosphorylated and unbound,
4. phosphorylated and bound to a phosphatase.

Therefore, there are $4^m + 2$ chemical species in the corresponding reaction network: free kinase and phosphatase, and 4^m states of the protein.

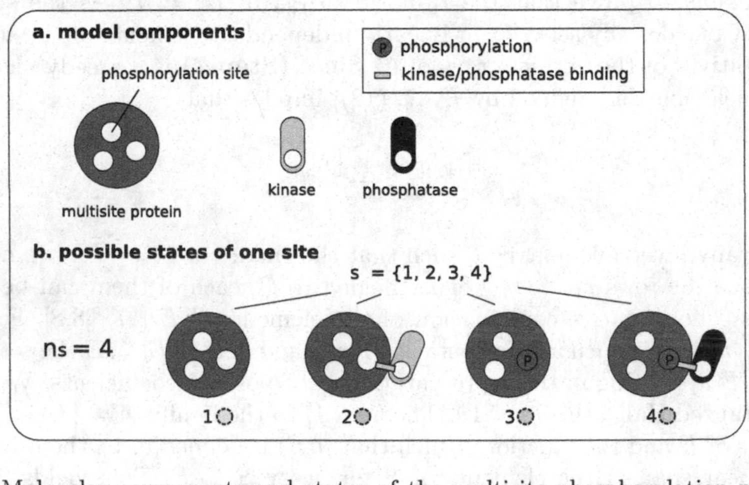

Fig. 1. Molecular components and states of the multisite phosphorylation model. a. Consists of multisite proteins and kinases. This example has $m = 3$ sites. b. There are 4 possible states for a single site: unphosphorylated and unbound, unphosphorylated and bound to a kinase, phosphorylated and unbound, phosphorylated and bound to a phosphatase.

Reductions by ERODE and CLUE. In the reduction computed by ERODE [5] (for $m = 2, \ldots, 8$), the concentrations of protein configurations are replaced by the sums of the concentrations of configurations differing by a permutation of the sites. Therefore, the number of macro-variables is equal to $\binom{m+3}{3} + 2$.

In contrast, the analysis performed by CLUE [16] always results in just six macro-variables. Two of them were always the concentrations of kinase and phosphatase as for ERODE. The other four were linear combinations with protein configurations. In [16, Section 4.2], for $m = 2$, interpretation was provided for the first three of them. However, for the last one, it was remarked that "the last macro-variable escaped physical intelligibility as it represents the difference between the free substrate with unphosphorylated sites and protein configurations that appear in the aforementioned lumps."

Our Results. We applied our algorithm to the cases $m = 2, 3, 4, 5$ and obtained new macro-variables, which have again included the concentrations of free kinase and phosphatase. Moreover, the three interpretable macro-variables from the analysis in [16] for $m = 2$ are kept. Each of the four our macro-variables involving the protein configurations corresponds to a state of a site (e.g., unbounded and unphosphorylated), and each protein configuration appears with a coefficient equal to the number of sites in it with this state. Examples of these new macro-variables are given on Fig. 2 for $m = 2$ and $m = 3$.

One way to interpret the result is that the constructed reduction replaces the concentration of the protein configurations with the "concentrations" of each of the four states of the sites (see also Example 1). From our interpretation, we expect that the models with larger m will have a reduction of the same form.

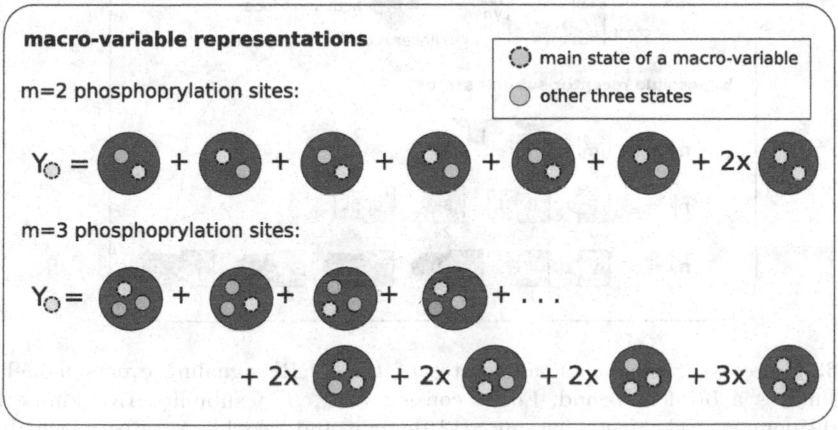

Fig. 2. New macro-variables for $m = 2, 3$. Each state of a binding site from Fig. 1-b can be the main state, yielding four macro-variables for each m. The coefficients are equal to the number of binding sites in a protein that are in the main state.

3.2 Fcϵ-RI Signaling Pathways

Setup. We consider a model for a different kind of multisite phosphorylation [8], a model for the early events in the signaling pathway of the high-affinity IgE receptor (FcϵRI) in mast cells and basophils.

The model details the rule-based interactions of FcϵRI receptor with a bivalent ligand (IgE dimer), the Src kinase Lyn, and the cytosolic protein tyrosine kinase Syk. The model is based on the following sequence of signaling events in FcϵRI [13,14] (the reactions are nicely summarized on [8, Figure 2]):

1. binding of IgE ligand and FcϵRI which aggregates at the plasma membrane,
2. transphosphorylation of tyrosine residues in the immunoreceptor tyrosine-based activation motifs (ITAMs) of the aggregated receptor by constitutively associated Lyn,
3. recruitment of an extra Lyn/Syk kinase to the phosphorylated ITAM sites,
4. transphosphorylation of Syk by Lyn and Syk on its linker region and activation loop, respectively.

For visualizing different chemical species occurring in the resulting reaction network, we use the representation [8, Figure 1] summarized in Fig. 3. In total, there are 354 of three types: monomers, dimers, and non-receptor states (free ligand/Lyn and Syk in each of 4 possible states of phosphorylation) (Fig. 4).

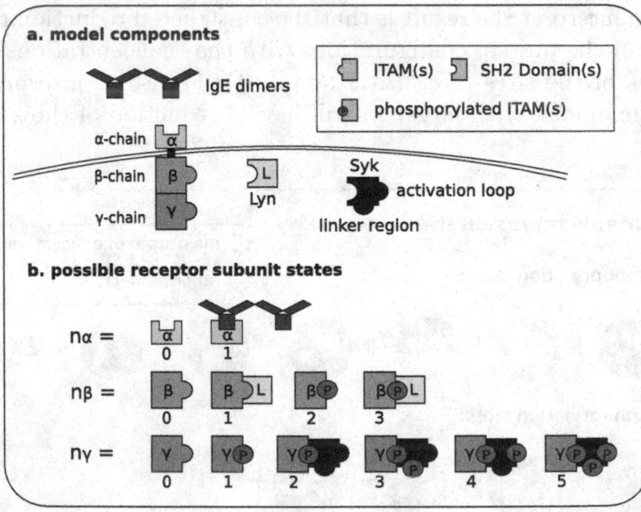

Fig. 3. Molecular components and states of the FcϵRI signaling events model. a. IgE dimer is a bivalent ligand. FcϵRI consists of α, β, γ subunits. Lyn kinase has an SH2 domain. Syk kinase has an SH2 domain and two ITAM sites which differ by the method of phosphorylation: Lyn at the linker region, and Syk at the activation loop. b. The α subunit can be unbound or bound to a ligand. β can be unphosphorylated/phosphorylated, with/without associated Lyn. γ can be unphosphorylated/phosphorylated, and the phosphorylated form binds to Syk in any of the four states of phosphorylation.

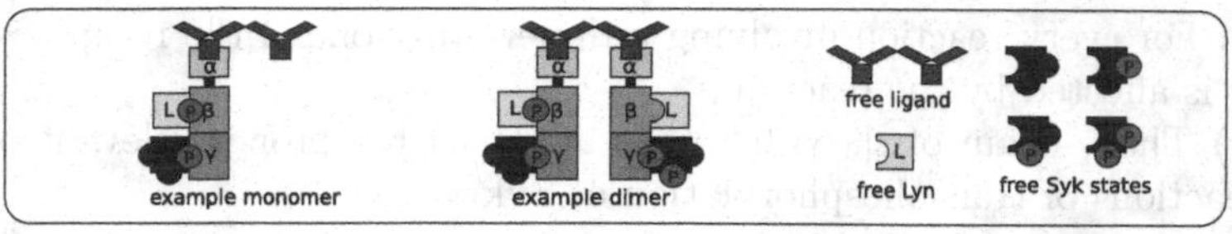

Fig. 4. Examples of a monomer, a dimer, and the free components.

Reductions by ERODE and CLUE. The reduction by ERODE [5] consists of 105 macro-variables, where all the complexes with the same configuration except for the phosphorylation state of the Syk units are summed up in a single macro-variable. We will refer to these macro-variables as *Syk-macro-variables.*

The model has been reduced using CLUE in [16, Section 4.2] with the observable being the total concentration of the free Syk (in all the four phosphorylation states). The reduced model had 69 macro-variables, and 51 of them were nontrivial. It has been observed in [16, Section 4.2] that some of these macro-variable carry a physical interpretation, but in some of them, negative elements were present, which hinder their physical intelligibility.

Our Results. We apply our algorithm to the reduced model. Among the new macro-variables, we have 51 nontrivial macro-variables which is the same as for the CLUE reduction. More precisely, **(Step 4)** produced 57 nontrivial macro-variables, and this number has been reduced to 51 when computing a linearly independent basis on **(Step 6)**. The resulting macro-variable refine the reduction by ERODE mentioned above in the sense that our new macro-variable are the sums of the Syk-macro-variables with non-negative coefficients. Therefore, in our reduction, all the complexes differing only by the phosphorylation state of the Syk units are in the same macro-variable. For the *monomers*, we obtain the same reduction: nontrivial macro-variable involving monomers are of the form described on Fig. 5.

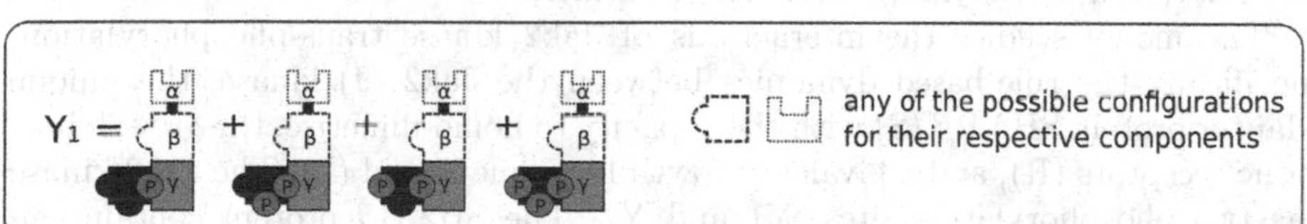

Fig. 5. The monomer macro-variables. In each of them, α and β are fixed, and we sum over all the phosphorylation patterns of Syk.

The macro-variable involving dimers are graphically described on Fig. 6. First, one can see that they are indeed linear combinations of the Syk-macro-variables. Our interpretation of these new macro-variables is based on two observations about the set of the reactions in the original model [8, Figure 2]:

(Obs. 1) For every reaction involving a dimer, only one of the receptors of the dimer is affected by the reaction.
(Obs. 2) The γ-chain of the other (not affected) receptor is relevant only for the reactions of transphosphorylation of Syk.

Since the complexes with different phosphorylation patterns are grouped together in the Syk-macro-variables, the second observation implies that the transphosphorylation reactions do not affect the values of the Syk-macro-variables at all. Therefore, the first observation suggests considering macro-variables as sums over all the dimer configurations in which one receptor is fixed (up to the phosphorylation of Syk), and for the other receptor, all the possible variants of the γ-chain are considered.

With this interpretation in mind, let us take a closer look at the Fig. 6:

- Each of the variables as on Fig. 6a is the sum over the configurations with the fixed left receptor not carrying Syk and the right receptor having each of the six possible γ-chains. If one of the complexes in the sum is fully symmetric, it appears with coefficient 2.
- Each of the variables as on Fig. 6b is a combination of complexes that have: the same β-chains and Syk on the left receptor, any phosphorylation pattern of the Syk on the left receptor, and any γ-chain on the right receptor. If the β-chains on the receptors are equal, the complexes with two Syk's (which are *symmetric up to* Syk phosphorylation) appear with coefficient 2.

Note that the coefficients 2 appearing in the presence of symmetry prevent ERODE [5] from finding this reduction.

3.3 Jak-Family Protein Tyrosine Kinase Activation

Setup. We study a simplified cellular model of a bipolar "clamp" mechanism for Jak-family kinase activation [2]. Kinases of the Janus kinase (JAK) family play an essential role in signal transduction mediated by cell surface receptors, which lack innate enzymatic activities to dimerize.

The model studies the interactions of Jak2 kinase trans-phosphorylation, specifically the rule-based dynamics between the Jak2 (J) kinase, the unique adaptor protein SH2-Bβ (S) with the capacity to homo-dimerize, the growth hormone receptors (R), and a bivalent growth hormone ligand (L). The Jak2 kinase has two phosphorylation sites, Y1 and Y2. The SH2-Bβ protein contains an N-terminal dimerization domain (DD) and a C-terminal Src homology-2 (SH2) domain.

The components can interact in the following ways:

1. binding of ligand and growth hormone receptors which aggregates at the plasma membrane,
2. constitutive binding of Jak2 kinase to the receptors, which autophosphorylates on the phosphorylation sites when two Jak2 kinases are bounded in the same complex,

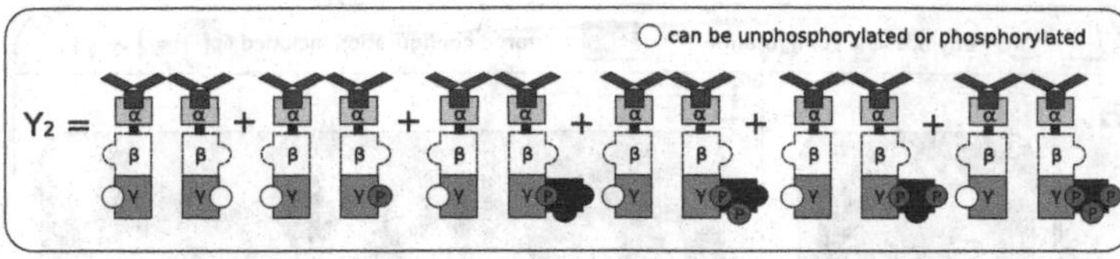

(a) No Syk on the left γ-chain, the right takes all the possible γ-configurations

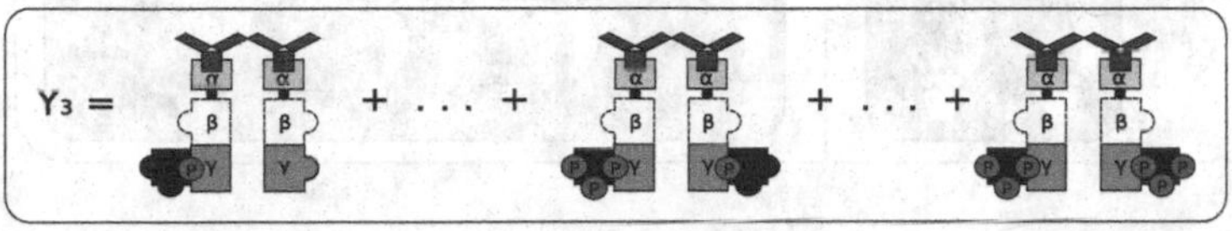

(b) Syk (with all possible phosphorylations) on the left γ-chain, the right takes all the possible γ-configurations.

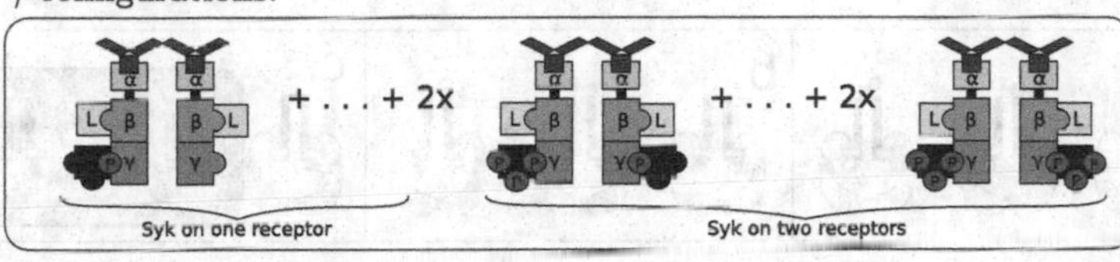

(c) An example configuration of Y_3 (b) with identical β-chains. A variable has coefficient 2 when Syk kinases are recruited on both receptors (regardless of Syk state).

Fig. 6. Macro-variables involving dimers

3. recruitment of SH2-Bβ protein at the SH2 domain by the Jak2's autophosphorylated Y1 site,
4. dimerization of SH2-Bβ protein through recruitment of an additional SH2-Bβ protein, engaged at the DD domains.

Receptors can undergo a process of *internalization*, in which the receptors can no longer associate with any Jak2, and the existing Jak2 and SH2-Bβ in the complex can dissociate at the normal rate (Fig. 7).

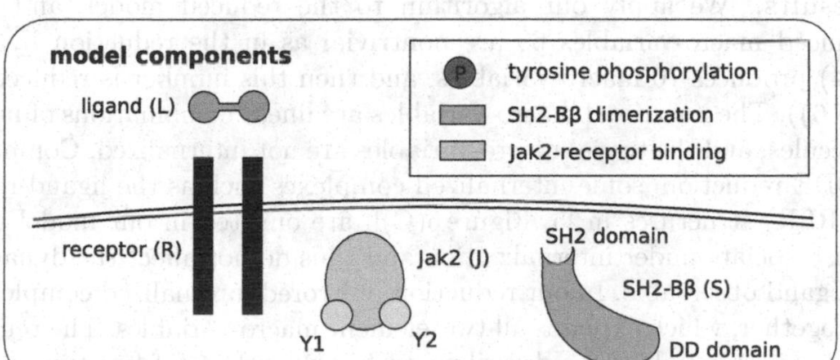

Fig. 7. Components of the Jak-family protein tyrosine kinase activation model.

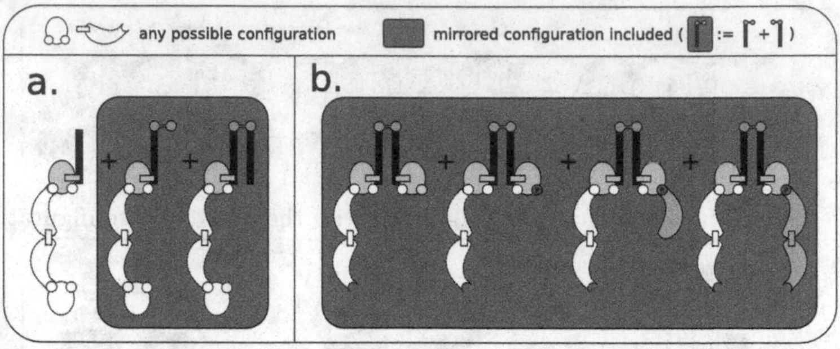

Fig. 8. Classes of internalized macro-variables. a. equivalent up to the connection between the ligand and the receptors. b. one receptor fixed.

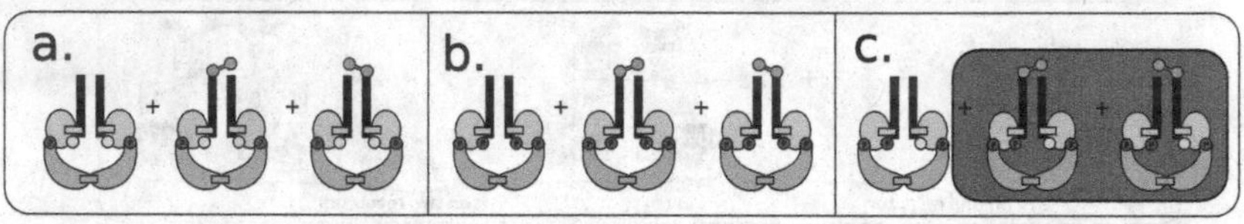

Fig. 9. Macro-variables for equivalent structures of the bipolar "clamp" mechanism

Reductions by ERODE and CLUE. The reduction obtained by ERODE in [5, Figure 5] contained 345 macro-variables. It grouped the internalized configurations, which differ by the connections between the receptors and the ligand, into macro-variables.

The model has been reduced by CLUE in [16], with the observable being the concentration of the free ligand. The reduced model had 322 macro-variables, and 69 of them were nontrivial. The model has been used in [16] for benchmarking purposes only, so the macro-variables have not been interpreted. The reduction included several macro-variables with negative coefficients, including one with 20 nonzero coefficients. We do not see any natural interpretation for them.

Our Results. We apply our algorithm to the reduced model, and among the produced macro-variables 69 are nontrivial as in the reduction by CLUE (**(Step 4)** produces 70 macro-variables, and then this number is reduced to 69 at **(Step 6)**). The nontrivial macro-variables are linear combinations of internalized molecules, and the trivial macro-variables are not internalized. Compared to the ERODE reduction, some internalized complexes such as the ligand-receptor (R, RL, RLR) structures in [5, Figure 5(C)], are omitted in our model as they do not disassociate under internalization and thus do not affect the dynamics of the free ligand observable. In our reduction, mirrored internalized complexes are lumped together, which explains all two-element macro-variables. The remaining nontrivial macro-variables are described on Fig. 8, and are of two types:

- *Configurations equivalent up to the connection between the ligand and the receptors (Fig. 8-a and 9).* The structures are equivalent under internalization

as the ligand and receptors cannot disassociate and were obtained also by ERODE [5, Figure 5(D, E)]. They are of two types: single-Jak2-Receptor case (Fig. 8-a) and "clamp" case (Fig. 9).
- *Configurations with one receptor fixed (Fig.* 8-b). These are similar to Fig. 6 from the case study in Sect. 3.2: since in the reactions with internalized complexes, only one receptor is affected, and this does not depend on the state of the other receptor, one can group together the complexes having one of the receptors the same.
When the receptors are symmetric, the element and its mirrored element are the same, so the corresponding configuration appears with coefficient 2.

4 Conclusion

We have hypothesized that the interpretability of the macro-variables in an exact linear reduction may be improved by a change of coordinates making the macro-variable nonnegative combinations of the original variables and minimizing the number of nonzero coefficients. We have designed and implemented an algorithm for performing such a transformation and applied it to three models (with hundreds of variables) for which the result of the reduction by CLUE [16] contained macro-variables without a clear physical interpretation. We have shown that the resulting macro-variables are interpretable, thus supporting the original hypothesis and demonstrating the usefulness of our algorithm.

Our results also give insight into the structure of reductions in which not all the coefficients are zeroes and ones. In particular, we can point out two different situations:

- The macro-variables are the "concentrations" of parts of molecules as in Sect. 3.1. The species having several identical pieces may appear with larger coefficients.
- Some of the molecules appearing in the macro-variable are symmetric (as in Sect. 3.3) or even partially symmetric (as in Sect. 3.2), and they appear with a coefficient accounting for the symmetries.

Acknowledgement. The authors are grateful to François Fages, Mathieu Hemery, Sylvain Soliman, and Mirco Tribastone for helpful discussions and to the referees for helpful suggestions. GP is grateful to Heather Harrington, Gregory Henselman-Petrusek, and Zvi Rosen for educating him about the matroid theory.

Appendix: Non-positivizable Reduction

As we have mentioned in Remark 4, we did not encounter examples from the literature for which Algorithm 1 would return NO. However, one can easily construct an artificial example with this property. Consider the system

$$\begin{cases} x_1' = x_1^2 + x_2^2, \\ x_2' = 2x_1x_2. \end{cases} \tag{7}$$

Then $y = x_1 - x_2$ yields a reduced system $y' = y^2$. However, since any change of macro-variables is a scaling of y, there is no equivalent lumping with nonnegative coefficients, so Algorithm 1 (with the input $L = (1\ -1)$) will return NO.

References

1. Antoulas, A.: Approximation of Large-Scale Dynamical Systems. Advances in Design and Control, SIAM (2005)
2. Barua, D., Faeder, J.R., Haugh, J.M.: A bipolar clamp mechanism for activation of Jak-family protein tyrosine kinases. PLoS Comput. Biol. **5**(4), e1000364 (2009). https://doi.org/10.1371/journal.pcbi.1000364
3. Borisov, N., Markevich, N., Hoek, J., Kholodenko, B.: Signaling through receptors and scaffolds: Independent interactions reduce combinatorial complexity. Biophys. J. **89**(2), 951–966 (2005). https://doi.org/10.1529/biophysj.105.060533
4. Cardelli, L., Tribastone, M., Tschaikowski, M., Vandin, A.: ERODE: a tool for the evaluation and reduction of ordinary differential equations. In: Legay, A., Margaria, T. (eds.) TACAS 2017. LNCS, vol. 10206, pp. 310–328. Springer, Heidelberg (2017). https://doi.org/10.1007/978-3-662-54580-5_19
5. Cardelli, L., Tribastone, M., Tschaikowski, M., Vandin, A.: Maximal aggregation of polynomial dynamical systems. Proc. Nat. Acad. Sci. **114**(38), 10029–10034 (2017). https://doi.org/10.1073/pnas.1702697114
6. Conzelmann, H., Fey, D., Gilles, E.: Exact model reduction of combinatorial reaction networks. BMC Syst. Biol. **2**(1), 78 (2008). https://doi.org/10.1186/1752-0509-2-78
7. Edmonds, J.: Matroids and the greedy algorithm. Math. Program. **1**(1), 127–136 (1971), https://doi.org/10.1007/bf01584082
8. Faeder, J.R., et al.: Investigation of early events in FcϵRI-mediated signaling using a detailed mathematical model. J. Immunol. **170**(7), 3769–3781 (2003). https://doi.org/10.4049/jimmunol.170.7.3769
9. Feret, J., Danos, V., Krivine, J., Harmer, R., Fontana, W.: Internal coarse-graining of molecular systems. Proc. Nat. Acad. Sci. **106**(16), 6453–6458 (2009). https://doi.org/10.1073/pnas.0809908106
10. Fieker, C., Hart, W., Hofmann, T., Johansson, F.: Nemo/Hecke: computer algebra and number theory packages for the Julia programming language. In: Proceedings of the 2017 ACM on International Symposium on Symbolic and Algebraic Computation, pp. 157–164. ISSAC 2017, ACM, New York (2017). https://doi.org/10.1145/3087604.3087611
11. Gawrilow, E., Joswig, M.: polymake: a framework for analyzing convex polytopes. In: Polytopes - Combinatorics and Computation, pp. 43–73. Birkhäuser Basel (2000). https://doi.org/10.1007/978-3-0348-8438-9_2
12. Gunawardena, J.: Multisite protein phosphorylation makes a good threshold but can be a poor switch. Proc. Nat. Acad. Sci. **102**(41), 14617–14622 (2005). https://doi.org/10.1073/pnas.0507322102
13. Metzger, H., Eglite, S., Haleem-Smith, H., Reischl, I., Torigoe, C.: Quantitative aspects of signal transduction by the receptor with high affinity for IgE. Mol. Immunol. **38**(16–18), 1207–1211 (2002). https://doi.org/10.1016/s0161-5890(02)00065-2

14. Nadler, M.J., Matthews, S.A., Turner, H., Kinet, J.P.: Signal transduction by the high-affinity immunoglobulin E receptor FcεRI: coupling form to function. Adv. Immunol. 76, 325–355 (2001). https://doi.org/10.1016/S0065-2776(01)76022-1. https://www.sciencedirect.com/science/article/pii/S0065277601760221
15. Okino, M., Mavrovouniotis, M.: Simplification of mathematical models of chemical reaction systems. Chem. Rev. **2**(98), 391–408 (1998). https://doi.org/10.1021/cr950223l
16. Ovchinnikov, A., Verona, I.P., Pogudin, G., Tribastone, M.: CLUE: exact maximal reduction of kinetic models by constrained lumping of differential equations. Bioinformatics (2021). https://doi.org/10.1093/bioinformatics/btab010
17. Schrijver, A.: Theory of Linear and Integer Programming. Wiley-Blackwell (1986)
18. Sneddon, M.W., Faeder, J.R., Emonet, T.: Efficient modeling, simulation and coarse-graining of biological complexity with NFsim. Nat. Methods **8**(2), 177–183 (2010). https://doi.org/10.1038/nmeth.1546
19. Welsh, D.: Generalized versions of Hall's theorem. J. Comb. The. Ser. B **10**(2), 95–101 (1971). https://doi.org/10.1016/0095-8956(71)90069-4
20. Ziegler, G.M.: Lectures on Polytopes. Springer, New York (1995). https://doi.org/10.1007/978-1-4613-8431-1

Explainable Artificial Neural Network for Recurrent Venous Thromboembolism Based on Plasma Proteomics

Misbah Razzaq[1,2(✉)], Louisa Goumidi[3], Maria-Jesus Iglesias[5,6], Gaëlle Munsch[1], Maria Bruzelius[7,8], Manal Ibrahim-Kosta[3,4], Lynn Butler[5,6], Jacob Odeberg[5,6], Pierre-Emmanuel Morange[3,4], and David Alexandre Tregouet[1,2(✉)]

[1] Univ. Bordeaux, Inserm, BPH, U1219, 33000 Bordeaux, France
misbah.razzaq@inserm.fr,
{gaelle.munsch,david-alexandre.tregouet}@u-bordeaux.fr
[2] Laboratory of Excellence GENMED (Medical Genomics), Bordeaux, France
[3] Aix Marseille Univ, Inserm, INRAE, C2VN, Marseille, France
{louisa.Goumidi,manal.ibrahim,pierre.morange}@ap-hm.fr
[4] Laboratory of Haematology, La Timone Hospital, Marseille, France
[5] Science for Life Laboratory, Department of Protein Science, CBH, KTH Royal Institute of Technology, Stockholm, Sweden
{mariajesus.iglesias,jacob.odeberg}@scilifelab.se
[6] Department of Clinical Medicine, Faculty of Health Science, The Arctic University of Tromsö, Tromsö, Norway
lynn.m.butler@uit.no
[7] Department of Medicine Solna, Karolinska Institute, Stockholm, Sweden
maria.bruzelius@sll.se
[8] Department of Hematology, Karolinska University Hospital, Stockholm, Sweden

Abstract. Venous thromboembolism (VTE) is the third most common cardiovascular disease, affecting ∼1,000,000 individuals each year in Europe. VTE is characterized by an annual recurrent rate of ∼6%, and ∼30% of patients with unprovoked VTE will face a recurrent event after a six-month course of anticoagulant treatment. Even if guidelines recommend life-long treatment for these patients, about ∼70% of them will never experience a recurrence and will receive unnecessary lifelong anti-coagulation that is associated with increased risk of bleeding and is highly costly for the society. There is then urgent need to identify biomarkers that could distinguish VTE patients with high risk of recurrence from low-risk patients.

Capitalizing on a sample of 913 patients followed up for the risk of VTE recurrence during a median of ∼10 years and profiled for 376 plasma proteomic antibodies, we here develop an artificial neural network (ANN) based strategy to identify a proteomic signature that helps discriminating patients at low and high risk of recurrence. In a first stage, we implemented a Repeated Editing Nearest Neighbors algorithm to select a homogeneous sub-sample of VTE patients. This sub-sample was then split in a training and a testing sets. The former was used for

E. Cinquemani and L. Paulevé (Eds.): CMSB 2021, LNBI 12881, pp. 108–121, 2021.
https://doi.org/10.1007/978-3-030-85633-5_7

training our ANN, the latter for testing its discriminatory properties. In the testing dataset, our ANN led to an accuracy of 0.86 that compared to an accuracy of 0.79 as provided by a random forest classifier. We then applied a Deep Learning Important FeaTures (DeepLIFT) – based approach to identify the variables that contribute the most to the ANN predictions. In addition to sex, the proposed DeepLIFT strategy identified 6 important proteins (DDX1, HTRA3, LRG1, MAST2, NFATC4 and STXBP5) whose exact roles in the etiology of VTE recurrence now deserve further experimental validations.

Keywords: Artificial neural network · Interpretation · Thrombosis · Proteomics · Imbalanced

1 Introduction

Venous thromboembolism (VTE) is the pathological result of a blood clot (thrombus) forming in the deep veins of the leg that can obstruct venous circulation (also known as deep vein thrombosis DVT) and that, in approximately 40% of cases, migrates to the lung artery, causing pulmonary embolism (PE). PE has a 6% mortality rate in the acute phase and a 20% mortality rate after one year. There is a high likelihood of thrombosis recurrence after the initial event. Even though the anti-coagulation treatment is effective, studies have shown that recurrent episodes of thrombosis occur in ∼20% to 30% of cases within the five years following the first event [11,17,37], and even in ∼4% of patients under anticoagulant therapy [8,9,43]. Therefore it is essential to understand which subgroup of patients are at higher risk of thrombosis recurrence. Our objective here is to identify novel molecular markers associated with VTE recurrence in order to get insights into the disease pathophysiology.

VTE recurrence is considered to be a multifactorial disease where genetic and non genetic factors act together to modulate the individual risk of recurrence. However, so far, limited risk factors of VTE recurrence have been identified [20]. To cite a few, the Factor V Leiden mutation has been proposed to be associated with ∼ 2-fold increased risk [48]. Increased plasma levels of D-dimer, factor VIII, and C-Reactive protein have also been shown to be associated with increased risk of recurrent VTE [44]. However, their ability to predict which VTE patients will develop a recurrence event are still limited, leaving room for identifying additional risk markers.

Plasma is an ideal source for discovering novel biomarkers for VTE and its complications. Indeed, plasma drawing is rather non-invasive, fast and cost-effective, and this intravascular compartment is the site of disease manifestation. In this work, we aim at identifying novel plasma proteins that could help discriminating patients at high or low risk of recurrence by deploying an Artificial Neural Network (ANN) strategy on 234 proteins measured in the plasma sample of 913 VTE patients followed up for VTE recurrence during 9.51 years. ANN is a class of artificial intelligence technique that is increasingly popular in the biomedical domain and that have the advantage, among others, to identify complex non linear relationships between variables [1,4,28,38]. However, their applications to thrombosis are still sparse. Calazans-Romano et al. [40] built an ANN model from 44 clinical and biological laboratory variables to predict the risk of recurrence in 240 DVT patients. More recently, Martins et al. [25] proposed different ANNs to predict VTE recurrence from 39 clinical variables in a sample of 261 VTE patients. While both studies demonstrated good accuracy for predicting VTE recurrence using ANN, they do not draw much attention about novel biological knowledge on molecular players involved in VTE recurrence.

In the present work, we propose an ANN based strategy to predict VTE recurrence from a larger number of features profiled in a much larger study of VTE patients. Besides, we implement a recently proposed methodology to investigate which features contribute the most to the ANN predictions with the objective to identify novel molecular determinants associated with the risk of VTE recurrence.

The remainder of the paper is structured as follows. In Sect. 2, we describe the datasets and the different methods implemented in the proposed workflow. Results are described in Sect. 3 and Sect. 4 provides general concluding remarks and perspectives.

2 Materials and Methods

2.1 MARTHA Study

The MARTHA population is composed of ~2,900 VTE patients recruited between 1994 and 2012 from the Thrombophilia center of La Timone hospital (Marseille, France) and free of any chronic conditions and of any well characterized genetic risk factors including antithrombin, protein C or protein S deficiency, homozygosity for FV Leiden or Factor II 20210A, and lupus anticoagulant. Detailed description of the MARTHA population has been provided elsewhere [10,34].

MARTHA Proteomics Sub-study. Out of ~2,900 patients, 913 patients were followed-up for recurrence information among which 162 experienced a recurrent event during the follow-up. At the time of inclusion in the study, before the start of the follow-up period, plasma samples were collected for these patients and were measured for 376 antibodies targeting 234 proteins using high-affinity bead array technology. The selected proteins were either reported to have an association to the cardiovascular trait or have shown an important role in the coagulation and fibrinolysis cascades. The 7 clinical variables, i.e., sex, age, bmi, smoking, type of first thrombosis, family history, and provoked or unprovoked thrombosis were selected as part of the learning dataset. These variables have previously shown to be involved in venous thrombosis. The proteomics data along with 7 clinical variables forms our initial dataset D. Detailed description of the technique used for plasma protein profiling can be found here [39]. All proteomic data were centered and standardized using the StandardScaler from scikit-learn [36] before further downstream analyses.

2.2 Proposed Workflow

Data Exploration. VTE recurrence is hypothesized to be a complex disorder where multiple players jointly contribute to the individual susceptibility to develop a recurrent thrombotic event, especially as the biological sources underlying the occurrence of the first event are also complex. As a consequence, we anticipated that the profiling of ~400 antibodies in the plasma of 913 VTE patients would not be able to easily discriminate between patients that will develop recurrence from those that will not. The application of different techniques such as principal component analysis (PCA), t-distributed stochastic neighbour embedding (t-SNE) [24], and uniform manifold approximation and projection (UMAP) [27] confirm that no clear separation between the recurrence and non-recurrence class could be obtained as shown in Fig. 1. This highlights the need for more complex non linear methodologies, such as ANN, to derive an accurate classification tool for recurrence.

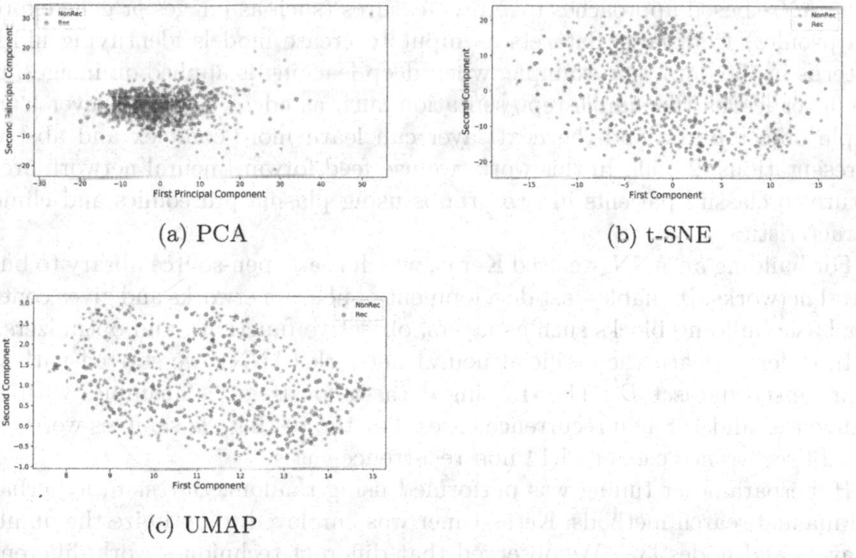

(a) PCA (b) t-SNE

(c) UMAP

Fig. 1. Visualization of the learning dataset. The class recurrence and non recurrence are represented by red and green colors, respectively. (Color figure online)

Repeated Edited Nearest Neighbors Algorithm. Classification algorithms are known to be very sensitive to unbalanced data when the aim is to derive classification/prediction tools for categorical classes. In general, algorithms will correctly classify the most frequent classes and lead to higher misclassification rates for the minority classes which is often the most interesting ones. In the

MARTHA sample, we have approximately 5 times more patients that did not experience VTE recurrence than patients that did. Therefore, before implementing an ANN for predicting VTE recurrence, we first deployed an under sampling algorithm, the repeated edited nearest neighbors algorithm (RENN) [15,47], to create a more balanced dataset $D^{'}$ of recurrent and non-recurrent patients, while keeping all recurrent patients and selecting a homogeneous, with respect to plasma proteomic data, sub-sample of non-recurrent patients. The RENN algorithm starts with the D dataset and iteratively removes the samples which contradicts with its k nearest neighbors where k is typically an odd number to avoid ties. These k nearest neighbors are calculated using an Euclidean distance function. By the completion of this algorithm, we were left with a $D^{'}$ dataset composed of 270 samples including 162 patients with recurrence and 108 without, that was then used to train and test an ANN model for VTE recurrence.

Artificial Neural Network. An ANN is a biologically inspired computing system composed of different units (neurons). Each unit takes real valued inputs and transforms it to a single real valued output. It can handle complex and noisy data, and, unlike classical machine learning methods, can also model non-linear relationships thanks to the layer-wise architecture of non-linear processing units. ANN based approaches take raw features (such as images or gene expression profiles) from large datasets as input to create models identifying hidden patterns in the data. For example, when deep learning is applied on images, the first layer learns the simple representation such as edges, the next layer learns simple object parts, and the next layer can learn more complex and abstract representations [23,50]. In this work, we use feed forward neural network architecture to classify patients in two groups using plasma proteomics and clinical characteristics.

For building an ANN, we used Keras, which is an open-source library to build neural networks. It enables fast development of neural networks and gives control over basic building blocks such as layers, objective functions, and optimizers.

In order to learn the artificial neural network (ANN), we started with the preprocessed dataset $D^{'}$. The training dataset consist of 242 samples with 162 recurrences and 108 non recurrence cases. For the testing, 28 samples were used with 14 recurrence cases and 14 non recurrence cases.

Hyperparameter tuning was performed using random, Bayesian, hyperband and manual search methods. Keras-tuner was employed to optimize the number of layers and nodes [32]. We observed that different techniques work differently on data as each data is unique [22]. In terms of computation time, the fastest method was hyperband followed by random search and Bayesian method. All three methods proposed to use 3 to 4 layers. Each method proposed different number of nodes in each layer ranging from 192 to 450. The best performing model was obtained by taking guidance from these methods and by following the rules of thumb for setting number of neurons. The general rule of thumbs states that number of neurons in the hidden layer [18,41,49]:

1. should be between the size of input and output neurons.

2. should not be greater than twice the number of neurons of input layer.
3. should be the 2/3 size of the input and output neurons.

We built a 3 layer artificial neural network. The ReLU activation function [16] was used to learn non-linear decision boundaries. The softmax activation function [3] was used to generate probabilities for recurrence and non recurrence class at the output layer. To avoid overfitting, regularization rate of 0.25 was used.

After fixing the number of nodes, layer and activation function, the process of training the neural network can start. Starting from random weights, forward propagation is used to generate the output of all nodes at all layers while moving from the input to the output layers. The generated final output is compared to the observed class phenotype and an error is calculated using the cross-entropy function [19]. Iteratively, this error was then backpropagated using a stochastic gradient descent method (Adam) [21] with learning rate of 0.004 and batch size of 32, to update weights according to their contribution to the error.

It is important to note that the model which we get at the end of the training may not be the best stable model. In order to reduce the over-fitting and get the best performing model, we used the callbacks feature proposed by the Keras [5].

We have performed the comparison with state of the art method, i.e., random forest classifier. We used scikit learn API to build a random forest classifier with preprocessed dataset $D^{'}$. The number of trees were set to 100 as suggested in literature [33]

DeepLIFT. DeepLIFT (Deep Learning Important FeaTures) is a model specific heuristic based method that has been proposed to identify important features contributing to ANN derived predictions. This method assigns contribution scores to the features of a particular observation according to a reference point (i.e. control value). More precisely, it explains the difference in the output from the reference output in terms of the difference in the input from the reference input. DeepLIFT basically tries to trace back contributions to the input features by backpropagating activated neurons. For each sample s, an importance score $IS_{s,i,j}$ is calculated, representing the contribution of input i to the output j in a sample s.

The choice of the reference value is crucial, requires a careful consideration, and relies on domain-specific knowledge. For example, for an image classification task, we can use the absence of pixels as a reference value. DeepLIFT uses a backpropagation-like algorithm that uses chain rules to compute importance scores. Note that if softmax function is used to generate probabilities at the output layer then the importance score should be mean normalized as recommended in [42].

DeepLIFT assigns positive scores to the features contributing toward the predicted class and negative scores to the features contributing against the predicted class. The magnitude of the importance score of particular feature represents the strength of the contribution.

A key element for applying DeepLIFT is the definition of the reference values that are used as control to define which features positively contribute to the recurrence class. As such, selecting as control/reference value, the clinical and proteomics characteristics of a VTE patient that did not experience any recurrent event during a long period of follow-up appear rather intuitive. However, as it has been shown that DeepLIFT [42], could be sensitive to the choice of reference values, the successive application of DeepLIFT with multiple different reference values has been proposed to obtain more stable results. As consequence, for our application, we selected 5 independent VTE patients that did not experience a recurrent VTE event despite a follow up time of more than 18 years. The clinical and proteomics characteristics of these 5 patients were then used a reference value for 5 independent run of DeepLIFT.

Most important variables identified by the DeepLIFT methodology were finally tested for association with VTE recurrence using a delay-entry Cox model [7,46] adjusted for sex, age at first thrombosis, provoked status and type of VTE (i.e. DVT vs PE) at the first thrombosis, smoking and anticoagulant therapy at the inclusion of the status.

3 Results

3.1 MARTHA Study

The original dataset D include 913 VTE patients with 2 different classes, i.e., recurrence (Rec) and non recurrence (NonRec). Of these 913 patients, 162 experienced recurrence during the follow-up and 751 remained free of new thrombotic events. Clinical characteristics of these individuals are shown in Table 1.

Table 1. Clinical characteristics of the MARTHA proteomics study.

	$D^{'}$ (Rec)	$D^{'}$ (NonRec)	$D - D^{'}$ (NonRec)
N	162	108	643
Age at sampling	44.22 (13.84)	50.26 (14.99)	47.25 (14.64)
Age at first VTE	40.10 (14.09)	42.75 (15.81)	42.00 (15.30)
Pulmonary embolism as first VTE	33 (20%)	29 (27%)	109 (17%)
Male Sex	65 (40%)	43 (40%)	220 (34%)
Oral contraception at sampling	16 (10%)	7 (6%)	38 (5%)
Anticoagulant therapy at sampling	49 (30%)	50 (46%)	220 (34%)
Smokers	30 (19%)	15 (14%)	113 (18%)
BMI	25.38 (4.30)	26.82 (4.87)	25.28 (4.63)
Follow up duration	6.19 (4.95)	10.09 (4.03)	10.25 (4.49)
First event provoked	147 (91%)	98 (91%)	602 (94%)

Data shown correspond to mean (standard deviation) and count (percentage) for continuous and categorical variables, respectively.

3.2 Constructing and Validation of the ANN

The application of the RENN algorithm led to the selection of a sub-sample $D^{'}$ of about 162 patients with recurrence and 108 patients without. The clinical characteristics of these patients are shown in Table 1. It can be noticed that non recurrent patients selected by the under-sampling algorithm are not completely representative of the whole set of patients that did not experience recurrence during the follow-up. Indeed, non recurrent patients in the $D^{'}$ sample tended to be, at inclusion in the study, slightly older, more frequently PE patients, more frequently under anticoagulant therapy and less frequently smokers, than non recurrent patients that were not selected by the RENN algorithm.

In order to learn our ANN, we started with the preprocessed dataset $D^{'}$. We divided this dataset into training and testing dataset. The training dataset consist of 242 samples with 148 VTE patients with recurrence and 94 non recurrent VTE patients. As a consequence, the testing dataset resulted in of 28 samples including 14 with recurrence and 14 without recurrence.

From the set of 383 features including 376 plasma antibodies and 7 clinical variables, we implemented a 3 hidden layers neural network. The first hidden layer contain 383 nodes. The second hidden layer contain 2/3 nodes of the input layer and finally 3rd layer contains 1/3 nodes of the input layer.

Once trained on the training dataset, the ANN performance was evaluated on the testing dataset. It led to a accuracy of 0.86 and to F1-scores of 0.86 for both recurrence/non recurrence classes. The resulting area under the operating curve (AUC) was 0.91 [0.82 – 0.97]. By comparison, the implementation of a Random-Forest classifier as described in the Materials and Methods led to an accuracy of 0.79 and to an AUC of 0.79 [0.63 – 0.93].

3.3 Post-hoc Explainability of ANN

Once a classification model is built, it is often desirable to understand the internal composition of such model and to identify the most important contributors to such classification tool. This is indeed a mandatory prerequisite in order to get insights into the underlying biological/clinical structure and to anticipate any clinical translation.

As mentioned in the Materials and Methods section, the DeepLIFT methodology was applied using 5 different references characterizing non recurrence cases to identify features (antibodies and clinical variables) that contribute the most to predict recurrence risk in the $D^{'}$ dataset. The top 20 variables with the highest average importance score over the 5 runs of DeepLIFT are shown in Fig. 2. The variable with the highest importance (~411) was plasma levels of NFATC4, while Sex, a known risk factor for VTE recurrence [45], ranked 12th in this list (average importance ~124). From Fig. 2, it can be deduced that in addition to the "Sex" variable, 6 antibodies targeting 6 different proteins, DDX1, HTRA3, LRG1, MAST2, NFACT4 and STXBP5, could be considered as important contributors of our ANN derived predictions for VTE recurrence.

By comparison, the average importance of the "smoking" and "type of first thrombosis" variables, two additional clinical factors known to associated with VTE recurrence, over the 5 runs of DeepLIFT were only ~15 and ~24.

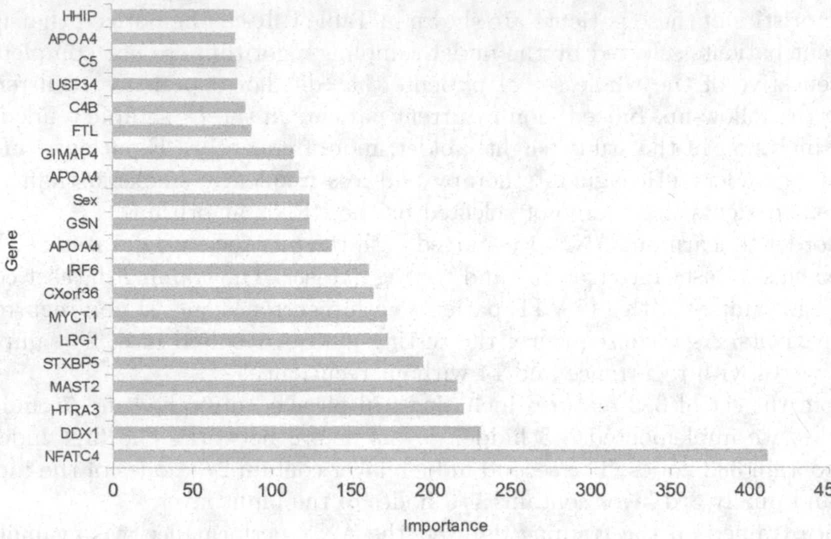

Fig. 2. The top 20 important features ordered by importance score.

As a final step, we were interested to assess whether the identified 6 proteins exert univariate effects on VTE recurrence or whether they were involved in more complex relationships. From the 6 candidate proteins, only DDX1 demonstrated significant and homogeneous (between the D' sub-sample and in the whole MARTHA study) association with recurrence, (Table 2). One standard deviation decrease of DDX1 plasma levels was associated with a Hazard Ratio (HR) for recurrence of 1.35 [1.11 − 1.64] ($p = 3.83\ 10^{-3}$). These observations strongly suggest that the 5 other candidate proteins would be involved in recurrence risk in a more complex pattern.

To explore this hypothesis further, we investigated whether these 5 proteins could interact with clinical factors mentioned above and that discriminate between non recurrent patients selected and non-selected by the undersampling strategy. Only one marginal interaction was observed and it relates to HTRA3 and smoking. While in non-smokers (number of recurrent events = 132 among 755 patients), plasma HTRA3 levels were not associated with recurrence ($HR = 1.02$ [0.86 − 1.21], $p = 0.84$), a strong association was observed in smokers (number of recurrent events = 30, among 158 patients). In smokers, one standard deviation decrease of HTRA3 levels was associated with an increased risk of $HR = 1.94$ [1.20 − 3.13] ($p = 0.007$), the test for interaction between these two HRs being significant at $p = 0.013$. Beyond this marginal interaction, no others were detected suggesting that the identified candidate proteins could interact with other unmeasured clinical factors and/or other proteins including some that were not in the top list identified by the DeepLIFT approach.

Table 2. Association of Top proteins identified by the DeepLIFT methodology with the risk of VTE recurrence in the D' sub-sample and whole MARTHA study.

	D'	D
Proteins	162 Rec among 270 VTE patients	162 Rec among 913 VTE patients
HTRA3	$0.627\ [0.543 - 0.723], p = 1.7\ 10^{-10}$	$0.903\ [0.765 - 1.065], p = 0.226$
STXBP5	$0.686\ [0.571 - 0.825], p = 5.98\ 10^{-5}$	$0.823\ [0.6895 - 0.982], p = 0.031$
LRG1	$1.177\ [1.007 - 1.373], p = 0.039$	$1.018\ [0.8650 - 1.200], p = 0.823$
NFATC4	$0.828\ [0.702 - 0.978], p = 0.026$	$0.872\ [0.733 - 1.038], p = 0.123$
DDX1	$0.717\ [0.586 - 0.878], p = 1.29\ 10^{-3}$	$0.740\ [0.608 - 0.902], p = 3.83\ 10^{-3}$
MAST2	$0.772\ [0.655 - 0.911], p = 2.27\ 10^{-3}$	$0.892\ [0.755 - 1.054], p = 0.181$

Values shown represent the Hazard Ratio (HR) [95levels. HRs were estimated using a delay-entry Cox model and were adjusted for adjusted for sex, age at rst thrombosis, provoked status and type (i.e DVT vs PE) of the rst thrombosis, smoking and anticoagulant therapy at the time of plasma sampling.

4 Conclusion

In this work, we applied an ANN-based strategy on proteomic and clinical dataset to identify novel molecular markers associated with the risk of recurrence. This study was based on a sample of 913 patients profiled for 383 clinical and proteomic markers, making our work, so far, the largest ANN based study for VTE recurrence, both in terms of number of patients and of number of assessed features. Building on a two-step approach, that includes first an under sampling strategy addressing both the unbalance aspect and the heterogeneity of the original dataset, and then a 3 layers ANN, our strategy was able to derive a prediction model for recurrence with good discriminatory ability characterized by an AUC of 0.91 $[0.82 - 0.97]$, greater than the one achieved by a standard Random Forest algorithm applied to the same data. To the best of our knowledge, two ANN based prediction models have been previously proposed to predict VTE recurrence [25,40]. Unfortunately, as all clinical and biological variables used to build these prediction models were not available in our study, it was not possible to compare these models to ours.

Besides, the search for proteomic variables that contribute the most to the ANN predictions identified a set of 6 candidate proteins, DDX1, HTRA3, LRG1, MAST2, NFACT4 and STXBP5. These proteins were included in our proteomic panel because they have been previously proposed to associate with cardiometabolic risk factors or the risk of VTE (e.g. MAST2 [29], STXBP5 [30]). However, their role in the etiology of VTE risk recurrence has never been reported so far and deserve further epidemiological and experimental validation. Among these 6 proteins, only DDX1 showed strong association with VTE recurrence in marginal analyses suggesting that other candidates would interact on the risk of recurrence in more complex pattern, or could tag for unmeasured molecular markers, both hypotheses also deserving further investigations. DDX1 was included in our proteomics analyses because its structural gene has been previously shown to harbor genetic variations associated with renal dysfunction [35], the latter being a risk factor for VTE risk [31]. Our observed association of DDX1 plasma levels with the risk of recurrence in VTE patients is consistent with recent findings reporting that chronic kidney disease is associated with the risk of VTE recurrence [14].

Nonetheless, few limitations of this work should be acknowledged. First, we started our general workflow by implementing an under sampling algorithm to overcome the unbalanced aspect of our original dataset with respect to the number of patients with and without recurrence. However, other methodologies such as ensemble modeling [12,13] and different performance metrics (Cohen's Kappa [6], Matthews Correlation Coefficient [26]) are available to tackle such issue and investigating how the application of these alternatives may impact the final results of our ANN workflow deserves further work. Second, our under sampling strategy led to a selection of a non-representative sample of the MARTHA population for training our ANN model. As a consequence, the identified candidate proteins may not be good biomarkers for VTE recurrence in the general population of VTE patients but only in subgroups of patients. This is illustrated by the HTRA3 protein that was selected by the ANN model but that was finally associated with recurrence only in the subgroup of smokers patients. Smoking was under represented in the non-recurrent patients used for training our ANN model and this may have generated an extreme selecting sampling design facilitating the identification of an interaction [2]. Third, the selection by the DeepLIFT method of the main features contributing to the ANN predictions was rather empirical and based on the observed distribution of average importance. We may have then missed additional proteins that could impact on VTE recurrence in non linear manner. In particular, we did not find any evidence for direct association of the top protein identified by DeepLIFT, NFATC4, with recurrence nor identified candidate interaction with clinical factors. This strongly suggests that NFATC4 could interact with other proteins to modulate the risk of recurrence and the identification of such interactions deserves further exploration.

In conclusion, using an original ANN based strategy applied to plasma proteomics data, we identify plasma levels of DDX1 as a new biomarker for VTE recurrence and propose HTRA3 as an additional candidate in smokers patients.

Acknowledgments. M.R was financially supported by the GENMED Laboratory of Excellence on Medical Genomics [ANR-10-LABX-0013], a research program managed by the National Research Agency (ANR) as part of the French Investment for the Future. DA.T was partially supported by the EPIDEMIOM-VTE Senior Chair from the Initiative of Excellence of the University of Bordeaux. The proteomics screening was financed by a grant from Stockholm County Council (SLL 2017-0842) and from Familjen Erling Perssons Foundation. G.M has benefited from training offered by the EUR DPH, a PhD program supported within the framework of the PIA3 (Investment for the future), project reference 17-EURE-0019.

Additional Information. The script to build the model can be found on the following link: https://github.com/misbahch6/paper_script.

References

1. Angermueller, C., Pärnamaa, T., Parts, L., Stegle, O.: Deep learning for computational biology. Mol. Syst. Biol. **12**(7), 878 (2016)
2. Boks, M.P.M., Schipper, M., Schubart, C.D., Sommer, I.E., Kahn, R.S., Ophoff, R.A.: Investigating gene-environment interaction in complex diseases: increasing power by selective sampling for environmental exposure. Int. J. Epidemiol. **36**(6), 1363–1369 (2007)
3. Bridle, J.S.: Probabilistic interpretation of feedforward classification network outputs, with relationships to statistical pattern recognition. In: Soulié, F.F., Hérault, J. (eds.) Neurocomputing. NATO ASI Series, vol. 68, pp. 227–236. Springer, Heidelberg (1990). https://doi.org/10.1007/978-3-642-76153-9_28
4. Ching, T., et al.: Opportunities and obstacles for deep learning in biology and medicine. J. R. Soc. Interface **15**(141), 20170387 (2018)
5. Chollet, F., et al.: Keras (2015). https://keras.io
6. Cohen, J.: A coefficient of agreement for nominal scales. Educ. Psychol. Meas. **20**(1), 37–46 (1960)
7. Commenges, D., Letenneur, L., Joly, P., Alioum, A., Dartigues, J.-F.: Modelling age-specific risk: application to dementia. Stat. Med. **17**(17), 1973–1988 (1998)
8. Douketis, J.D., Crowther, M.A., Foster, G.A., Ginsberg, J.S.: Does the location of thrombosis determine the risk of disease recurrence in patients with proximal deep vein thrombosis? Am. J. Med. **110**(7), 515–519 (2001)
9. Douketis, J.D., Kearon, C., Bates, S., Duku, E.K., Ginsberg, J.S.: Risk of fatal pulmonary embolism in patients with treated venous thromboembolism. Jama **279**(6), 458–462 (1998)
10. Drobin, K., Nilsson, P., Schwenk, J.M.: Highly multiplexed antibody suspension bead arrays for plasma protein profiling. In: Bäckvall, H., Lehtiö, J. (eds.) The Low Molecular Weight Proteome. MIMB, vol. 1023, pp. 137–145. Springer, New York (2013). https://doi.org/10.1007/978-1-4614-7209-4_8
11. Farzamnia, H., Rabiei, K., Sadeghi, M., Roghani, F.: The predictive factors of recurrent deep vein thrombosis. ARYA Atherosclerosis **7**(3), 123 (2011)
12. Feng, W., Huang, W., Ren, J.: Class imbalance ensemble learning based on the margin theory. Appl. Sci. **8**(5), 815 (2018)
13. Galar, M., Fernandez, A., Barrenechea, E., Bustince, H., Herrera, F.: A review on ensembles for the class imbalance problem: bagging-, boosting-, and hybrid-based approaches. IEEE Trans. Syst. Man Cybern. Part C (Appl. Rev.) **42**(4), 463–484 (2011)
14. Goto, S., et al.: Assessment of outcomes among patients with venous thromboembolism with and without chronic kidney disease. JAMA Netw. Open **3**(10), e2022886–e2022886 (2020)
15. Guan, D., Yuan, W., Lee, Y.-K., Lee, S.: Nearest neighbor editing aided by unlabeled data. Inf. Sci. **179**(13), 2273–2282 (2009)
16. Hahnloser, R.H.R., Sarpeshkar, R., Mahowald, M.A., Douglas, R.J., Seung, H.S.: Digital selection and analogue amplification coexist in a cortex-inspired silicon circuit. Nature **405**(6789), 947–951 (2000)
17. Hansson, P.-O., Sörbo, J., Eriksson, H.: Recurrent venous thromboembolism after deep vein thrombosis: incidence and risk factors. Arch. Intern. Med. **160**(6), 769–774 (2000)
18. Heaton, J.: AIFH, volume 3: deep learning and neural networks. J. Chem. Inf. Model. **3**, Heaton Research Inc (2015)

19. Hinton, G.E., Dayan, P., Frey, B.J., Neal, R.M.: The "wake-sleep" algorithm for unsupervised neural networks. Science **268**(5214), 1158–1161 (1995)
20. Jensen, S.B., et al.: Discovery of novel plasma biomarkers for future incident venous thromboembolism by untargeted synchronous precursor selection mass spectrometry proteomics. J. Thromb. Haemost. **16**(9), 1763–1774 (2018)
21. Kingma, D.P., Ba, J.: Adam: a method for stochastic optimization. arXiv preprint arXiv:1412.6980 (2014)
22. Kong, J., Kowalczyk, W., Nguyen, D.A., Bäck, T., Menzel, S.: Hyperparameter optimisation for improving classification under class imbalance. In: 2019 IEEE Symposium Series on Computational Intelligence (SSCI), pp. 3072–3078. IEEE (2019)
23. LeCun, Y., Bengio, Y., Hinton, G.: Deep learning. Nature **521**(7553), 436–444 (2015)
24. van der Maaten, L., Hinton, G.: Visualizing data using T-SNE. J. Mach. Learn. Res. **9**, 2579–2605 (2008)
25. Martins, T.D., Annichino-Bizzacchi, J.M., Romano, A.V.C., Filho, R.M.: Artificial neural networks for prediction of recurrent venous thromboembolism. Int. J. Med. Inform. **141**, 104221 (2020)
26. Matthews, B.W.: Comparison of the predicted and observed secondary structure of t4 phage lysozyme. Biochimica et Biophysica Acta (BBA)-Protein Struct. **405**(2), 442–451 (1975)
27. McInnes, L., Healy, J., Melville, J.: UMAP: uniform manifold approximation and projection for dimension reduction. arXiv preprint arXiv:1802.03426 (2018)
28. Min, S., Lee, B., Yoon, S.: Deep learning in bioinformatics. Brief. Bioinform. **18**(5), 851–869 (2017)
29. Morange, P.-E., et al.: A rare coding mutation in the MAST2 gene causes venous thrombosis in a French family with unexplained thrombophilia: the Breizh MAST2 Arg89Gln variant. PLoS Genet. **17**(1), e1009284 (2021)
30. Morange, P.-E., Suchon, P., Trégouët, D.-A.: Genetics of venous thrombosis: update in 2015. Thromb. Haemost. **114**(11), 910–919 (2015)
31. Ocak, G., et al.: Risk of venous thrombosis in patients with chronic kidney disease: identification of high-risk groups. J Thromb. Haemost. **11**(4), 627–633 (2013)
32. O'Malley, T., Bursztein, E., Long, J., Chollet, F., Jin, H., Invernizzi, L., et al.: Keras Tuner (2019). https://github.com/keras-team/keras-tuner
33. Oshiro, T.M., Perez, P.S., Baranauskas, J.A.: How many trees in a random forest? In: Perner, P. (ed.) MLDM 2012. LNCS (LNAI), vol. 7376, pp. 154–168. Springer, Heidelberg (2012). https://doi.org/10.1007/978-3-642-31537-4_13
34. Oudot-Mellakh, T., et al.: Genome wide association study for plasma levels of natural anticoagulant inhibitors and protein C anticoagulant pathway: the MARTHA project. Br. J. Haematol. **157**(2), 230–239 (2012)
35. Pattaro, C., et al.: Genome-wide association and functional follow-up reveals new loci for kidney function. PLoS Genet. **8**(3), e1002584 (2012)
36. Pedregosa, F., et al.: Scikit-learn: machine learning in Python. J. Mach. Learn. Res. **12**, 2825–2830 (2011)
37. Prandoni, P., et al.: The long-term clinical course of acute deep venous thrombosis. Ann. Intern. Med. **125**(1), 1–7 (1996)
38. Razzak, M.I., Naz, S., Zaib, A.: Deep learning for medical image processing: overview, challenges and the future. In: Dey, N., Ashour, A.S., Borra, S. (eds.) Classification in BioApps. LNCVB, vol. 26, pp. 323–350. Springer, Cham (2018). https://doi.org/10.1007/978-3-319-65981-7_12

39. Razzaq, M., et al.: An artificial neural network approach integrating plasma proteomics and genetic data identifies PLXNA4 as a new susceptibility locus for pulmonary embolism. medRxiv (2020)
40. Romano, A.V.C., Martins, T.D., Maciel, R., De Paula, E.V., Annichino-Bizzacchi, J.M.: Artificial neural network for prediction of venous thrombosis recurrence. Blood **128**(22), 3771 (2016). ISSN 0006-4971
41. Gnana Sheela, K., Deepa, S.N.: Review on methods to fix number of hidden neurons in neural networks. Math. Probl. Eng. **2013** (2013)
42. Shrikumar, A., Greenside, P., Kundaje, A.: Learning important features through propagating activation differences. arXiv preprint arXiv:1704.02685 (2017)
43. Siragusa, S., Cosmi, B., Piovella, F., Hirsh, J., Ginsberg, J.S.: Low-molecular-weight heparins and unfractionated heparin in the treatment of patients with acute venous thromboembolism: results of a meta-analysis. Am. J. Med. **100**(3), 269–277 (1996)
44. Stevens, H., Peter, K., Tran, H., McFadyen, J.: Predicting the risk of recurrent venous thromboembolism: current challenges and future opportunities. J. Clin. Med. **9**(5), 1582 (2020)
45. Tagalakis, V., et al.: Men had a higher risk of recurrent venous thromboembolism than women: a large population study. Gender Med. **9**(1), 33–43 (2012)
46. Thiébaut, A.C.M., Bénichou, J.: Choice of time-scale in Cox's model analysis of epidemiologic cohort data: a simulation study. Stat. Med. **23**(24), 3803–3820 (2004)
47. Tomek, I., et al.: An experiment with the edited nearest-neighbor rule. IEEE Trans. Syst. Man Cybern. **SMC-6**(6), 448–452 (1976)
48. van Hylckama Vlieg, A., et al.: Genetic variations associated with recurrent venous thrombosis. Circ. Cardiovasc. Genet. **7**(6), 806–813 (2014)
49. Xu, S., Chen, L.: A novel approach for determining the optimal number of hidden layer neurons for FNN's and its application in data mining. In: 5th International Conference on Information Technology and Applications (ICITA) (2008)
50. Zeiler, M.D., Fergus, R.: Visualizing and understanding convolutional networks. In: Fleet, D., Pajdla, T., Schiele, B., Tuytelaars, T. (eds.) ECCV 2014. LNCS, vol. 8689, pp. 818–833. Springer, Cham (2014). https://doi.org/10.1007/978-3-319-10590-1_53

Neural Networks to Predict Survival from RNA-seq Data in Oncology

Mathilde Sautreuil(✉), Sarah Lemler, and Paul-Henry Cournède

Laboratoire MICS, CentraleSupélec, Université Paris-Saclay, 9 rue Joliot Curie, 91190 Gif-sur-Yvette, France
mathilde.sautreuil@centralesupelec.fr

Abstract. Survival analysis consists of studying the elapsed time until an event of interest, such as the death or recovery of a patient in medical studies. This work explores the potential of neural networks in survival analysis from clinical and RNA-seq data. If the neural network approach is not recent in survival analysis, methods were classically considered for low-dimensional input data. But with the emergence of high-throughput sequencing data, the number of covariates of interest has become very large, with new statistical issues to consider. We present and test a few recent neural network approaches for survival analysis adapted to high-dimensional inputs.

Keywords: Survival analysis · Neural networks · High-dimension · Cancer · Transcriptomics data

1 Introduction

Survival analysis consists of studying the elapsed time until an event of interest, such as the death or recovery of a patient in medical studies. This paper aims to compare methods to predict a patient's survival from clinical and gene expression data.

The Cox model [8] is the reference model in the field of survival analysis. It relates the survival duration of an individual to the set of explanatory covariates. It also enables to take into account censored data that are often present in clinical studies. With high-throughput sequencing techniques, transcriptomics data are more and more often used as covariates in survival analysis. Adding these covariates raise issues of high-dimensional statistics, when we have more covariates than individuals in the sample. Methods based on regularization or screening [10,31] have been developed and used to solve this issue.

The Cox model relies on the proportional hazard hypothesis, and in its classical version, does not account for nonlinear effects or interactions, which proves limited in some real situations. Therefore, in this paper, we focus on another type of methods: neural networks. Deep learning methods are more and more popular, notably due to their flexibility and their ability to handle interactions and nonlinear effects, including in the biomedical field [23,26,30]. The use of

E. Cinquemani and L. Paulevé (Eds.): CMSB 2021, LNBI 12881, pp. 122–140, 2021.
https://doi.org/10.1007/978-3-030-85633-5_8

neural networks for survival analysis is not recent, since it dates back to the 90's ([3,11]), but it began being widely used only recently. We can differentiate two strategies. The first one relies on the use of a neural network based on the Cox partial log-likelihood as those developed by [6,11,21,22]. The second strategy consists of using a neural network based on a discrete-time survival model, as introduced by [3]. [3] have studied this neural network only in low-dimension. In this paper, our objective is to study and adapt this model to the high-dimensional cases, and compare its performances to two other methods: the two-step procedure with the classical estimation of the parameters of the Cox model with a Lasso penalty to estimate the regression parameter and a kernel estimator of the baseline function (as in [16]) and the Cox-nnet neural network [6] based on the partial likelihood of the Cox model. Section 2 recalls the different notations used in survival analysis and presents the different models. Then, we introduce the simulation plan created to compare the models. Finally, we underline the results to conclude with the potential of neural networks in survival analysis.

2 Models

First, we introduce the following notations:

- Y_i the survival time
- C_i the censorship time
- $T_i = \min(Y_i, C_i)$ the observed time
- δ_i the censorship indicator (which will be equal to 1 if the interest event occurs and else to 0).

2.1 The Cox Model

The Cox model [8] predicts the survival probability of an individual from explanatory covariates $X_{i.} = (X_{i1}, \ldots, X_{ip})^T \in \mathbb{R}^p$. The hazard function λ is given by:

$$\lambda(t|X_{i.}) = \alpha_0(t)\exp(\beta^T X_{i.}), \tag{1}$$

where $\alpha_0(t)$ corresponds to the baseline hazard and $\beta = (\beta_1, \ldots, \beta_p)^T \in \mathbb{R}^p$ is the vector of regression coefficients. A benefit of this model is that only $\alpha_0(t)$ depends on time while the second term of the right hand side of (1) depends only on the covariates (proportional hazard model). The Cox model structure can be helpful when we are interested in the prognostic factors because β can be estimated without knowing the function α_0. It is possible thanks to the Cox partial log-likelihood, which is the part of the total log-likelihood that does not depend on $\alpha_0(t)$, and is defined by:

$$\mathcal{L}(\beta) = \sum_{i=1}^{n} (\beta^T X_{i.}) - \sum_{i=1}^{n} \delta_i \log\left(\sum_{l \in R_i} \exp(\beta^T X_{l.})\right),$$

with R_i the individuals at risk at the observation time T_i of individual i, and δ_i the censorship indicator of individual i. The Lasso procedure was proposed by [31] for the estimation of β in the high-dimensional setting. The non-relevant variables are set to zero thanks to the L_1- penalty added to the Cox partial likelihood: $\mathcal{L}(\beta) + \lambda||\beta||_1$. However, to predict the survival function S, we need to fully estimate the hazard risk $\lambda(s|X_{i.})$ since:

$$S(t) = \mathbb{P}(T_i > t|X_{i.}) = \exp\left(-\int_0^t \alpha_0(s)\exp(\beta^T X_{i.})ds\right).$$

We follow the two-step procedure of [16]: first, we estimate β from the penalized Cox partial likelihood, and then we estimate $\alpha_0(t)$ from the kernel estimator introduced by [27], in which we have plugged the Lasso estimate of β.

2.2 Neural Networks

The studied neural networks in this paper are fully-connected multi-layer perceptrons. Several layers constitute this network with at least one input layer, one output layer, and one or several hidden layers.

Cox-nnet. In 1995, [11] developed a neural network based on the proportional hazards model. The idea of [11] was to replace the linear prediction of the Cox regression with the neural network's hidden layer's output. [11] only applied their neural network to survival analysis from clinical data, in low dimension. More recently, some authors revisited this method [6,21,22]. However, only Cox-nnet [6] was applied in a high-dimensional setting. We will thus use this model as benchmark in our study.

The principle of Cox-nnet is that its output layer corresponds to a Cox regression: the output of the hidden layer replaces the linear function of the covariates in the exponential of the Cox model equation.

To estimate the neural network weights, [6] uses the Cox partial log-likelihood as the neural network loss:

$$\mathcal{L}(\beta, W, b) = \sum_{i=1}^{n} \theta_i - \sum_{i=1}^{n} \delta_i \log\left(\sum_{l \in R_i} \exp(\theta_l)\right) \tag{2}$$

with δ_i the censoring indicator and $\theta_i = \beta^T G(W^T X_{i.} + b)$, where G is the activation function of the hidden layer, $W = (w_{dh})_{1\leq d\leq p, 1\leq h\leq H}$ with H the number of neurons in the hidden layer, and $\beta = (\beta_1, \ldots, \beta_H)^T$ the weights and b the biases of the neural network to be estimated. In this network, the activation function *tanh* is used. To the partial log-likelihood, [6] adds a ridge penalty in L_2-norm of the parameters. Thus, the final cost function for this neural network is:

$$Loss(\beta, W, b) = \mathcal{L}(\beta, W, b) + \lambda(\|\beta\|_2 + \|W\|_2 + \|b\|_2). \tag{3}$$

We maximize this loss function to deduce estimators of β, W and b. The principle in this neural network is that the activation function for the output layer is a Cox regression, so that we have:

$$\hat{h}_i = \exp\left(\underbrace{\sum_{h=1}^{H} \hat{\beta}_h G\left(\hat{b}_h + \hat{W}^T X_{i.}\right)}_{\hat{\theta}_i = \hat{\beta}^T G(\hat{W}^T X_{i.} + \hat{b})} \right). \tag{4}$$

The output of the neural network $\hat{h}_i$ corresponds to the part of the Cox regression that does not depend on time. [6] only used $\hat{h}_i$, but in our study, we are interested in the complete survival function, and thus we need to estimate the complete hazard function $\hat{h}(x_i, t)$. For that purpose, we estimate the baseline risk $\alpha_0(t)$, with the kernel estimator introduced by [27]. As for the Cox model, we estimate $\alpha_0(t)$ with the two-steps procedure of [16] and this estimator is defined by:

$$\widehat{\alpha}_m(t) = \frac{1}{nm} \sum_{i=1}^{n} K\left(\frac{t-u}{m}\right) \frac{\delta_i}{\sum_{l \in R_i} \widehat{h}_l}, \tag{5}$$

with $\hat{h}_l$ the estimator defined by (4), $K : \mathbb{R} \to \mathbb{R}$ a kernel (a positive function with integral equal to 1), m the bandwidth, which is a strictly positive real parameter. m can be obtained by cross-validation or by the Goldenshluger & Lepski method [14] for instance, and we choose the latter. We can finally derive an estimator of the survival function for individual i:

$$\widehat{S}(t|X_{i.}) = \exp\left(-\int_0^t \ddot{\alpha}_m(s)\hat{h}_i ds\right). \tag{6}$$

Discrete Time Neural Network. [3] has proposed a neural network based on a discrete-time model. They introduced L time intervals $A_l =]t_{l-1}, t_l]$, and build a model predicting in which interval, the failure event occurs. We write the discrete hazard as the conditional probability of survival:

$$h_{il} = P(Y_i \in A_l | Y_i > t_{l-1}), \tag{7}$$

with Y_i the survival time of individual i. [3] duplicates the individuals as input of the neural network. The duplication of individuals gives it a more original structure than that of a classical multi-layer perceptron. The [3]'s neural network takes as input the set of variables of the individual and an additional variable corresponding to the mid-point of each interval. Due to the addition of this variable, the p variables of each individual are repeated for each time interval. The output is thus the estimated hazard $h_{il} = h_l(X_i, a_l)$ for the individual i at time a_l. We schematize the structure of this neural network on Fig. 1. [3] initially used a 3-layers neural network with a logistic function as the activation function for both the hidden and output layers. The output of the neural network with H neurons in the hidden layer and $p+1$ input variables is given by:

$$h_{il} = h(x_i, t_l) = f_2\left(a + \beta^T f_1\left(b + W^T X_{i.}\right)\right),$$

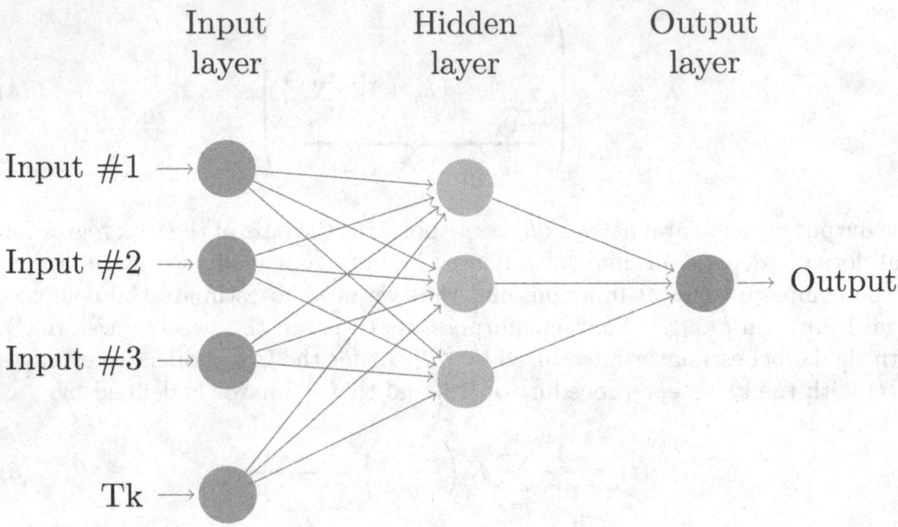

Fig. 1. Structure of the neural network based on the discrete-time model of [3]

where $W = (w_{dh})_{1\le d\le p+1, 1\le h\le H}$, and $\beta = (\beta_1, \ldots, \beta_H)^T$ are the weights of the neural network, a and b are the biases of the neural network to be estimated, and f_1 and f_2 the sigmoid activation functions. The target of this neural network is the death indicator d_{il}, which will indicate if the individual i dies in the interval A_l. We introduce $l_i \le L$ the number of intervals in which the individual i is observed, $d_{i0}, \ldots, d_{i(l_i-1)} = 0$ whatever the status of the individual i and d_{il_i} is equal to 0 if the individual i is censored and 1 otherwise. The cost function used by [3] is the cross-entropy function and the weights of the neural network can be estimated by minimizing it:

$$\mathcal{L}(\beta, W, a, b) = -\sum_{i=1}^{n}\sum_{l=1}^{l_i} d_{il}\log(h_{il}) + (1-d_{il})\log(1-h_{il}). \tag{8}$$

The duplication of the individuals for each time interval increases the sample size in the neural network, it is an advantage in a high-dimensional framework. Moreover, [3] added a ridge penalty to their cross-entropy function (8):

$$Loss(\beta, W, a, b) = \mathcal{L}(\beta, W, a, b) + \lambda(\|\beta\|_2 + \|W\|_2 + \|a\|_2 + \|b\|_2), \tag{9}$$

In [3], λ was chosen by deriving an Information Criteria. We choose instead to use cross-validation since it improved model the predictive capacity.

After estimating the parameters of the neural network by minimizing the loss function (9), the output obtained is the estimate of the discrete risk $\widehat{h}_{il}$ for each individual i and the survival function of individual i is estimated using:

$$\widehat{S}(T_{l_i}) = \prod_{i=1}^{l_i}(1-\widehat{h}_{il}). \tag{10}$$

This model was only applied for low-dimensional inputs, and this paper investigates its performance and capacity to adapt to high-dimensional settings. We denote this network NNsurv. We noticed an improvement of the performance when using a ReLU activation function for the hidden layers and thus used it instead of the original sigmoid functions. Moreover, original neural network only has one hidden layer. We propose to add one supplementary hidden layer to study if a deeper structure could improve the neural network prediction capacity. We call the deeper version NNsurv-deep. Its structure is similar to the one schematized in Fig. 1, but with two hidden layers instead of one. The input layer does not change, and the individuals are always duplicated at the input of the neural network. The output layer also has a single neuron corresponding to the discrete hazard estimate. These neural networks are implemented in a package available on https://github.com/mathildesautreuil/NNsurv.

We will compare the performances of these four models (Cox-Lasso, Cox-nnet, NNsurv, NNsurv-deep) on simulated data and then to a real dataset.

3 Simulations

We create a simulation design to compare different neural network approaches to predict survival time in high-dimension. We divide the simulation plan into two parts. The first part concerns a simulation study based on [2] which proposes to generate the survival data from a Cox model. Data simulated with this model naturally favors the two methods based on the Cox model. We also consider a model with a more complex behavior: the Accelerated Hazards (AH) model [5]. In the AH model, variables will accelerate or decelerate the hazard risk. The survival curves of the AH model can therefore cross each other. Other choices of models were also possible, and in the Appendix A, we also present the results for the Accelerated Failure Time (AFT) model [19] which does not satisfy the proportional risk assumption either, but does not allow the intersection of survival curves of different patients.

The models' baseline risk function is assumed known and follows a particular probability distribution in all cases. We use the Weibull distribution for the Cox model and the log-normal distribution for the AH model. Several simulations are considered, by varying the sample size, the total number of explanatory variables, and the number of relevant explanatory variables considered in the model to be the closest to the RNA-seq real datasets in oncology. First, we considered a low number of covariates (10 covariates), representing the case with only clinical data in survival studies. Then, we considered some simulation cases to look like RNA-seq data with filtering. The first case corresponds to important filtering with only 100 covariates, and the second case is a filtering case more realistic with 1000 covariates. We use the package that we have developed called `survMS` and available on CRAN or https://github.com/mathildesautreuil/survMS.

3.1 Generation of Survival Times

Considering the survival models (Cox, AFT, and AH models), the survival function $S(t|X)$ can be written as:

$$S(t|X) = \exp(-H_0(\psi_1(X)t)\psi_2(X) \text{ with} \tag{11}$$

H_0 is the cumulative hazard and

$$(\psi_1(X), \psi_2(X)) = \begin{cases} (1, \exp(\beta^T X)) & \text{for the Cox model} \\ (\exp(\beta^T X), \exp(-\beta^T X)) & \text{for the AH model} \\ (\exp(\beta^T X), 1) & \text{for the AFT model.} \end{cases}$$

The distribution function is deduced from the survival function from the following formula:

$$F(t|X) = 1 - S(t|X). \tag{12}$$

For data generation, if Y is a random variable that follows a probability distribution F, then $U = F(Y)$ follows a uniform distribution on the interval $[0, 1]$, and $(1 - U)$ also follows a uniform distribution $\mathcal{U}[0, 1]$. From Eq. (12), we finally obtain that:

$$1 - U = S(t|X) \tag{13}$$
$$= \exp(-H_0(\psi_1(X)t)\psi_2(X)). \tag{14}$$

If $\alpha_0(t)$ is positive for all t, then $H_0(t)$ can be inverted, and we can express the survival time of each of the models considered (Cox, AFT and AH) from $H_0^{-1}(u)$. We write in a general form the expression of the random survival times for each of the survival models:

$$T = \frac{1}{\psi_1(X)} H_0^{-1}\left(\frac{\log(1-U)}{\psi_2(X)}\right). \tag{15}$$

Two distributions are used for the cumulative hazard function $H_0(t)$ to generate the survival data. If the survival times are distributed according to a Weibull distribution $\mathcal{W}(a, \lambda)$, the baseline hazard is of the form :

$$\alpha_0(t) = a\lambda t^{a-1}, \lambda > 0, a > 0. \tag{16}$$

The inverse of the cumulative risk function is expressed as follows:

$$H_0^{-1}(u) = \left(\frac{u}{\lambda}\right)^{1/a}. \tag{17}$$

For survival times following a log-normal distribution $\mathcal{LN}(\mu, \sigma)$ with mean μ and standard deviation σ, the basic risk function is therefore written:

$$\alpha_0(t) = \frac{\frac{1}{\sigma\sqrt{2\pi t}} \exp\left[-\frac{(\log t - \mu)^2}{2\sigma^2}\right]}{1 - \Phi\left[\frac{\log t - \mu}{\sigma}\right]}, \tag{18}$$

with $\Phi(t)$ the distribution function of a standard Normal distribution. The inverse of the cumulative hazard function is expressed by:

$$H_0^{-1}(u) = \exp(\sigma\Phi^{-1}(1 - \exp(-u)) + \mu), \tag{19}$$

with $\Phi^{-1}(t)$ the inverse of the distribution function of a centered and reduced normal distribution.

3.2 Simulation with the Cox - Weibull Model

Survival Times and Baseline Function: Generating survival times from a variety of parametric distributions were described by [2]. In the case of a Cox model with a baseline function distributed from a Weibull distribution, the inverse cumulative hazard function is $H_0^{-1}(t) = (\frac{t}{\lambda})^{\frac{1}{a}}$ and the survival time T of the Cox model is expressed as:

$$T = \left(-\frac{1}{\lambda}\log(1 - U)\exp(-X_{i.}\beta)\right)^{\frac{1}{a}}, \tag{20}$$

where U is a random variable with $U \sim \mathcal{U}[0, 1]$.

Choice of Parameters of the Weibull Distribution: We chose the Weibull distribution parameters so that our design of simulation is close to real datasets. The mean and the standard deviation of Breast cancer real dataset (available on www.ncbi.nlm.nih.gov/geo/query/acc.cgi?acc=GSE6532) is around 2325 days and 1304 days respectively. As the survival times follow a Weibull distribution, the mean and the variance of T write as:

$$\mathbb{E}(T) = \frac{1}{\sqrt[a]{\lambda}}\Gamma\left(\frac{1}{a} + 1\right) \text{ and } \mathbb{V}(T) = \frac{1}{\sqrt[a]{\lambda^2}}\left[\Gamma\left(\frac{2}{a} + 1\right) - \Gamma^2\left(\frac{1}{a} + 1\right)\right],$$

where Γ is the Gamma function. We set $a = 2$ and $\lambda = 1.3e^{-7}$ to have a mean and variance of our simulated datasets close to those of the Breast cancer real dataset.

3.3 Simulation with the AH - Log-Normal Model

Survival Times and Baseline Function: Building on the work of [2], we also simulate the survival data from the AH model. We perform this simulation to generate data whose survival curves will intersect. For this simulation, we consider that the survival times follow a log-normal distribution $\mathcal{LN}(\mu, \sigma)$. In this case, the inverse of the cumulative hazard function is expressed as (19), and we have:

$$T = \frac{1}{\exp(\beta^T X_{i.})}\sigma\Phi^{-1}\left(\frac{\log(1 - U)}{\exp(-\beta^T X_{i.})} + \mu\right) \tag{21}$$

with $\Phi^{-1}(t)$ the inverse of the distribution function of a centered and reduced normal distribution.

Choice of Parameters of the Log-Normal Distribution: As in the previous simulation, we ensure that the distribution of the simulated data is close to that of the real ones and we use the formulas:

$$\mu = \ln(\mathrm{E}(T)) - \frac{1}{2}\sigma^2 \text{ and } \sigma^2 = \ln\left(1 + \frac{\mathrm{Var}(T)}{(\mathrm{E}(T))^2}\right). \tag{22}$$

Since, the expectation and the standard deviation are respectively 2325 and 1304, the values of μ and σ used for the simulation of the survival data should be $\mu = 7.73$ and $\sigma = 0.1760$. However, to have survival curves crossing rapidly, we take a higher value of σ: $\sigma = 0.7$.

3.4 Metrics

To assess the performance of the survival models, we use two classical metrics, the Concordance Index (CI) and the Integrated Brier Score (IBS).

Concordance Index. The index measures whether the prediction of the model under study matches the rank of the survival data. If the event time of an individual i is shorter than that of an individual j, a good model will predict a higher probability of survival for individual j. This metric takes into account censored data, and it takes a value between 0 and 1. If the C-index is equal to 0.5, the model is equivalent to random guessing. The time-dependent C-index proposed by [1], adapted to non-proportional hazard models, is chosen in this study.

Consider n individuals, and for $1 \leq i \leq n$, T_i their observation times (either survival or censoring times) and δ_i their censorship indicators. For $i, j = 1, \ldots, n \; i \neq j$, we define the indicators:

$$comp_{ij} = \mathbb{1}_{\{(T_i < T_j; \delta_i = 1) \cup (T_i = T_j; \delta_i = 1, \delta_j = 0)\}}$$

and

$$conc_{ij}^{td} = \mathbb{1}_{\{S(T_i | X_{i.}) < S(T_j | X_{j.})\}} comp_{ij},$$

The estimate of the time-dependent C-index for survival models is equal to:

$$\widehat{C}_{td} = \frac{\sum_{i=1}^{n} \sum_{j \neq i} conc_{ij}^{td}}{\sum_{i=1}^{n} \sum_{j \neq i} comp_{ij}} \tag{23}$$

If we are in the proportional hazards or linear transformation models' case, the metric $\widehat{C}_{td}$ of the Eq. (23) is equivalent to the usual C-index [12].

Integrated Brier Score. The Brier score measures the squared error between the indicator function of surviving at time t, $\mathbb{1}_{\{T_i \geq t\}}$, and its prediction by the model $\widehat{S}(t|X_{i.})$. [15] adapted the Brier score [4] for censored survival data using the inverse probability of censoring weights (IPCW) and [13] subsequently proposed a consistent estimator of the Brier score in the presence of censored data. The Brier score is defined by:

$$BS(t,\widehat{S}) = \mathbb{E}\left[\left(Y_i(t) - \widehat{S}(t|X_{i.})\right)^2\right], \tag{24}$$

where $Y_i(t) = \mathbb{1}_{\{T_i \geq t\}}$ is the status of individual i at time t and $\widehat{S}(t|X_{i.})$ is the predicted survival probability at time t for individual i. Unlike the C-index, a lower value of this score shows a better predictive ability of the model.

As mentioned above, [13] gave an estimate of the Brier score in the presence of censored survival data. The estimate of the Brier score under right censoring is:

$$\widehat{BS}(t,\widehat{S}) = \frac{1}{n}\sum_{i=1}^{n} \widehat{W}_i(t)(Y_i(t) - \widehat{S}(t|X_{i.}))^2, \tag{25}$$

with n the number of individuals in the test set. Moreover, in the presence of censored data it is necessary to adjust the score by weighting it by the inverse probability of censoring weights (IPCW). This weighting is defined by:

$$\widehat{W}_i(t) = \frac{(1 - Y_i(t))\delta_i}{\widehat{G}(T_i|X_{i.})} + \frac{Y_i(t)}{\widehat{G}(t|X_{i.})}, \tag{26}$$

where δ_i is the censored indicator equal to 1 if we observe the survival time and equal to 0 if the survival time is censored, and $\widehat{G}(t|x)$ is the Kaplan-Meier [20] estimator of the censored time survival function at time t.

The integrated Brier score [25] summarizes the predictive performance estimated by the Brier score [4]:

$$\widehat{IBS} = \frac{1}{\tau}\int_0^{\tau} \widehat{BS}(t,\widehat{S})dt, \tag{27}$$

where $\widehat{BS}(t,\widehat{S})$ is the estimated Brier score and $\tau > 0$. We take $\tau > 0$ as the maximum of the observed times and the Brier score is averaged over the interval $[0,\tau]$. As for the Brier score, a lower value of the IBS indicates a better predictive ability of the model.

4 Results

In this section, we compare the performances of the Cox model with Lasso (denoted CoxL1) [31], the neural network based on the Cox partial log-likelihood Cox-nnet [6] presented in Sect. 2, and discrete-time neural networks (NNsurv and NNsurv-deep), adapted from [3] and also presented in Sect. 2. The performances are compared on simulated data (with the Cox and AH models, and for several parametric configurations) and on a real data case presented below. The discrete-time C-index (C_{td}) and Integrated Brier Score (IBS) are used for this purpose. We can calculate the reference C_{td} and IBS values from our simulations based on the exact model used for the simulation. Note however, that the models under comparison can sometimes "beat" these reference values by chance (due to the random generation of survival times).

4.1 Simulation Study

n is the number of samples, $n \in 200; 1000$, and p is the number of covariates, $p \in 10; 100; 1000$. Note that even if our objective is to apply our models to predict survival from RNA-seq data, we present simulation results up to 1000 covariates (instead of the potential several tens of thousands usually available with RNA-seq). Indeed, when we performed tests with 10,000 inputs, none of the model were able to perform well, thus underlining the necessity of a preliminary filtering as classically done when handling RNA-seq data [7].

Results for the Cox - Weibull Simulation. The Cox-Weibull simulation corresponds to a Cox model's data with a baseline risk modeled by a Weibull distribution. In this simulation, the model satisfies the proportional hazards assumption. The results of this simulation in Table 1 show that Cox-nnet performs best concerning the C_{td} in all settings (regardless of the number of variables or sample size) and most settings for the IBS. The best IBS values for Cox-nnet, as we can see from Table 1, are for sample size equal to 200 and number of variables to 10 and 100 or sample size worth to 1000 and number of variables is to 100 and 1000. CoxL1 also has the best IBS (*i.e.* the lowest) for a sample size of 1000 and 10 variables. These good results of CoxL1 and Cox-nnet are not surprising because we simulated the data from a Cox model. We can observe in Table 1 that NNsurv-deep obtains the lowest IBS value for 200 individuals and 1000 variables. We can also see that the IBS values of NNsurv and NNsurv-deep are very close to the reference IBS values. This phenomenon is also true when the sample size is equal to 1000, and the number of variables is equal to 100. Moreover, we can observe in Table 1 that some of the values of C_{td} obtained for NNsurv and NNsurv-deep are close to those of Cox-nnet. We notice notably this case when the sample size is equal to 200, and the number of variables is equal to 10, and when the number of samples is 1000 and the number of variables is of 10 and 100. We can see that some of the values of the C_{td} for the discrete-time neural networks are better than those obtained from the Cox model, for example, for a sample size equal to 200 and number of variables worth to 100 or for a sample size worth to 1000 and whatever the number of variables is.

Synthesis: Not surprisingly, Cox-nnet has the best results on this dataset simulated from a Cox model with a Weibull distribution. However, the neural networks based on a discrete-time model (NNsurv and NNsurv-deep) have very comparable performances, and clearly outperforms the CoxL1 model when the number of variables increases.

Results for the AH - Log-Normal Simulation. The results presented in Table 2 are those obtained on the AH simulation with the baseline hazard following a log-normal distribution. In this simulation, the risks are not proportional, and the survival functions of different individuals can cross.

We can observe that the neural networks based on a discrete-time model have the best performances concerning the C_{td} and the IBS, and their values

Table 1. Results of predicting methods on Cox-Weibull simulation

	n	200			1000		
Method	p	10	100	1000	10	100	1000
Reference	$C^{\star}_{td}$	*0.7442*	*0.7428*	*0.7309*	*0.7442*	*0.7428*	*0.7309*
	$IBS^{\star}$	*0.0471*	*0.0549*	*0.0582*	*0.0471*	*0.0549*	*0.0582*
NNsurv	C_{td}	0.7137	0.6224	0.5036	0.7398	0.7282	0.5700
	IBS	0.0980	0.0646	0.1359	0.0759	0.0537	0.1007
NNsurv	C_{td}	0.7225	0.5982	0.5054	0.7424	0.7236	0.5741
deep	IBS	0.0878	0.0689	**0.1080**	0.0591	0.0555	0.1185
Cox	C_{td}	**0.7313**	**0.6481**	**0.5351**	**0.7427**	**0.7309**	**0.6110**
-nnet	IBS	**0.0688**	**0.0622**	0.1402	0.0640	**0.0498**	**0.0710**
CoxL1	C_{td}	0.7292	0.5330	0.5011	0.7419	0.7243	0.5
	IBS	0.0715	0.0672	0.1175	**0.0541**	0.0509	0.0770

are close to the reference C_{td} and IBS. This phenomenon is particularly correct for the IBS when the sample size is equal to 1000, the IBS values of NNsurv and NNsurv-deep are lower than those of the reference IBS. On the other hand, the methods based on the Cox partial likelihood have the highest C_{td} values for a small sample size ($n = 200$) and a small number of variables ($p = 10$) or, on the contrary, for a large sample size ($n = 1000$) and a large number of variables ($p = 1000$). For a sample size equal to 200, neural networks based on a discrete-time model have higher C_{td} values than those obtained by CoxL1 and Cox-nnet. The values obtained for the IBS by the two methods using the Cox partial likelihood are good. For a small number of individuals ($n = 200$), the IBS values of CoxL1 and Cox-nnet are very high. For example, Cox-nnet obtains IBS values equal to 0.2243 and 0.1609 respectively for 10 and 100 variables, and CoxL1 gets IBS values equal to 0.2278 and 0.1614, respectively. These values are very high compared to the baseline IBS. CoxL1 and Cox-nnet, therefore, have more difficulty with a small number of samples. The predictions of these two methods are not as good as those given by discrete-time neural networks.

Synthesis: On the dataset simulated from an AH model with a log-normal distribution, neural networks based on the discrete-time model have the best performances in most situations. The deep version of the model is also better than the one with only one hidden layer. In this simulation, the data do not check the proportional hazards assumption, and survival curves exhibit complex patterns for which the more versatile NNsurv-deep appears more adapted.

4.2 Application on Real Datasets

Breast Cancer Dataset

Description of Data: The METABRIC data (for Molecular Taxonomy of Breast Cancer International Consortiulm) [9] include 2509 patients with early

Table 2. Results of predicting methods on AH/Log-normal simulation

	n	200			1000		
Method	p	10	100	1000	10	100	1000
Reference	$C^{\star}_{td}$	*0.7225*	*0.6857*	*0.7070*	*0.7225*	*0.6867*	*0.7070*
	$IBS^{\star}$	*0.0755*	*0.0316*	*0.0651*	*0.0755*	*0.0316*	*0.0651*
NNsurv	C_{td}	0.6863	**0.5971**	**0.5358**	0.7084	0.6088	0.5654
	IBS	**0.1247**	**0.0780**	**0.0859**	0.0699	0.0347	0.0533
NNsurv	C_{td}	0.7042	0.5793	0.5325	**0.7155**	**0.6450**	0.5702
deep	IBS	0.1789	0.2529	0.1554	**0.0602**	**0.0303**	**0.0484**
Cox	C_{td}	**0.7128**	0.5812	0.5356	0.7097	0.6047	**0.5720**
-nnet	IBS	0.1342	0.2243	0.1609	0.0843	0.0875	0.0553
CoxL1	C_{td}	0.7042	0.5219	0.5112	0.7088	0.5597	0.5
	IBS	0.1350	0.2278	0.1614	0.0608	0.0408	0.0553

breast cancer. These data are available at https://www.synapse.org/#!Synapse:syn1688369/wiki/27311. Survival time, clinical variables, and expression data were present for 1981 patients, with six clinical variables (age, tumor size, hormone therapy, chemotherapy, tumor grades and number of invaded lymph nodes), and 863 genes (pre-filtered). The percentage of censored individuals is high, equal to 55%.

Results: The comparison results of the METABRIC dataset are summarized in Table 3. NNsurv-deep manages to get the highest value of C_{td}. The C_{td} of NNsurv is equivalent to that of Cox, but Cox-nnet has a lower value. The integrated Brier score is very close for NNsurv-deep, Cox-nnet, and CoxL1, although the latter has the lowest IBS value.

On this real dataset, the differences between the models are not striking, despite the small superiority of NNsurv-deep.

Table 3. Results of different methods on the breast dataset (METABRIC)

		CoxL1	Cox-nnet	NNsurv-deep	NNsurv
Metabric	C_{td}	0.6757	0.6676	**0.6853**	0.6728
	IBS	**0.1937**	0.1965	0.1972	0.2038

5 Discussions

This work is a study of neural networks for the prediction of survival in high-dimension. In this context, usual methods such as the estimation in a Cox model with the Cox partial likelihood can no longer be performed. Several methods (such as dimension reduction or machine learning methods, like Random Survival Forests [18]) have been proposed, but our interest in this study has been directed towards neural networks and their potential for survival analysis from RNA-seq data. Two neural-network based approaches have been proposed. The first one is based on the Cox model but introduces a neural network for risk determination [11]. The second approach is based on a discrete-time model [3] and its adaptation to the high-dimensional setting was the main contribution of our work. In Sect. 4, we compared the standard Cox model with Lasso penalty and a neural network based on the Cox model (Cox-nnet) with those based on a discrete-time model adapted to the high dimension (NNsurv, and NNsurv-deep). To evaluate this comparison rigorously, we created a design of simulations. We simulated data from different models (Cox, AH, and AFT in appendix) with varying numbers of variables and sample sizes, allowing diverse levels of complexity. The variation of parameter numbers enables simulated datasets to be closer to real datasets (Clinical dataset or RNA-seq dataset with filtering).

We concluded from this study that the best neural network in most situations is Cox-nnet. It can handle nonlinear effects as well as interactions. However, the neural network based on discrete-time modelling, which directly predicts the hazard risk, with several hidden layers (NNsurv-deep), has shown its superiority in the most complex situations, especially in the presence of non-proportional risks and intersecting survival curves. On the Metabric data, NNsurv-deep performs the best, but only marginally better than the Cox partial log-likelihood-based Lasso estimation procedure, suggesting slight non-linearity and interactions.

The Neural networks seem to be interesting methods to predict survival in high-dimension and, in particular, in the presence of complex data. The effect of censoring in these models was not studied in this work, but [28] evaluated several methods to cope with censoring in neural networks models for survival analysis. For practical applications, a disadvantage of neural networks is the interpretation difficulty. The Cox model associated with the Lasso procedure enables identifying the linear factor prognostics and is privileged by the domain's users nowadays. The interpretability issue of neural networks is more and more studied [17,29] and is an exciting research avenue to explore. For example, [29] propose a neural network, called DeepLIFT, enabling to explain the output from some 'reference' output. They compute the 'reference' output from some 'reference' input chosen according to the knowledge of the domain's experts. [17] propose to use sparse coding, making the connections between layers sparse for the interpretation ease. The connections within the network not contributing to minimizing the loss are removed. Further investigation based on these works will allow us to have better comprehensible survival prediction output and identify nonlinear factor prognostics.

A Appendix: Supplementary Results

A.1 Simulation from the AFT - Log-Normal Model

Survival Times and Baseline Function: To simulate the data from the AFT/Log-normal model, we relied on [24]. We chose to perform this simulation to generate survival data that do not respect the proportional hazards assumption. For this simulation, we consider that the survival times follow a log-normal distribution $\mathcal{LN}(\mu, \sigma)$. In this case, the inverse of the cumulative hazard function is expressed as (19). Survival times can therefore be simulated from:

$$T = \frac{1}{\exp(\beta^T X_{i.})} \exp(\sigma\phi^{-1}(U) + \mu). \tag{28}$$

Choice of Parameters of Log-Normal Distribution: We wish the distribution of the simulated data is close to the real data. We follow the same approach to choose the parameters σ and μ of the survival time distribution as for the Cox/Weibull simulation presented above. The value of the parameters is obtained from the explicit formulas:

$$\mu = \ln(\mathrm{E}(T)) - \frac{1}{2}\sigma^2 \text{ and } \sigma^2 = \ln\left(1 + \frac{\mathrm{Var}(T)}{(\mathrm{E}(T))^2}\right). \tag{29}$$

Given the expectation and the standard deviation are respectively 2325 and 1304, the values of μ and σ used for the simulation of the survival data should be $\mu = 7.73$ and $\sigma = 0.1760$.

A.2 Simulation Study

Results for the AFT - Log-Normal Simulation. This section presents the results for data simulated from an AFT model with a baseline risk modeled by a log-normal distribution. The specificity of these simulated data is that they do not satisfy the proportional hazards assumption, but the survival curves do not cross.

Table 4 shows that CoxL1 and Cox-nnet have the best results in most configurations considering C_{td} or IBS. This good result for C_{td} is particularly right when the sample size is equal to 200 or when the sample size is equal to 1000, and the number of variables is equal to 10 and 100. The C_{td} obtained by the CoxL1 model is equal to 0.9867 for 200 individuals and ten variables, and the C_{td} obtained for the Cox-nnet model is equal to 0.9060 for 1000 individuals and 100 variables. We can see in Table 4 that the C_{td} obtained for the neural networks based on a discrete-time model is very close to those obtained by CoxL1 and Cox-nnet and is either higher than the reference one or slightly below. For example, for a sample size equal to 200 and a number of variables equal to 10, the C_{td} of NNsurv is equal to 0.9832, that of Cox-nnet is equal to 0.9867, and the reference one is equal to 0.9203. We have the same behavior for 100 variables and the same sample size or 100 variables and a sample size of 1000.

Moreover, the *IBS* values are the lowest for the methods based on Cox modeling in most situations. But the *IBS* values for NNsurv and NNsurv-deep are also excellent. They are lower than the reference *IBS* in many cases and are very close to CoxL1 and Cox-nnet. We can observe these results when the number of variables is less than or equal to 100 regardless of the sample size. The good results of CoxL1 and Cox-nnet might seem surprising, but we can explain it because we simulate these data from an AFT model whose survival curves do not cross. A method based on a Cox model will predict survival functions that do not cross. For this simulation, the survival function prediction obtained by CoxL1 and Cox-nnet is not cross and is undoubtedly closer to the survival function of the AFT simulation compared to discrete-time neural networks.

Table 4. Results of predicting methods on AFT/Log-normal simulation

	n	200			1000		
Method	p	10	100	1000	10	100	1000
Reference	$C^{\star}_{td}$	*0.9203*	*0.9136*	*0.9037*	*0.9203*	*0.9136*	*0.9037*
	$IBS^{\star}$	*0.0504*	*0.0604*	*0.0417*	*0.0504*	*0.0604*	*0.0417*
NNsurv	C_{td}	0.9832	0.8349	0.5425	0.9851	0.9038	0.7426
	IBS	0.0265	**0.0560**	0.2577	0.0247	0.0188	0.0642
NNsurv deep	C_{td}	0.9786	0.8275	0.5576	**0.9857**	**0.9060**	**0.7500**
	IBS	0.0295	0.0561	0.1886	0.0261	0.0207	**0.0631**
Cox -nnet	C_{td}	0.9825	**0.8558**	**0.5979**	0.9844	**0.9060**	0.7085
	IBS	**0.0122**	0.0906	**0.0959**	0.0126	0.0374	0.0808
CoxL1	C_{td}	**0.9867**	0.7827	0.5091	0.9856	0.9028	0.5349
	IBS	0.0146	0.0965	0.0960	**0.0077**	**0.0182**	0.0827

Synthesis: For data simulated from an AFT model with a log-normal distribution, Cox-nnet is the neural network with the best results in most situations when the sample size is small. When the sample size increases, NNsurv-deep is the best model considering the C_{td} in most situations. Moreover, NNsurv and NNsurv-deep also seem to perform well when the number of variables is less than or equal to 100. We assume that the good results of Cox-nnet are due to the low level of complexity of the data. Indeed, the survival curves of the individuals in this dataset never cross.

References

1. Antolini, L., Boracchi, P., Biganzoli, E.: A time-dependent discrimination index for survival data. Stat. Med. **24**(24), 3927–3944 (2005). https://doi.org/10.1002/sim.2427, https://onlinelibrary.wiley.com/doi/abs/10.1002/sim.2427

2. Bender, R., Augustin, T., Blettner, M.: Generating survival times to simulate Cox proportional hazards models. Stat. Med. **24**(11), 1713–1723 (2005). https://doi.org/10.1002/sim.2059, http://doi.wiley.com/10.1002/sim.2059
3. Biganzoli, E., Boracchi, P., Mariani, L., Marubini, E.: Feed forward neural networks for the analysis of censored survival data: a partial logistic regression approach. Stat. Med. **17**(10), 1169–1186 (1998). https://doi.org/10.1002/(SICI)1097-0258(19980530)17:10⟨1169::AID-SIM796⟩3.0.CO;2-D
4. Brier, G.W.: Verification of forecasts expressed in terms of probability. Monthly Weather Rev. **78**(1), 1–3 (1950). https://doi.org/10.1175/1520-0493(1950)078⟨0001:VOFEIT⟩2.0.CO;2, https://journals.ametsoc.org/mwr/article/78/1/1/96424/VERIFICATION-OF-FORECASTS-EXPRESSED-IN-TERMSOF
5. Chen, Y.Q., Wang, M.C.: Analysis of accelerated hazards models. J. Am. Stat. Associ. **95**(450), 608–618 (2000). https://doi.org/10.1080/01621459.2000.10474236, https://www.tandfonline.com/doi/abs/10.1080/01621459.2000.10474236
6. Ching, T., Zhu, X., Garmire, L.X.: Cox-nnet: an artificial neural network method for prognosis prediction of high-throughput omics data. PLOS Computational Biology **14**(4), e1006076 (2018). https://doi.org/10.1371/journal.pcbi.1006076, https://dx.plos.org/10.1371/journal.pcbi.1006076
7. Conesa, A., et al.: A survey of best practices for RNA-seq data analysis. Genome Biol. **17**(1), 1–19 (2016)
8. Cox, D.R.: Regression models and life-tables. J. Royal Stat. Soc. Ser. B (Methodol.) 34(2), 187–220 (1972). https://www.jstor.org/stable/2985181
9. Curtis, C., et al.: The genomic and transcriptomic architecture of 2,000 breast tumours reveals novel subgroups. Nature **486**(7403), 346–352 (2012). https://doi.org/10.1038/nature10983
10. Fan, J., Feng, Y., Wu, Y.: High-dimensional variable selection for Cox's proportional hazards model. Borrowing Strength: Theory Powering Applications — A Festschrift for Lawrence D. Brown **6**, 70–86 (2010). https://doi.org/10.1214/10-IMSCOLL606, https://projecteuclid.org/ebooks/institute-of-mathematical-statistics-collections/Borrowing-Strength--Theory-Powering-Applications--A-Festschrift-for/chapter/High-dimensional-variable-selection-for-Coxs-proportional-hazards-model/10.1214/10-IMSCOLL606
11. Faraggi, D., Simon, R.: A neural network model for survival data. Stat. Med. **14**(1), 73–82 (1995). https://doi.org/10.1002/sim.4780140108, https://onlinelibrary.wiley.com/doi/abs/10.1002/sim.4780140108
12. Gerds, T.A., Kattan, M.W., Schumacher, M., Yu, C.: Estimating a time-dependent concordance index for survival prediction models with covariate dependent censoring. Stat. Med. **32**(13), 2173–2184 (2013). https://doi.org/10.1002/sim.5681, https://onlinelibrary.wiley.com/doi/abs/10.1002/sim.5681
13. Gerds, T.A., Schumacher, M.: Consistent estimation of the expected brier score in general survival models with right-censored event times. Biometrical J. **48**(6), 1029–1040 (2006). https://doi.org/10.1002/bimj.200610301, https://onlinelibrary.wiley.com/doi/abs/10.1002/bimj.200610301
14. Goldenshluger, A., Lepski, O.: Bandwidth selection in Kernel density estimation: oracle inequalities and adaptive minimax optimality. Ann. Stat. **39**(3), 1608–1632 (2011). https://doi.org/10.1214/11-AOS883, https://projecteuclid.org/euclid.aos/1307452130

15. Graf, E., Schmoor, C., Sauerbrei, W., Schumacher, M.: Assessment and comparison of prognostic classification schemes for survival data. Stat. Med. **18**(17–18), 2529–2545 (1999). https://doi.org/10.1002/(SICI)1097-0258(19990915/30)18:17/18⟨2529::AID-SIM274⟩3.0.CO;2-5, https://onlinelibrary.wiley.com/doi/abs/10.1002/%28SICI%291097-0258%2819990915/30%2918%3A17/18%3C2529%3A%3AAID-SIM274%3E3.0.CO%3B2-5
16. Guilloux, A., Lemler, S., Taupin, M.L.: Adaptive Kernel estimation of the baseline function in the cox model with high-dimensional covariates. J. Multivar. Anal **148**, 141–159 (2016)
17. Hao, J., Kim, Y., Mallavarapu, T., Oh, J.H., Kang, M.: Interpretable deep neural network for cancer survival analysis by integrating genomic and clinical data. BMC Med. genomics **12**(10), 1–13 (2019)
18. Ishwaran, H., Kogalur, U.B., Blackstone, E.H., Lauer, M.S., et al.: Random survival forests. Ann. Appl. Stat. **2**(3), 841–860 (2008)
19. Kalbfleisch, J.D., Prentice, R.L.: The Statistical Analysis of Failure Time Data: Kalbfleisch/The Statistical. Wiley Series in Probability and Statistics, John Wiley & Sons, Inc., Hoboken, NJ, USA, August 2002. https://doi.org/10.1002/9781118032985, http://doi.wiley.com/10.1002/9781118032985
20. Kaplan, E.L., Meier, P.: Nonparametric estimation from incomplete observations. J. Am. Stat. Assoc. **53**(282), 457–481 (1958). https://doi.org/10.1080/01621459.1958.10501452, https://www.tandfonline.com/doi/abs/10.1080/01621459.1958.10501452
21. Katzman, J.L., Shaham, U., Cloninger, A., Bates, J., Jiang, T., Kluger, Y.: DeepSurv: personalized treatment recommender system using a Cox proportional hazards deep neural network. BMC Med. Res. Methodolo. **18**(1), 24 (2018). https://doi.org/10.1186/s12874-018-0482-1, https://doi.org/10.1186/s12874-018-0482-1
22. Kvamme, H., Borgan, Ø., Scheel, I.: Time-to-event prediction with neural networks and cox regression. J. Mach. Learn. Res. **20**(129), 1–30 (2019). http://jmlr.org/papers/v20/18-424.html
23. Kwong, C., Ling, A.Y., Crawford, M.H., Zhao, S.X., Shah, N.H.: A clinical score for predicting atrial fibrillation in patients with cryptogenic stroke or transient ischemic attack. Cardiology **138**(3), 133–140 (2017). https://doi.org/10.1159/000476030, https://www.ncbi.nlm.nih.gov/pmc/articles/PMC5683906/
24. Leemis, L.M., Shih, L.H., Reynertson, K.: Variate generation for accelerated life and proportional hazards models with time dependent covariates. Stat. Probab. Lett. **10**(4), 335–339 (1990). https://doi.org/10.1016/0167-7152(90)90052-9, https://linkinghub.elsevier.com/retrieve/pii/0167715290900529
25. Mogensen, U.B., Ishwaran, H., Gerds, T.A.: Evaluating random forests for survival analysis using prediction error curves. J. Stat. Softw. **50**(11), 1–23 (2012). https://www.ncbi.nlm.nih.gov/pmc/articles/PMC4194196/
26. Rajkomar, A., et al.: Scalable and accurate deep learning with electronic health records. npj Digital Med. **1**(1), 18 (2018). https://doi.org/10.1038/s41746-018-0029-1, https://www.nature.com/articles/s41746-018-0029-1
27. Ramlau-Hansen, H.: smoothing counting process intensities by means of kernel functions. Ann. Stat. **11**(2), 453–466 (1983). https://www.jstor.org/stable/2240560
28. Roblin, E., Cournede, P.-H., Michiels, S.: On the use of neural networks with censored time-to-event data. In: Bebis, G., Alekseyev, M., Cho, H., Gevertz, J., Rodriguez Martinez, M. (eds.) ISMCO 2020. LNCS, vol. 12508, pp. 56–67. Springer, Cham (2020). https://doi.org/10.1007/978-3-030-64511-3_6

29. Shrikumar, A., Greenside, P., Kundaje, A.: Learning important features through propagating activation differences. In: International Conference on Machine Learning, pp. 3145–3153. PMLR (2017)
30. Suo, Q., et al.: Deep patient similarity learning for personalized healthcare. IEEE Trans. NanoBiosci. **17**(3), 219–227 (2018). https://doi.org/10.1109/TNB.2018.2837622
31. Tibshirani, R.: The Lasso method for variable selection in the Cox model. Stat. Med. **16**(4), 385–395 (1997). https://doi.org/10.1002/(SICI)1097-0258(19970228)16:4⟨385::AID-SIM380⟩3.0.CO;2-3

Microbial Community Decision Making Models in Batch and Chemostat Cultures

Axel Theorell and Jörg Stelling(✉)

Department of Biosystems Science and Engineering, and SIB Swiss Institute of Bioinformatics, ETH Zurich, 4058 Basel, Switzerland
joerg.stelling@bsse.ethz.ch

Abstract. Microbial community simulations using genome scale metabolic networks (GSMs) are relevant for many application areas, such as the analysis of the human microbiome. Such simulations rely on assumptions about the culturing environment, affecting if the culture may reach a metabolically stationary state with constant microbial concentrations. They also require assumptions on decision making by the microbes: metabolic strategies can be in the interest of individual community members or of the whole community. However, the impact of such common assumptions on community simulation results has not been investigated systematically. Here, we investigate four combinations of assumptions, elucidate how they are applied in literature, provide novel mathematical formulations for their simulation, and show how the resulting predictions differ qualitatively. Crucially, our results stress that different assumption combinations give qualitatively different predictions on microbial coexistence by differential substrate utilization. This fundamental mechanism is critically under explored in the steady state GSM literature with its strong focus on coexistence states due to crossfeeding (division of labor).

Keywords: Microbial communities · Flux balance analysis · Game theory

1 Introduction

Microbial communities perform essential functions in diverse environments such as the soil [11] and the human gut [13]. While the experimental characterization of community composition is relatively easy with metagenomics methods, this is not true for the analysis of functional metabolic interactions between community members [10]. The paradigm of constraint based modelling of metabolism with genome scale models (GSMs) [4] has therefore become increasingly popular for the analysis of microbial communities [1,3]. For example, a recent GSM-based study stipulated that whether a microbial community is cooperative or competitive correlates strongly with the nutrient abundance in its natural habitat [20].

E. Cinquemani and L. Paulevé (Eds.): CMSB 2021, LNBI 12881, pp. 141–158, 2021.
https://doi.org/10.1007/978-3-030-85633-5_9

Approaching community functions with GSMs requires two key ingredients: models and simulation methods. Models are no longer a main limitation because of the ease with which large, organism-specific and relatively predictive GSMs can be derived automatically from genome sequences [15]. However, the main simulation methods for GSMs such as flux balance analysis (FBA) [16] and stochastic sampling [14] were originally developed for single species, not communities.

In single-species FBA, a key assumption is that the simulated species optimizes its fitness (e.g., growth). This can be interpreted as a decision making problem where the organism needs to optimally control its (evolved) metabolic network. However, in co-culture, the degree to which one species reaches its objective may depend on the metabolic activity of all species, for example, when species compete for nutrients. Dynamic FBA (dFBA) explicitly accounts for nutrient concentrations and thereby for such interactions; it combines the FBA principle with iterations over time to reflect changing environmental conditions [28]. Recently, also scalable methods for dFBA simulation of communities have been proposed [23]. Yet, a drawback of dFBA is that it requires reliable knowledge on the form and parameters of uptake kinetics. These are hard to obtain and without them, the simulation results can be unreliable [3].

Incomplete information on uptake kinetics raises a new frontier in decision making for the simulation of interacting microbes in co-culture: the presence of multiple decision making entities with potentially conflicting objectives. For example, in d-OptCom, an influential method for dFBA of a community of GSMs, decisions are based on a community objective (high community biomass production) as well as individual objectives (high growth rate) [29]. In other methods, the emergence of multiple decision makers has stimulated the use of game theory for the analysis of microbial interactions [24].

To alleviate the dependence on uptake kinetics, community analyses with GSMs are often restricted to metabolically stationary states (that is, metabolic fluxes are constant over time) [7,31]. The long term behavior, represented by the steady states, is the primary interest of most investigations. This makes investment of computational resources into predicting transient behavior less attractive. Furthermore, dynamics make the conceptualization of microbial decision making more complex (raising questions about when a consortium strives to achieve an objective).

However, as we will detail in Sect. 2, going from one to several microbial species, the interpretation of the metabolically stationary state assumption in constraint-based modeling suddenly depends on the type of cultivation environment. Moreover, it turns out that differences in the environment also have implications for models of *decision making* in FBA-type analyses. In particular, assumptions on environment and decision making have fundamental impact on whether organisms in a community of GSMs can coexist or not. These dependencies have not yet been investigated systematically.

Here, we formulate four methods for simulating metabolically stationary states, corresponding to combinations of two different environments, batch and chemostat cultivation, and two different modes of microbial decision making, distributed (rational agent) and centralized (rational community). In these formulations, we put a novel emphasis on what information (local/global) the decision makers have access to. The combination steady state batch/rational community resembles the SteadyCom formulation [8]; the chemostat formulations applicable to GSMs are new. We demonstrate the qualitative differences between the approaches on two toy-examples, a prisoners dilemma (PD) model for decision making and a nutrient limitation model for coexistence. As expected, switching from rational agent to rational community, PD switches from defection to cooperation. For nutrient limitation, the four models yield qualitatively different results. We argue that, which model to apply in a practical scenario should be considered carefully and has to reflect both the chemical environment and whether the community can be expected to have developed community strategies. We believe that the chemostat formulations are of particular value because important microbial environments such as the human gut resemble a chemostat [9].

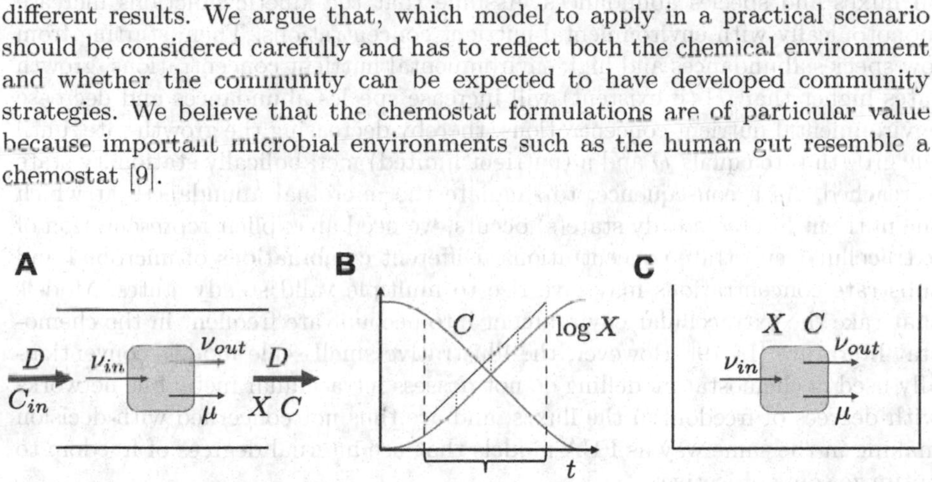

Fig. 1. Cultivation systems and their implications for metabolically stationary state conditions. For definitions of mathematical variables, see Sect. 2.1. (A) Chemostat as an open system in steady state. Black: time-constant entities; bold arrows: flows; normal arrows: metabolic fluxes; rounded rectangle: cell. (B) Dynamics in batch cultivation of cells with a phase of metabolically stationary state (constant growth rate, implying linear increase of the logarithm of the species concentration, insensitive to external concentrations) between dashed vertical lines. (C) Metabolically stationary state in the closed batch system (time-constant entities in black).

2 Concepts

The two main environments for cultivating microbes are (assumed) chemostat and batch processes. For FBA-based analysis, they imply different concepts of metabolically stationary state, leading to different forms of microbial coexistence and of models for decision making.

2.1 Chemostat vs Batch Environment

In a *chemostat* as an open system, a fluid flow (dilution rate D) adds nutrients (inflow concentrations C_{in}) and flushes out parts of the cultivation medium, keeping the cultivation volume constant (see Fig. 1A). A metabolically stationary state requires that metabolic fluxes (ν), species abundances (X), and environmental nutrient concentrations (C) are constant over time (t). For the (non-zero) absolute microbial species abundances to be constant, the growth rates (μ) must be equal to the dilution rate D (henceforth called D-growth).

Assuming growth maximization, the growth rate depends on the environmental nutrient concentrations via uptake kinetic functions that determine the upper bounds of uptake fluxes. In turn, environmental nutrient concentrations depend on fluxes and species abundances. Assume that the kinetic functions increase monotonically with environmental nutrient concentrations. Then, starting from low species abundances and high environmental nutrient concentrations, growth rates higher than D (if existent) will increase species abundances and decrease environmental nutrient concentrations, thereby decreasing the growth rate, until the growth rate equals D and a (nutrient limited) metabolically stationary state is reached. As a consequence, to simulate the microbial abundances at which the nutrient limited steady state(s) occurs, we need an explicit representation of extracellular substrate concentrations. Different combinations of microbial and substrate concentrations may give rise to multiple valid steady states. Models that take the extracellular environment into account are frequent in the chemostat literature [18,19]. However, the illustrative small scale models conventionally used in chemostat modelling do not possess intracellular metabolic networks with degrees of freedom in the fluxes, and are thus not concerned with decision making in the same way as FBA-models that use internal degrees of freedom to optimize some objective.

In a *batch* process as a closed system, all nutrients are provided at the beginning of the cultivation and nothing is flushed out (Fig. 1B, C). Here, a modeled metabolically stationary state refers to the condition that metabolic fluxes as well as growth rates are time-constant. This can hold, for example, during exponential growth. It has two important implications: First, relevant environmental nutrient concentrations are assumed to be in a regime where the kinetic functions determining the upper bounds of the growth limiting uptake fluxes are insensitive to the nutrient concentrations. This allows for community models without a representation of environmental nutrient concentrations. Second, the relative species concentrations must be constant, implying that all species with non-zero abundance grow at the same rate averaged over time (henceforth called balanced growth). This allows to properly model inter species crossfeeding of compounds. Some GSM-based studies of communities apply balanced growth [8,17]. However, others do not [6,7,30], thus assuming a non-closed system. Throughout this manuscript, any system operating a metabolically stationary state under non-limiting extracellular nutrient conditions will be called a steady state batch. Though such a system may not have to be a batch cultivation, we will use the name batch throughout the manuscript, since batch cultivation is the model system addressed in this manuscript.

2.2 Implications for Coexistence

As demonstrated, in the context of GSMs, chemostat and batch imply distinct conditions on metabolically stationary states. These distinctions have crucial consequences for the possibility of co-existence of microbes. For non-interacting microbes in a batch, balanced growth will only occur if all concerned species have the exact same growth rate by chance, a situation that never happens in practice. Therefore, to simulate coexistence in a consortium, an explicit interaction between microbes, such as crossfeeding [8] or some form of *agreement* to grow at the same rate is mandatory. In contrast, for a chemostat operated with constant nutrient concentrations in the feed, competing species may coexist under D-growth, if they are limited by different nutrients [2]. Indeed, this enables models with coexistence states originating from both crossfeeding and differential nutrient limitations [22].

2.3 Implications for Decision Making

The assumptions on the environment—implying observability of nutrient concentrations or lack of observability—also have implications for models of *decision making* in FBA-type analyses. As mentioned for the GSM community simulation method d-OptCom [29], as well as for its metabolically stationary state sibling OptCom, [30], decision making is modeled as a bi-level optimization problem. On one level, the community strives towards a fitness goal (high community biomass production) and on the other level each microbial species optimizes its own fitness (growth rate). Abstractly, there are two types of decision makers, one making community decisions and one making decisions for individuals. The existence of an apparent community decision maker is hypothetical—it could result from species co-evolution [27,31]. Generally, it has been shown that cooperative (generous) strategies are evolutionarily robust in repeated PD games in simulations [25].

Because community and individual decision makers may follow contradictory strategies, a principle for conflict resolution is needed. Some possibilities used for GSMs are: the community strategy takes precedence over individual decision makers [30], a community strategy must be Pareto optimal for the individual decision makers [6], and a community strategy must be a Nash equilibrium for the individual decision makers [7].

In particular, Cai et al. [7] makes the differences in conflict resolution mechanisms concrete by converting the so-called called Prisoners Dilemma (PD) [12] game theory example to a metabolic network setting. PD is a two player symmetric game with payoffs shown in Table 1. Mutual cooperation generates the largest overall benefit, but defection by one player yields a higher payoff for this player if the other player cooperates. Figure 2 shows a metabolic community version of PD, where species 1 and 2 both have the capacity to produce metabolites

Table 1. Generic prisoners dilemma payoff matrix (numbers unrelated to Fig. 2). The first and second number in the round brackets denote the payoffs for player 1 and 2, respectively.

Player 1	Player 2	
	Cooperate	Defect
Cooperate	(3, 3)	(1, 4)
Defect	(4, 1)	(2, 2)

A and B and need both to grow, but where species 1 produces A and species 2 produces B at lower yield than the other. Thus, for the community, mutual cooperation (crossfeeding) will lead to the highest biomass yield, whereas for the individual species, the highest yield is obtained by not secreting anything, while still being fed by the other species. PD is a good testing ground for conflict resolution: it pits the community and individual decision makers against each other. As expected, the Nash equilibrium mechanism suggested in [7] results in no crossfeeding, whereas giving the community decision maker precedence [30] yields crossfeeding. Yeast cells feeding off sucrose may be a biological PD. The sucrose is hydrolyzed to glucose and fructose extracellularly by the enzyme invertase. It is expected that producing and secreting invertase comes at a metabolic cost. However, it may also give a growth benefit, if being an invertase producer means that more sugars will be hydrolyzed close to the producer. If the cost is relatively high and the benefit relatively low, cheating by producing no invertase becomes a desirable strategy [7].

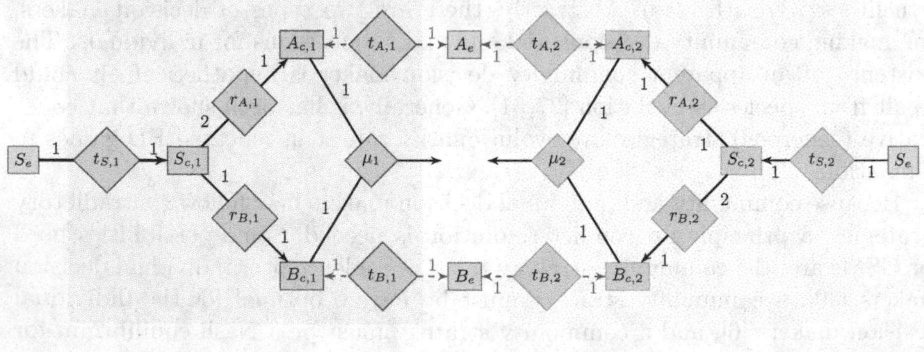

Fig. 2. A PD microbial consortium [7]. Rectangles are metabolites and diamonds are reactions. Red rectangles are extracellular metabolites. Numbers next to lines are stoichiometric coefficients. The subscripts c and e denote intra- and extracellular compounds, respectively. Species 1 and 2 (blue and brown symbols) can choose to crossfeed the compounds A and B to increase their yields by activating the reactions with the red dashed lines. (Color figure online)

3 Community Models

To cover the two principal dimensions environment (chemostat vs batch) and decision making model (rational agent vs rational community), we developed four models of microbial community growth at metabolically stationary state using metabolic networks. We first introduce the general system of equations and differential equations governing the metabolite and species concentrations that the models are based on. Then we impose assumptions about steady state conditions and decision making that lead to the models.

3.1 General Consortium Models

We are interested in the time development of the extracellular compound concentrations $C \geq 0 \in \mathcal{R}^{n_C}$ (we denote dimensionalities of variable x by n_x) and the organism concentrations $X \geq 0 \in \mathcal{R}^{n_X}$ (see also Fig. 1). We consider a system with inflow rate D_{in} and outflow rate D_{out}. The inflow has nutrient concentrations $C_{in} \in \mathcal{R}^{n_C}$. The vector of metabolic fluxes (reaction rates) of microbial species i is denoted $\nu_i \in \mathcal{R}^{n_{\nu_i}}$. One element of each flux vector ν_i is the biomass production (growth) rate $\nu_{\mu,i}$. The matrix $T_i \in \mathcal{R}^{n_C \times n_{\nu_i}}$ maps reactions to exchanges of extracellular compounds. Assuming that compounds and cells are flushed out at a rate proportional to their concentrations, the dynamics of C and X are described by:

$$\frac{dC}{dt} = D_{in}C_{in} - D_{out}C - \sum_i T_i \nu_i X_i \qquad (1)$$

$$\frac{dX_i}{dt} = X_i(\nu_{\mu,i} - D_{out}), \ \forall i \,. \qquad (2)$$

For steady state, left-hand sides of the system of ordinary differential equations (ODEs) Eqs. (1–2) are zero.

Common in both FBA and dFBA, as well as used here, is the assumption of intracellular (metabolically) stationary state [16,21]. Modelling reactions between n_S intracellular compounds at constant concentrations, intracellular stationary state introduces a stoichiometric matrix $S_i \in \mathcal{R}^{n_S \times n_{\nu_i}}$ for which holds that

$$S_i \nu_i = 0, \ \forall i \,. \qquad (3)$$

Furthermore, for some matrix $A_i \in \mathcal{R}^{n_A \times n_{\nu_i}}$ and a vector $b_i \in \mathcal{R}^{n_A}$, the fluxes have capacity constraints

$$A_i \nu_i \leq b_i, \ \forall i \,. \qquad (4)$$

The *steady state models* of interest here are chemostat and steady state batch. In a chemostat at steady state, $D_{in} = D_{out} = D$ and the steady state algebraic relations from Eqs. (1)–(4) apply directly.

In contrast, in steady state batch, the extracellular compound concentrations are assumed to have no influence on the fluxes. To avoid infinite uptakes, flux exchanges with the environment, modeled by changes in C in Eq. (1), are captured by a vector of culture uptake bounds, $u \in \mathcal{R}^{n_C}$. The species concentrations X are exchanged for the relative species concentrations x. The change to relative species concentrations allows them to stay constant over time under balanced growth. To represent balanced growth, a community growth rate $\nu_\mu^\star$ is introduced. In combination, the steady state batch system is then:

$$\begin{aligned} u - \sum_i T_i \nu_i \cdot x_i \geq 0 \\ S_i \nu_i = 0,\ \forall i \\ A_i \nu_i \leq b_i,\ \forall i \\ x_i(\nu_\mu^\star - \nu_{\mu,i}) = 0, \forall i \\ \sum_i x_i = 1\ . \end{aligned}$$

We use these formulations of chemostat and batch in metabolically stationary state to introduce four models. For ease of comparison, all model equations, plus extra information such as Karush Kuhn Tucker (KKT) [26] conditions used for solving the models, can be seen side-by-side in Table A.1.

3.2 Rational Agents

We assume that each cell is a decision making entity, using the extracellular concentrations as information to maximize its growth rate. As foundation for its decision making, each cell uses local information, in this case the extracellular compound concentrations, as well as its own flux constraints Eqs. (3–4). This assumption seems intuitive for microbial species that do not share an evolutionary history of interactions.

Assuming a *chemostat environment* and denoting variables resulting from an optimization problem with hat notation $\hat{\nu}_i$, the rational agent (CA, where 'C' stands for chemostat and 'A' for agent) model becomes:

$$\begin{aligned} D(C_{in} - C) - \sum_i T_i \hat{\nu}_i(C) X_i = 0 \\ X_i(D - \hat{\nu}_{\mu,i}(C)) = 0,\ \forall i \\ C, X \geq 0 \\ \hat{\nu}_i(C) = \underset{\nu_i \in \mathcal{R}^{n_{\nu_i}}}{\operatorname{argmax}}\, \nu_{\mu,i}, \forall i \\ s.t.\ S_i \nu_i = 0,\ \forall i \\ A_i \nu_i \leq b_i(C),\ \forall i\ . \end{aligned} \tag{5}$$

Importantly, here the right hand side of the capacity constraints $b(C)$ depends on the extracellular concentrations C through uptake kinetics. Without implementing a concentration dependency, the optimization problem is independent of the substrate and organism concentrations. This means that the modeled cells do not adapt their growth to changes in extracellular nutrient concentrations. In most cases, this will imply that no solution will fulfill the D-growth requirement and only the trivial solution $X = 0$ will be feasible.

Correspondingly, the *steady state batch* rational agent (BA) system is:

$$\begin{aligned} u - \sum_i T_i \hat{\nu}_i \cdot x_i &\geq 0 \\ x_i(\nu_\mu^\star - \hat{\nu}_{\mu,i}) &= 0, \forall i \\ \sum_i x_i &= 1 \\ x &\geq 0 \\ \hat{\nu}_i = \underset{\nu_i \in \mathcal{R}^{n_{\nu_i}}}{\operatorname{argmax}}\, &\nu_{\mu,i}, \forall i \\ s.t.\ S_i \nu_i &= 0, \forall i \\ A_i \nu_i &\leq b_i, \forall i\ . \end{aligned} \tag{6}$$

Contrary to practice in the CSM consortium literature [3], the rational agent assumption does not include the global equations (first lines) in the optimization problem. Furthermore, since the extracellular substrate concentrations are assumed to be constant, the optimization problem is independent of the first two lines of Eq. (6). Combined with the balanced growth assumption, this means that coexistence is only possible for organisms that independently developed the exact same growth rate, a situation that will never occur in practice. The steady state batch rational agent model is therefore of little practical relevance and we included it only for completeness.

3.3 Rational Community

The rational community model assumes that, through a time of coexistence, a community has learnt to optimize its (D- or balanced-) growth rate while cooperating to create a favourable nutrient environment.

To model a *chemostat environment* and a rational community (CC), note that what the community wants to achieve through cooperation, and with it the formal community objective function, may vary. A biologically relevant community objective, so far not formulated as FBA objective, is resistance to invasion by pathogenic species [5]. Here, for simplicity and in line with the literature [3], we consider only the community objective of maximizing total biomass production. Using a concatenated flux variable $\nu = [\nu_1, , , \nu_{n_X}] \in \mathcal{R}^{n_\nu}$, the CC model reads:

$$
\begin{aligned}
X_i(D - \hat{\nu}_{\mu,i}(X)) &= 0, \forall i \\
X &\geq 0 \\
\hat{\nu}(X) = \underset{\nu \in \mathcal{R}^{n_\nu}, C \in \mathcal{R}^{n_C}}{\text{argmax}} & \sum_i \nu_{\mu,i} X_i \\
s.t.\ D(C_{in} - C) - \sum_i T_i \nu_i X_i &= 0 \\
S_i \nu_i &= 0, \ \forall i \\
A_i \nu_i &\leq b_i(C), \ \forall i \\
C &\geq 0 \, .
\end{aligned}
\tag{7}
$$

Contrary to the rational agent models, Eq. (1) is now inside the optimization problem, and therefore, so are also all instances of the global variables C. This encodes our assumption that the community has knowledge of and power over the global cellular exchanges of compounds.

An important detail of the CC model is that the abundances X do not enter the optimization problem as optimization variables. Since different community decisions may benefit different organisms (in terms of species abundances and other factors), having a range of community optimal strategies in terms of fluxes and extracellular concentrations, but given different species abundances, it is not possible to know which strategy the community will settle for without detailed knowledge of the "negotiation" process leading up to a decision. Thus, the step from rational agent to rational community is not about assuming full knowledge of how the community decides, but that actively influencing C is taken into account in its decision, while optimizing some assumed objective.

In the *steady state batch* rational community (BC) case, having no explicit representation of C, the community decision maker cannot take C into account in the optimization problem. Thus, the difference to the rational agent steady state batch model is that the community takes the macroscopic equation $u + \sum_i T_i \nu_i \cdot x_i \geq 0$ into account in the decision making process, leading to the system:

$$
\begin{aligned}
x_i(\nu_\mu^\star - \hat{\nu}_{\mu,i}(x)) &= 0, \forall i \\
\sum_i x_i &= 1 \\
x &\geq 0 \\
\hat{\nu}(x) = \underset{\nu \in \mathcal{R}^{n_\nu}}{\text{argmax}} & \sum_i \nu_{\mu,i} x_i \\
s.t.\ u - \sum_i T_i \nu_i \cdot x_i &\geq 0 \\
S_i \nu_i &= 0, \ \forall i \\
A_i \nu_i &\leq b_i, \ \forall i \, .
\end{aligned}
\tag{8}
$$

4 Applications

We tested the CA, CC and BC models on two toy examples: PD (Fig. 2) for exploring decision making and a competition scenario for exploring coexistence. For named reactions and compounds in Figs. 2 and 3, such as $t_{S,1}$ and S_e, the corresponding fluxes and concentrations will be referred to as $\nu_{t_S,1}$ and C_{S_e}. All examples were solved symbolically in Mathematica 9 using the KKT formulations in Table A.1. Notice that the optimization problems in Eqs. (5)–(8) are all linear in the optimization arguments, implying that KKT provides sufficient conditions for global optimality [26]. By solving the respective systems symbolically, we are confident that all solutions are found.

4.1 Prisoners Dilemma

For PD (Fig. 2), we are interested in whether, using a specific simulation model, the community achieves a fitness bonus by utilizing crossfeeding (using reactions with dashed red arrows in Fig. 2) or whether the organisms refuse to cooperate. For CA and CC simulations, we set the inflow nutrient concentration mixture to $C_{in,A_e} = 0$, $C_{in,B_e} = 0$ and $C_{in,S_e} = 10$. The flow rate was set to $D = 0.5$. Similarly for BC, we used the culture uptake bounds $u_{A_e} = 0$, $u_{B_e} = 0$ and $u_{S_e} = 10$.

Quantitative simulation results are shown in Table 2. As expected, without a joint objective for the organisms, CA finds no crossfeeding. CC and BC find crossfeeding solutions, but these solutions differ. In CC, the secretion fluxes are greater than the uptake fluxes because some of the secreted material will be flushed out of the chemostat, rather than taken up by another organism. This generally makes crossfeeding in chemostats less attractive. For example when increasing the flow rate D from 0.5 to 1.2, the benefit of crossfeeding vanishes and CC switches to a solution without crossfeeding. In BC, void of an active out flush mechanism, all secreted material is taken up. Apart from the symmetric (non-zero) solutions, non-symmetric solutions where one species has zero abundance occur. These solutions, with only one participating species and thereby no potential for cooperation, are not considered here.

4.2 Coexistence Microbial Consortium

In a chemostat with two supplied nutrients A_e and B_e, coexistence of two distinct species may emerge if, depending on the supply concentrations, the species reach a state in which they are limited by different nutrients [2,19]. This (potentially competitive) coexistence does not rely on direct interactions such as crossfeeding. For the CA, CC and BC models, we investigated under what circumstances coexistence emerges for the non-crossfeeding metabolic network models in Fig. 3. There, species 1 needs more of compound A_c to grow and species 2 needs more of compound B_c to grow.

For CA and CC, we varied the nutrient composition of the inflow, $(C_{in,A_e}/C_{in,B_e})$, linearly from (0/10) to (10/0). We set the flow rate $D = 1$

Table 2. Flux values of PD simulations for CA, CC and BC. The variable names correspond to the reactions in Fig. 2. In the last column, CC was run with a higher flow rate, $D = 1.2$.

Variable	CA	CC	BC	CC $D = 1.2$
C_{A_e}	0	0.5		0
C_{B_e}	0	0.5		0
C_{S_e}	1.5	1.13		3.6
X_1	1.42	1.97	0.5	1.07
X_2	1.42	1.97	0.5	1.07
$\nu_{t_S,1}$	1.5	1.13	10	3.6
$\nu_{t_A,1}$	0	0.5	5	0
$\nu_{t_B,1}$	0	−0.627	−5	0
$\nu_{r_A,1}$	0.5	0	0	1.2
$\nu_{r_B,1}$	0.5	1.13	10	1.2
$\nu_{\mu,1}$	0.5	0.5	5	1.2
$\nu_{t_S,2}$	1.5	1.13	10	3.6
$\nu_{t_A,2}$	0	−0.627	−5	0
$\nu_{t_B,2}$	0	0.5	5	0
$\nu_{r_A,2}$	0.5	1.13	10	1.2
$\nu_{r_B,2}$	0.5	0	0	1.2
$\nu_{\mu,2}$	0.5	0.5	5	1.2

and the uptake flux limitations $\nu_{t_A,i} \leq 2 \cdot C_{A_e}$ and $\nu_{t_B,i} \leq 2 \cdot C_{B_e}$ (symbols defined in Fig. 3). Lacking a potential to crossfeed, CA and CC generated identical results. Figure 4A shows identical, horizontally mirrored, zero solutions (ZS), that is, solutions where only one species exists. The ZS of species 1 starts flat at zero, which is a regime where the concentration of C_{A_e} is so low that species 1 cannot grow at the flow rate ($D = 1$); it is flushed out of the chemostat. After the zero regime comes a regime in which the growth rate of species 1 is limited by C_{A_e} and the concentration of species 1 increases linearly with C_{in,A_e}. This continues with increasing C_{in,A_e} and decreasing C_{in,B_e}, until C_{B_e} becomes growth limiting, and the species concentration goes down linearly. A coexistence solution (CS) exists in one central regime, throughout which species 1 is limited by C_{A_e} and species 2 is limited by C_{B_e}. At the concentration mixture where the dark blue curve (CS) goes to zero and ends, the light blue curve (CS) touches the light green curve (ZS). At this point, where the lower CS goes to zero, the upper CS becomes a ZS.

For BC, we varied the culture uptake bounds u_{A_e} and u_{B_e} linearly from (0/1) to (1/0). The uptake bounds were $\nu_{t_A,i} \leq 2$ and $\nu_{t_B,i} \leq 2$. The main distinction from the chemostat scenario is that in BC, the ZSs are identically one (Fig. 4B), due to the relative species concentrations.

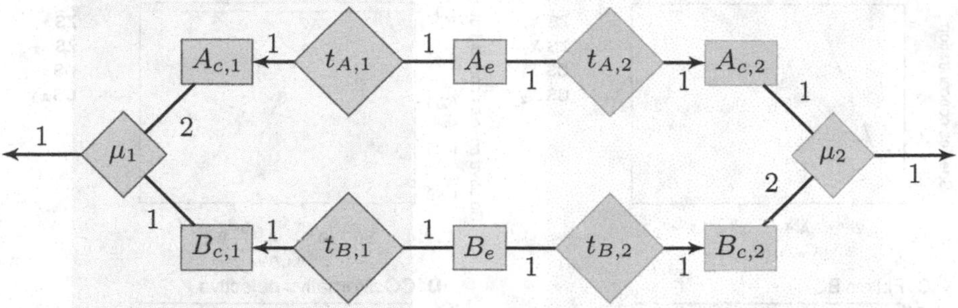

Fig. 3. A coexistence microbial consortium. Rectangles are metabolites and diamonds are reactions. Red rectangles are extracellular metabolites. The subscripts c and e denotes intra- and extracellular compounds, respectively. Numbers next to lines are stoichiometric coefficients. (Color figure online)

Despite the apparent similarity between CA and BC, the interpretation of the coexistence solutions (CSs) differs. For CA, a CS emerges without interspecies *communication*, simply because, at the species level, the growth rates of species 1 and 2 are limited by the uptake rates of A_e and B_e, respectively. This is a known result from chemostat modelling [18]. Thus, at their steady state concentrations, the species reach a self stabilizing equilibrium, where neither species can grow faster than $D = 1$.

In contrast, the growth rates of the species in the CS in BC (Fig. 4B) are not restricted by individual species uptake fluxes. Figure 4C shows for species 1 (a horizontal mirror image of species 2) that the uptake flux $\nu_{t_A,1}$ always remains below its upper bound of 2. Instead, the growth rates are restricted by the global nutrient restrictions $u - \sum_i T_i \nu_i \cdot x_i \geq 0$. With regard to the global nutrient restrictions, the balanced growth solutions, where the species grow at the same rate are not the only solutions. As shown in Fig. 4C, in the CS, species 1 voluntarily grows at a rate that is lower than its maximal growth rate (CS max $\nu_{\mu,1}$ exceeds the take-all solution ZS $\nu_{\mu,1}$ since it is operating at a lower relative species concentration). If the species did not *communicate* that growing at the same rate maximizes community biomass production, single species would claim more resources for themselves and break the metabolically stationary state. Thus, the CS solution we see is a result of the objective function.

To elucidate the dependence of the CSs of CC and BC on the community objective function, we changed the objective of CC and BC to maximizing the sum of growth rates, $\sum_i \nu_{\mu,i}\delta(X_i > 0)$, rather than total community biomass production, $\sum_i \nu_{\mu,i} X_i$ (X_i is replaced by x_i in BC). Figure 4D-E shows that the changed objective function results in CSs that differ from the ones in Fig. 4A-B.

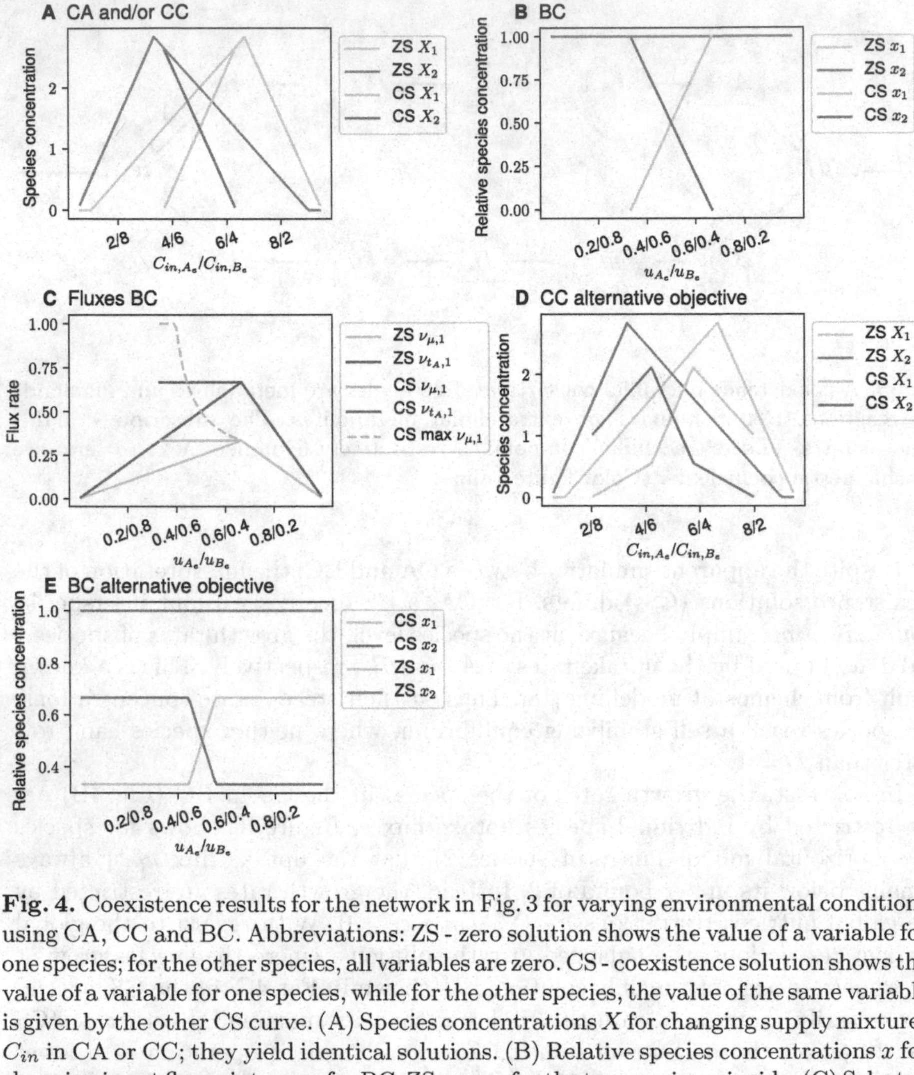

Fig. 4. Coexistence results for the network in Fig. 3 for varying environmental conditions using CA, CC and BC. Abbreviations: ZS - zero solution shows the value of a variable for one species; for the other species, all variables are zero. CS - coexistence solution shows the value of a variable for one species, while for the other species, the value of the same variable is given by the other CS curve. (A) Species concentrations X for changing supply mixtures C_{in} in CA or CC; they yield identical solutions. (B) Relative species concentrations x for changing input flux mixtures u for BC. ZS curves for the two species coincide. (C) Selected fluxes of species 1 for changing input flux mixtures u for BC. (D) Same information as in (A), but only using CC and with an alternative objective function. (E) Same information as in (B), but with an alternative objective function.

5 Discussion

Our study draws heavily on the long tradition of chemostat community models [19]. We followed in the same spirit: to keep models small and to use them to demonstrate general system properties, rather than detailed properties of cells with specific genomes. On the contrary, GSMs facilitate a detailed, species-specific analysis of intracellular fluxes and related properties. We consider the work presented here as an early-stage attempt to combine the two worlds.

To incorporate FBA models in the chemostat community model frame work, due to the internal degrees of freedom of FBA models, the fundamentally game theoretic problem of multiple decision makers has to be taken into account [24]. Here, and in line we previous proposals for community modeling, we therefore explored two flavors: rational agent and rational community. For our rational community models, we allowed the community to optimize both its shared metabolism and the environmental nutrient concentrations to achieve a community objective. However, we did not explicitly optimize the species concentration variables. This acknowledges that, if different species concentrations favor different species, and thereby yield multiple optima in terms of fluxes and nutrient concentrations, we do not know which optimum the community would choose.

One would expect that the decision a community takes depends on the overarching frame work (here: rational agent and rational community) and on the particular objective imposed. For example, by maximizing biomass production of the community, crossfeeding emerges in the PD scenario. However, our community models demonstrates that also environmental variables play a role. For example, by increasing the flow rate in the chemostat, the benefit of crossfeeding decreased, so that CC switched from crossfeeding to no crossfeeding. This phenomenon might be relevant for the gut microbiome, where the significance of other aspects of flow has been investigated [9].

More specifically, we believe that the qualitative results of PD are relatively robust to changes in the community objective, such as switching to a sum of growth rates objective. Contrarily, for the coexistence example, we saw that changing the community objective function gave a new set of coexistence states.

Our models also suggest that coexistence in batch (BC) relies on a different mechanism than in chemostat (CA and CC). In BC, the community steady state is not a consequence of nutrient limitations caused by community growth. Coexistence requires *agreement* to coexist in the community, without any external enforcement mechanism, contrary to the chemostat models. Agreement without enforcement may amount to *forced altruism*, a modelling artifact discussed in detail in the context of PD by Chan et al. [8]. The emergence of forced altruism in terms of coexistence at balanced growth, rather than in terms of crossfeeding, is to our knowledge a new perspective that may be relevant for future community simulation methods.

A limitation of the present work is that we considered only toy metabolic networks. This choice was intentional to provide general insights. It also allows us to solve the problems symbolically, which certified that all roots to the equations were found. However, this approach is of course not scalable. Before any real applications can be considered, an efficient numerical solution scheme needs to be developed. As alternative to solving the KKT equations, one could directly run corresponding dFBA simulations until stationarity. However, also this approach would need to be complemented with a mechanism for finding multiple solutions.

Lastly, in the chemostat literature [19], stability of stationary solutions of ODEs is a central topic, which we did not address. If we assume that the microbial species can make decisions and actively uphold a state or an equilibrium, exactly what stability means in this scenario may need additional theoretical attention. Such concepts may be important to evaluate resistance of microbial communities to invasion by pathogens.

Acknowledgments. We thank Jakob Vanhoever, Mattia Gollub and Charlotte Ramon for support and stimulating discussions. This work was supported by the Swiss National Science Foundation Sinergia project with grant #177164.

Appendix

Table A.1. Models, Lagrangians and KKT formulations. For BA and CA, Lagrangian multipliers (inequality, λ_1, and equality, λ_2, multipliers) are introduced on a per species level (subscript i), whereas for BC and CC, global multipliers are introduced. Dimensionalities of multipliers vary between formulations. Rows #var and #EQ confirm that the numbers of unknowns and equations are equal.

	BA	CA	BC	CC
Eqs	$u - \sum_i T_i \hat{\nu}_i \cdot x_i \geq 0$ $x_i(\nu^*_\mu - \hat{\nu}_{i,\mu}) = 0, \forall i$ $\sum_i x_i = 1$ $x \geq 0$ $\hat{\nu}_i = \underset{\nu_i \in \mathcal{R}^{n_{\nu_i}}}{\text{argmax}}\, \nu_{\mu,i}, \forall i$ $s.t.\ S_i \nu_i = 0,\ \forall i$ $A_i \nu_i \leq b_i,\ \forall i$	$D(C_{in} - C)$ $- \sum_i T_i \hat{\nu}_i(C) X_i = 0$ $X_i(D - \hat{\nu}_{i,\mu}(C)) = 0, \forall i$ $C, X \geq 0$ $\hat{\nu}_i(C) = \underset{\nu_i \in \mathcal{R}^{n_{\nu_i}}}{\text{argmax}}\, \nu_{\mu,i}, \forall i$ $s.t.\ S_i \nu_i = 0,\ \forall i$ $A_i \nu_i \leq b_i(C),\ \forall i$	$x_i(\nu^*_\mu - \hat{\nu}_{i,\mu}(x)) = 0, \forall i,$ $\sum_i x_i = 1,$ $x \geq 0,$ $\hat{\nu}(x) = \underset{\nu \in \mathcal{R}^{n_\nu}}{\text{argmax}} \sum_i \nu_{\mu,i} x_i,$ $s.t.\ u - \sum_i T_i \nu_i \cdot x_i \geq 0,$ $S_i \nu_i = 0,\ \forall i,$ $A_i \nu_i \leq b_i,\ \forall i$	$X_i(D - \hat{\nu}_{i,\mu}(X)) = 0, \forall i$ $X \geq 0$ $\hat{\nu}(X) =$ $\underset{\nu \in \mathcal{R}^{n_\nu}, C \in \mathcal{R}^{n_C}}{\text{argmax}} \sum_i \nu_{\mu,i} X_i$ $s.t.\ D(C_{in} - C)$ $- \sum_i T_i \nu_i X_i = 0$ $S_i \nu_i = 0,\ \forall i$ $A_i \nu_i \leq b_i(C),\ \forall i$ $C \geq 0$
Lgr	$L_i(\nu_i) =$ $-\nu_{\mu,i} + \lambda^T_{i,1}(A_i \nu_i - b_i)$ $+\lambda^T_{i,2}(S_i \nu_i), \forall i$	$L_i(\nu_i(C)) =$ $-\nu_{\mu,i} + \lambda^T_{i,1}(A_i \nu_i - b_i(C))$ $+\lambda^T_{i,2}(S_i \nu_i), \forall i$	$L(\nu) =$ $- \sum_i \nu_{\mu,i} x_i$ $+\lambda^T_1 \begin{bmatrix} A\nu - b \\ -u + \sum_i T_i \nu_i x_i \end{bmatrix}$ $+\lambda^T_{i,2} S_i$	$L([\nu, C]) =$ $- \sum_i \nu_{\mu,i} X_i$ $+\lambda^T_1 \begin{bmatrix} (A\nu - b(C)) \\ -C \end{bmatrix}$ $+\lambda^T_2 \begin{bmatrix} S \\ D(C_{supply} - C) \\ - \sum_i T_i \nu_i X_i \end{bmatrix}$
KKT	$\begin{bmatrix} 0 \\ \vdots \\ -1 \end{bmatrix}^T + \lambda^T_{i,1} A_i$ $+\lambda^T_{i,2} S_i = 0, \forall i,$ $\lambda_{i,1} \geq 0, \forall i,$ $\lambda^T_{i,1}(A_i \nu_i - b_i) = 0, \forall i$	$\begin{bmatrix} 0 \\ \vdots \\ -1 \end{bmatrix}^T + \lambda^T_{i,1} A_i$ $+\lambda^T_{i,2} S_i = 0, \forall i,$ $\lambda_{i,1} \geq 0, \forall i,$ $\lambda^T_{i,1}(A_i \nu_i - b_i(C)) = 0, \forall i$	$\begin{bmatrix} 0 \\ \vdots \\ -x_1 \\ 0 \\ \vdots \\ -x_{n_x} \end{bmatrix}^T$ $+\lambda^T_1 \begin{bmatrix} A \\ \sum_i T_i x_i \end{bmatrix}$ $+\lambda^T_2 S = 0, \forall i,$ $\lambda_1 \geq 0,$ $\lambda^T_1 \begin{bmatrix} A\nu - b \\ -u + \sum_i T_i \nu_i x_i \end{bmatrix} = 0$	$\begin{bmatrix} 0 \\ \vdots \\ -X_1 \\ 0 \\ \vdots \\ -X_{n_X} \\ 0 \\ \vdots \end{bmatrix}^T$ $+\lambda^T_1 \begin{bmatrix} A & -\frac{db(C)}{dC} \\ 0 & -I \end{bmatrix}$ $+\lambda^T_2 \begin{bmatrix} S & 0 \\ -\sum_i T_i X_i & -ID \end{bmatrix}$ $= 0, \forall i,$ $\lambda_1 \geq 0,$ $\lambda^T_1 \begin{bmatrix} (A\nu - b(C)) \\ -C \end{bmatrix} = 0$
#var	$1 + n_x + n_\nu + n_S + n_A$	$n_C + n_X + n_\nu + n_S + n_A$	$1 + n_x + n_C + n_\nu + n_S + n_A$	$3n_C + n_X + n_\nu + n_S + n_A$
#EQ	$n_x + 1 + n_S + n_\nu + n_A$	$n_C + n_X + n_S + n_\nu + n_A$	$n_x + 1 + n_S + n_\nu + n_C + n_A$	$n_X + n_C + n_S + n_\nu + n_C + n_A + n_C$

References

1. Altamirano, Á., Saa, P.A., Garrido, D.: Inferring composition and function of the human gut microbiome in time and space: a review of genome-scale metabolic modelling tools. Comput. Struct. Biotechnol. J. **18**, 3897–3904 (2020)
2. Armstrong, R.A., McGehee, R.: Competitive exclusion. Am. Nat. **115**(2), 151–170 (1980)
3. Biggs, M.B., Medlock, G.L., Kolling, G.L., Papin, J.A.: Metabolic network modeling of microbial communities. Wiley Interdiscip. Rev. Syst. Biol. Med. **7**(5), 317–334 (2015)
4. Bordbar, A., Monk, J.M., King, Z.A., Palsson, B.O.: Constraint-based models predict metabolic and associated cellular functions. Nat. Rev. Genet. **15**(2), 107–120 (2014)
5. Brugiroux, S., et al.: Genome-guided design of a defined mouse microbiota that confers colonization resistance against salmonella enterica serovar typhimurium. Nat. Microbiol. **2**(2), 1–12 (2016)
6. Budinich, M., Bourdon, J., Larhlimi, A., Eveillard, D.: A multi-objective constraint-based approach for modeling genome-scale microbial ecosystems. PloS One **12**(2), e0171744 (2017)
7. Cai, J., Tan, T., Joshua Chan, S.: Predicting Nash equilibria for microbial metabolic interactions. Bioinformatics **36**, 5649–5655 (2020)
8. Chan, S.H.J., Simons, M.N., Maranas, C.D.: SteadyCom: predicting microbial abundances while ensuring community stability. PLoS Comput. Biol. **13**(5), e1005539 (2017)
9. Cremer, J., Arnoldini, M., Hwa, T.: Effect of water flow and chemical environment on microbiota growth and composition in the human colon. Proc. Natl. Acad. Sci. **114**(25), 6438–6443 (2017)
10. Aguirre de Cárcer, D.: Experimental and computational approaches to unravel microbial community assembly. Comput. Struct. Biotechnol. J. **18**, 4071–4081 (2020). https://doi.org/10.1016/j.csbj.2020.11.031
11. Fierer, N.: Embracing the unknown: disentangling the complexities of the soil microbiome. Nat. Rev. Microbiol. **15**, 579–590 (2017). https://doi.org/10.1038/nrmicro.2017.87
12. Frey, E.: Evolutionary game theory: theoretical concepts and applications to microbial communities. Physica A **389**(20), 4265–4298 (2010)
13. Gilbert, J.A., Blaser, M.J., Caporaso, J.G., Jansson, J.K., Lynch, S.V., Knight, R.: Current understanding of the human microbiome. Nat. Med. **24**, 392–400 (2018). https://doi.org/10.1038/nm.4517
14. Gollub, M.G., Kaltenbach, H.M., Stelling, J.: Probabilistic thermodynamic analysis of metabolic networks. Bioinformatics btab194 (2021). https://doi.org/10.1093/bioinformatics/btab194
15. Gu, C., Kim, G.B., Kim, W.J., Kim, H.U., Lee, S.Y.: Current status and applications of genome-scale metabolic models. Genome Biol. **20**(1), 1–18 (2019)
16. Kauffman, K.J., Prakash, P., Edwards, J.S.: Advances in flux balance analysis. Curr. Opin. Biotechnol. **14**(5), 491–496 (2003)
17. Khandelwal, R.A., Olivier, B.G., Röling, W.F., Teusink, B., Bruggeman, F.J.: Community flux balance analysis for microbial consortia at balanced growth. PloS One **8**(5), e64567 (2013)
18. Li, Z., Liu, B., Li, S.H.J., King, C.G., Gitai, Z., Wingreen, N.S.: Modeling microbial metabolic trade-offs in a chemostat. PLoS Comput. Biol. **16**(8), e1008156 (2020)

19. Lobry, C.: The Chemostat. Wiley Online Library (2017)
20. Machado, D., et al.: Polarization of microbial communities between competitive and cooperative metabolism. Nat. Ecol. Evol. **5**, 195–203 (2021). https://doi.org/10.1038/s41559-020-01353-4
21. Mahadevan, R., Edwards, J.S., Doyle III, F.J.: Dynamic flux balance analysis of diauxic growth in Escherichia coli. Biophys. J. **83**(3), 1331–1340 (2002)
22. Nakaoka, S., Takeuchi, Y.: Two types of coexistence in cross-feeding microbial consortia. In: AIP Conference Proceedings, vol. 1028, pp. 233–260. American Institute of Physics (2008)
23. Popp, D., Centler, F.: μBialSim: constraint-based dynamic simulation of complex microbiomes. Front. Bioeng. Biotechnol. **8**, 574 (2020)
24. Pusa, T., Wannagat, M., Sagot, M.F.: Metabolic games. Front. Appl. Math. Stat. **5**, 18 (2019)
25. Stewart, A.J., Plotkin, J.B.: From extortion to generosity, evolution in the iterated prisoner's dilemma. Proc. Natl. Acad. Sci. **110**(38), 15348–15353 (2013)
26. Sun, W., Yuan, Y.X.: Optimization Theory and Methods: Nonlinear Programming. Springer Optimization and Its Applications, vol. 1. Springer, Heidelberg (2006). https://doi.org/10.1007/b106451
27. Van Hoek, M.J., Merks, R.M.: Emergence of microbial diversity due to cross-feeding interactions in a spatial model of gut microbial metabolism. BMC Syst. Biol. **11**(1), 1–18 (2017)
28. Zhuang, K., et al.: Genome-scale dynamic modeling of the competition between Rhodoferax and Geobacter in anoxic subsurface environments. ISME J. **5**(2), 305–316 (2011)
29. Zomorrodi, A.R., Islam, M.M., Maranas, C.D.: d-OptCom: dynamic multi-level and multi-objective metabolic modeling of microbial communities. ACS Synth. Biol. **3**(4), 247–257 (2014)
30. Zomorrodi, A.R., Maranas, C.D.: OptCom: a multi-level optimization framework for the metabolic modeling and analysis of microbial communities. PLoS Comput. Biol. **8**(2), e1002363 (2012)
31. Zomorrodi, A.R., Segrè, D.: Genome-driven evolutionary game theory helps understand the rise of metabolic interdependencies in microbial communities. Nat. Commun. **8**(1), 1–12 (2017)

Learning Boolean Controls in Regulated Metabolic Networks: A Case-Study

Kerian Thuillier[1], Caroline Baroukh[2], Alexander Bockmayr[3], Ludovic Cottret[2], Loïc Paulevé[4(✉)], and Anne Siegel[1]

[1] University of Rennes, Inria, CNRS, IRISA, 35000 Rennes, France
[2] LIPME, INRAE, CNRS, Université de Toulouse, Castanet–Tolosan, France
[3] Freie Universität Berlin, Institute of Mathematics, 14195 Berlin, Germany
[4] University of Bordeaux, Bordeaux INP, CNRS, LaBRI, UMR5800, 33400 Talence, France
loic.pauleve@labri.fr

Abstract. Many techniques have been developed to infer Boolean regulations from a prior knowledge network and experimental data. Existing methods are able to reverse-engineer Boolean regulations for transcriptional and signaling networks, but they fail to infer regulations that control metabolic networks. This paper provides a formalisation of the inference of regulations for metabolic networks as a satisfiability problem with two levels of quantifiers, and introduces a method based on Answer Set Programming to solve this problem on a small-scale example.

Keywords: Inference · Regulated metabolism · Satisfiability problem

1 Introduction

During the last twenty years, both the amount and the type of available data have allowed scientists to consider intracellular processes as a whole. Boolean networks have been refined to include non-deterministic dynamics in order to model the response of regulatory interactions [2,5,16]. Similarly, the study of metabolism at steady state has led to various constraint-based approaches [17,19], which usually assume that internal metabolites are in a quasi-steady-state (QSS). The classical approach to analyze metabolic networks at steady state is flux balance analysis (FBA) [19]. In this approach, a linear function, e.g. biomass production, is optimized with respect to stoichiometric and thermodynamic constraints, resulting in a linear programming problem (LP).

However, both the Boolean approach for regulation and the QSS approximation for metabolism are often developed *"in solo"*, without considering that cellular biology is multi-layered in the sense that the metabolic layer interacts through feed-forward and feedback loops with the regulatory layer [4,9,21,27]. Indeed, cellular metabolism transforms nutrients into biomass constituents. Metabolic reactions are catalysed by enzymes, which themselves are controlled by a cascade of regulations involving other proteins, metabolites and abiotic factors,

E. Cinquemani and L. Paulevé (Eds.): CMSB 2021, LNBI 12881, pp. 159–180, 2021.
https://doi.org/10.1007/978-3-030-85633-5_10

such as temperature and pH. A biological system thus has several layers of control, which mutually depend on each other. It cannot be simply viewed as a purely hierarchical system because there are regulatory feed-forward and feedback mechanisms to inform each layer on the state of the other ones. In concrete terms, some compounds produced by the metabolic layer have the capability to block or induce signaling regulation cascades, which themselves can block or induce transcription of genes leading to changes in the control of the initial metabolic process.

To figure out how gene expression triggers specific phenotypes depending on the environmental constraints [3], several constraint-based approaches for integrating metabolic and regulatory networks have been developed that combine Boolean dynamics for the regulatory layer with quasi-steady-state approximations of the metabolic layer (see [17] for an overview), one of them being FlexFlux [18], which implements the rFBA framework [9]. A major limitation when using such frameworks to analyse regulated metabolic models is that they require a precise description of the regulatory and signaling layers in the form of Boolean rules. A noticeable exception is [24], where RBA is used to deduce regulations according to perturbations of the environment. However, to induce regulations, the authors assume that no feedback from metabolism to regulation occurs, which does not correspond to the functioning of most systems. In practice, these rules are manually curated from the literature or experimental data. This has been done for example in the case of *E. coli* [7,8] and a few other organisms. But, the need for a manual curation of Boolean rules of regulated metabolism is a strong limitation to the use of these frameworks.

Signaling and regulatory rules can be identified from transcriptomic or phosphoproteomics data by solving combinatorial or MILP problems in order to optimize data-fitting and parsimony hypotheses [20,22,23,25,26]. In this direction, the caspoTS and the BoNesis approaches [6,20,22,26] were developed for inferring Boolean rules to model the response of regulatory and signaling networks from multiple time-series data. The goal of this paper is to lay foundation for the extension of these approaches to the inference of regulatory rules driving metabolism. This is done by discretizing both the rFBA framework (especially the QSS approximation) and the metabolic data used as input of the inference procedure.

This paper is structured as follows. Section 2 gives the background on the dynamic rFBA framework for the simulation of coupled metabolic and regulatory networks. In Sect. 3, we define a formal Boolean abstraction of dynamic rFBA simulations. Then, in Sect. 4, we build on this Boolean abstraction to express the inference of the logic of metabolic regulations as a satisfiability problem. Finally, in Sect. 5, we apply the obtained inference framework on a case study of simplified core carbon metabolism.

Notations. The cardinality of a finite set X is denoted by $|X|$. Given a vector $x \in D^n$ and a set of indices $I \subseteq \{1, \cdots, n\}$, x_I denotes the vector of dimension $|I|$ equal to $(x_i)_{i \in I}$. The Boolean domain is denoted by $\mathbb{B} = \{0, 1\}$. Given two Boolean vectors $x, y \in \mathbb{B}^n$, we write $x \preceq y$ iff $\forall i \in \{1, \cdots, n\}, x_i \leq y_i$.

Finally, given a non-negative real vector $s \in \mathbb{R}^n_{\geq 0}$, we denote by $\beta(s) \in \mathbb{B}^n$ its *binarization, i.e.* $\forall i \in \{1, \ldots, n\}, \beta(s)_i = 1$, if $s_i > 0$, and $\beta(s)_i = 0$, if $s_i = 0$.

2 Background: Regulated Metabolic Networks

2.1 Coupling Metabolic and Regulatory Networks

A *regulated metabolic network* consists of two layers. The regulatory layer is modelled by a Boolean network, which controls the metabolites and fluxes of the metabolic layer, which is characterized by linear equations. Feedbacks are provided by the components of the metabolic network, which are involved in the Boolean functions associated with the regulatory layer.

Formally, a metabolic network is given by a set of biochemical reactions linked together by the metabolites that they consume and produce.

Definition 1. *A* metabolic network *is a tuple* $\mathcal{N} = (\mathrm{Int}, \mathrm{Ext}, \mathcal{R}, S)$ *with a set of internal metabolites* Int, *a set of external metabolites* Ext, *a set* $\mathcal{R}$ *of irreversible reactions, and a stoichiometric matrix* $S \in \mathbb{R}^{(|\mathrm{Int}|+|\mathrm{Ext}|)\times|\mathcal{R}|}$.

Given flux bounds $l_r, u_r \in \mathbb{R}, 0 \leq l_r \leq u_r$, *for each* $r \in \mathcal{R}$, *a* metabolic steady state *is a flux vector* $v \in \mathbb{R}^{|\mathcal{R}|}$ *with* $S_{\mathrm{Int},\mathcal{R}} \cdot v = 0$ *and* $l_r \leq v_r \leq u_r$, *for all* $r \in \mathcal{R}$. *Here* $S_{\mathrm{Int},\mathcal{R}}$ *denotes the submatrix of* S *whose rows correspond to the internal metabolites.*

For the sake of simplicity, we assume that all reactions are irreversible. Reversible reactions may be split into a forward and backward reaction if necessary.

Definition 2 (Input and output metabolites). *For an external metabolite* $m \in \mathrm{Ext}$, *we denote by* $w_m = w_m(t) \in \mathbb{R}_{\geq 0}$ *the concentration of* m *at time* $t \geq 0$.

An external metabolite $m \in \mathrm{Ext}$ *is called an* input (resp. output) metabolite *if there exists a reaction* $r \in \mathcal{R}$ *with* $S_{mr} < 0$ *(resp.* $S_{mr} > 0$*). Here* S_{mr} *denotes the stoichiometric coefficient of metabolite* m *in reaction* r. *The set of all input metabolites is denoted by* $\mathrm{Inp} \subseteq \mathrm{Ext}$.

A regulatory network is a set of biological entities (e.g. genes, reactions, metabolites) or even abiotic entities (e.g. temperature, pH) that are linked by causal effects: the activity of some nodes can affect positively or negatively the activity of other nodes. This activity can be represented by a Boolean network.

Definition 3. *A* Boolean network *(BN) of dimension* n *is a function* $f : \mathbb{B}^n \to \mathbb{B}^n$. *For each* $i \in \{1, \ldots, n\}$, *the* i*-th component* $f_i : \mathbb{B}^n \to \mathbb{B}$ *is called the* local function *of* i.

The influence graph $G(f)$ *of* f *is a signed digraph* (V, E) *with* $V = \{1, \ldots, n\}$ *and* $E \subseteq V \times \{-, +\} \times V$ *such that* $(i, s, j) \in E$ *if and only if there exists* $x \in \mathbb{B}^n$ *with* $x_i = 0$ *such that* $s \cdot f_j(x) < s \cdot f_j(x_1, \cdots, x_{i-1}, 1, x_{i+1}, \cdots, x_n)$. *In the following we will slightly abuse notation by identifying* $G(f)$ *with its edge set,* i.e. $G(f) = E$.

A BN f is locally monotone *whenever for each influence $(i, s, j) \in G(f)$, there is no influence with opposite sign,* i.e. $(i, -s, j) \notin G(f)$.

We assume here that the fluxes of a metabolic network can be controlled by the activity of the input metabolites and additional regulatory proteins. More precisely, the activity of some reactions can be blocked (forced to have a zero flux) whenever certain conditions on the activity of input metabolites and regulatory proteins are met. Moreover, we assume that the activity of regulatory proteins is mediated by the metabolic network only. The resulting model is then supposed to run on two time scales: the metabolic network is a fast system, which, depending on the activity of input metabolites and regulatory proteins will converge to a steady state of the reactions fluxes; the regulatory network is a slow system, which gets updated once the metabolic network is in steady state.

Definition 4 (Regulated metabolic network). *A* regulated metabolic network *is a triplet $(\mathcal{N}, \mathcal{P}, f)$ composed of:*

- *a metabolic network $\mathcal{N} = (\text{Int}, \text{Ext}, \mathcal{R}, S)$ with k input metabolites* $\text{Inp} = \{e_1, \cdots, e_k\} \subseteq \text{Ext}$ *and m reactions $\mathcal{R} = \{r_1, \cdots, r_m\}$;*
- *a set of d regulatory proteins $\mathcal{P} = \{p_1, \dots, p_d\}$*
- *a BN f of dimension $n = |\text{Inp}| + |\mathcal{R}| + |\mathcal{P}|$ where $\{1, \dots, n\} = \text{Inp} \cup \mathcal{R} \cup \mathcal{P}$ such that $G(f)$ is a bipartite graph between $\mathcal{P}$ and $\text{Inp} \cup \mathcal{R}$.*

In this work, local functions for input metabolites in the BN f are never used (although the local functions of reactions may depend on them). Therefore we set arbitrarily $f_e = 0, \forall e \in \text{Inp}$.

The BN f models the regulation of the fluxes in the metabolic network $\mathcal{N}$. This regulation is always in one direction: either a flux v_r is only restricted by the flux bounds $l_r \leq v_r \leq u_r$, whenever $f_r(x) = 1$, or it is blocked, $v_r = 0$, whenever $f_r(x) = 0$. Following this convention, a reaction $r \in \mathcal{R}$ is never regulated whenever $f_r(x) = 1$. As we will define formally in the next section, the regulations impact the steady states of the metabolic network.

An example of regulated metabolic network is shown in Fig. 1. This example is based on a highly simplified model of core carbon metabolism, originally proposed in [9]. At the metabolic level (Fig. 1a), there are 9 metabolites and $m = 9$ reactions. The internal metabolites are Int = {A, D, E, O2, ATP, NADH}, the external metabolites are Ext = {Carbon1, Carbon2, Oxygen}. All the $k = 3$ external metabolites are input metabolites, Ext = Inp. The set of irreversible reactions is $\mathcal{R}$ = {Tc1, Tc2, To2, Td, Te, Growth, Rres, R6, R7}. The stoichiometric coefficients are also given in Fig. 1a. By default, they are set to 1, except for the reactions $R6$ and $R7$.

The regulatory level (Fig. 1b) of the regulated metabolism introduces $d = 2$ regulatory proteins: $\mathcal{P}$ = {RPcl, RPO2}. Thus, the Boolean network f is of dimension $n = k+m+d = 14$. It consists of 14 functions (see Fig. 1b) which map a Boolean vector $x = (x_{\text{Carbon1}}, x_{\text{Carbon2}}, x_{\text{Oxygen}}, x_{\text{RPcl}}, x_{\text{RPO2}}, x_{\text{Tc1}}, x_{\text{Tc2}}, x_{\text{To2}}, x_{\text{Td}}, x_{\text{Te}}, x_{\text{Growth}}, x_{\text{Rres}}, x_{\text{R6}}, x_{\text{R7}}) \in \mathbb{B}^n$ to a Boolean value in $\mathbb{B}$. The local functions associated with regulatory proteins in $\mathcal{P}$ involve only external

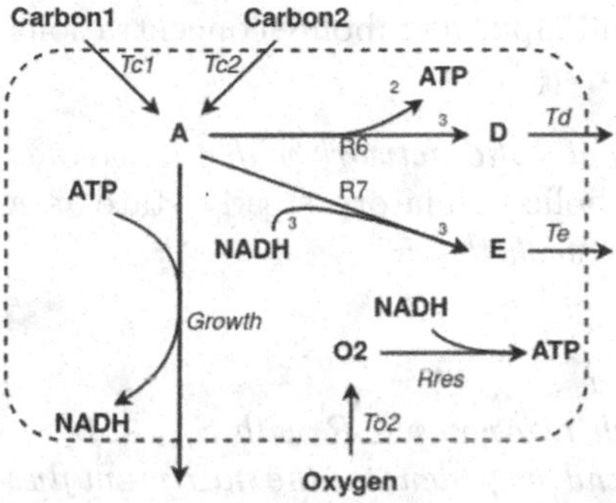

(a) *Metabolic Network*

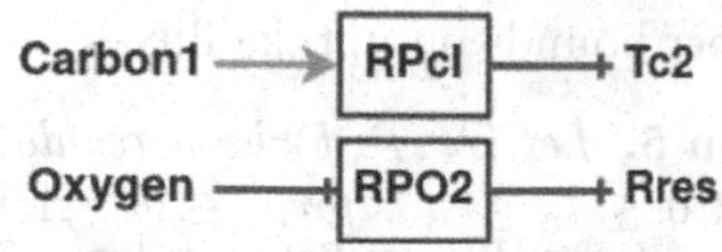

(c) Influence graph $G(f)$ of the regulatory Boolean network f. Nodes without in-going or out-going edges are not represented. Positive edges are drawn in green with a regular tipping arrow, negative edges are drawn in red with a bar arrow.

	Regulatory proteins		Input metabolites		
Local function	$f_{RPO2}(x)$	$f_{RPcl}(x)$	$f_{Carbon1}(x)$	$f_{Carbon2}(x)$	$f_{Oxygen}(x)$
Boolean rule	$\neg x_{Oxygen}$	$x_{Carbon1}$	0	0	0

	Reactions								
Local function	$f_{Tc1}(x)$	$f_{Tc2}(x)$	$f_{To2}(x)$	$f_{Td}(x)$	$f_{Te}(x)$	$f_{Growth}(x)$	$f_{Rres}(x)$	$f_{R6}(x)$	$f_{R7}(x)$
Boolean rule	1	$\neg x_{RPcl}$	1	1	1	1	$\neg x_{RPO2}$	1	1

(b) *Boolean Network.* All Boolean functions equal to 1 correspond to reactions which are not regulated by the Boolean network.

Fig. 1. Example of regulated metabolic network. In the metabolic network **(a)**, each node represents a metabolite and each hyperedge a reaction. For instance, the hyperedge R7 linking {A; NADH} to {E} models the reaction A + 3 NADH → 3 E. Integer values over hyperedges are stoichiometric coefficients, the default value is 1. **(b)** defines the Boolean network regulating the metabolic network in (a), with $x \in \mathbb{B}^n$ and $n = 14$. **(c)** shows the influence (or regulatory) graph of the Boolean network in (b), with square nodes denoting the regulatory proteins.

metabolite variables. Among the 9 functions associated with reactions, only two (Tc2, Rres) are non-constant: they involve the two regulatory proteins.

The influence graph of the network is shown in Fig. 1c. Only the shown nodes (RPcl, RPO2, Tc2, Rres) have a non-constant local function or are used in the local function of another node (Carbon1, Oxygen). The influence graph shows the multi-layered regulations of the network: external input metabolites (Carbon1, Oxygen) regulate regulatory proteins (RPcl, RPO2), which regulate reactions (Tc2, Rres).

2.2 Dynamic rFBA

Flux Balance Analysis (FBA) [19] returns an *optimal* metabolic steady state, according to a given linear objective function in the reaction fluxes. In the following, we assume that the objective function is to maximize the flux through a reaction *Growth*. For regulated metabolic networks, the rFBA framework [9] allows defining a discrete time series of optimal steady states, where regulatory variables can force reaction fluxes to be zero and input metabolite concentrations define upper bounds on uptake fluxes.

Definition 5. *Let $(\mathcal{N}, \mathcal{P}, f)$ be a regulated metabolic network with flux bounds $l_r, u_r \in \mathbb{R}, 0 \leq l_r \leq u_r$, for $r \in \mathcal{R}$. A* metabolic-regulatory steady state *is a triple $(v, w, x) \in \mathbb{R}^{|\mathcal{R}|} \times \mathbb{R}^{|\text{Ext}|} \times \mathbb{B}^{|\text{Inp}|+|\mathcal{R}|+|\mathcal{P}|}$ such that*

- $S_{\text{Int},\mathcal{R}} \cdot v = 0$,
- *for each reaction $r \in \mathcal{R}$, $l_r \cdot x_r \leq v_r \leq u_r \cdot x_r$,*
- *for each input metabolite $m \in \text{Inp}$ and each reaction $r \in \mathcal{R}$ with $S_{mr} < 0$, $v_r \leq uptake\_bound(w_m)$, where $uptake\_bound(w_m)$ denotes the maximum flux through uptake reaction r, given the input metabolite concentration w_m.*

Two successive metabolic-regulatory steady states (v^k, w^k, x^k) at time t^k, and $(v^{k+1}, w^{k+1}, x^{k+1})$ at time t^{k+1}, are linked by the following relations:

1. The external metabolite concentrations w^{k+1} are obtained from the previous concentrations w^k by assuming the constant uptake/secretion fluxes v^k for the whole time period $[t^k, t^{k+1}]$.
2. The Boolean state x^{k+1} is obtained by applying the regulatory function f to the binarized input metabolites concentrations $x'_{\text{Inp}} = \beta(w^{k+1}_{\text{Inp}})$ at time t^{k+1}, together with the binarized reaction fluxes $x'_{\mathcal{R}} = \beta(v^k)$ and the Boolean values $x'_{\mathcal{P}} = x^k_{\mathcal{P}}$ of the regulatory proteins at time t^k, *i.e.*,
$$x^{k+1} = f(x')$$
3. $(v^{k+1}, w^{k+1}, x^{k+1})$ is a metabolic-regulatory steady state maximizing the flux through the *Growth* reaction, i.e., there is no metabolic-regulatory steady state (v', w^{k+1}, x^{k+1}) such that $v'_{Growth} > v^{k+1}_{Growth}$.

In this paper, we rely on the FlexFlux implementation of rFBA [18], which assumes a fixed time step τ between successive metabolic-regulatory steady states ($t^{k+1} - t^k = \tau$ for any k). The *Growth* reaction is assumed to reflect the growth of the cell. FlexFlux computes the evolution of the total biomass of the cell as $\text{biomass}^{k+1} = \text{biomass}^k \cdot e^{v^k_{Growth} \cdot \tau}$ (from a given initial biomass^0). The maximum uptake fluxes of input metabolites $m \in \text{Inp}$ at step k are defined as
$$uptake\_bound(w_m) = w_m / (\text{biomass}^k \cdot \tau).$$
Finally, the update of the external metabolite concentrations is computed as
$$w^{k+1}_m = w^k_m - (S_{mr} v^k_r / v^k_{Growth}) \cdot (\text{biomass}^k - \text{biomass}^{k+1}),$$
where $r \in \mathcal{R}$ is the uptake/secretion reaction for the external metabolite m ($S_{mr} < 0$ or $S_{mr} > 0$), which is assumed to be unique.

An example of a dynamic rFBA simulation using FlexFlux of the regulated metabolic network of Fig. 1 is shown in Fig. 2. It uses a time step of 0.01 h and is initialized with 100 mM of Oxygen, 20 mM of Carbon1 and 20 mM of Carbon2.

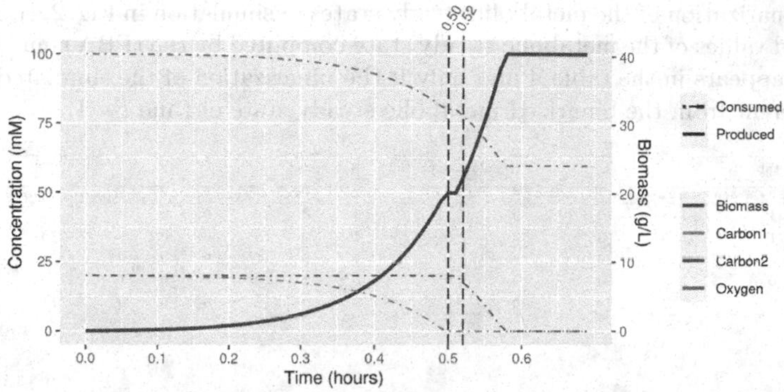

(a) Simulation showing the evolution of the concentrations of the external metabolites (Oxygen, Carbon1, Carbon2) and the production of biomass by the Growth reaction.

		External metabolites			Regulatory proteins		Reaction flows								
Time	w_{biomass}	w_{Carbon1}	w_{Carbon2}	w_{Oxygen}	x_{RPO2}	x_{RPcl}	v_{Tc1}	v_{Tc2}	v_{To2}	v_{Td}	v_{Te}	v_{Growth}	v_{Rres}	v_{R6}	v_{R7}
0.49	17.05	2.95	20.0	82.95	0	1	10.5	0.0	10.5	0.0	0.0	10.5	10.5	0.0	0.0
0.50	18.05	1.05	20.0	81.05	0	1	6.15	0.0	6.15	0.0	0.0	6.15	6.15	0.0	0.0
0.51	20.10	0.0	20.0	79.90	0	0	0.0	0.0	0.0	0.0	0.0	0.0	0.0	0.0	0.0
0.52	20.10	0.0	20.0	79.90	0	0	0.0	10.5	10.5	0.0	0.0	10.5	10.5	0.0	0.0
0.53	22.35	0.0	17.76	77.65	0	0	0.0	10.5	10.5	0.0	0.0	10.5	10.5	0.0	0.0

(b) Focus on the times from $0.49h$ to $0.53h$ in the simulation, showing the switch from Carbon1 to Carbon2 for biomass production.

Fig. 2. Dynamic rFBA simulation of the regulated metabolic network in Fig. 1. The simulation is done with FLEXFLUX and is initialized with 100 mM of Oxygen, 20 mM of Carbon1, and 20 mM Carbon2. Tke time step is set to $0.01\,h$. The flux bounds are $\forall r \in \{\text{Tc1, Tc2}\}$, $(l_r, u_r) = (0, 10.5)$, $\forall r \in \{\text{Td, Te}\}$, $(l_r, u_r) = (0, 12.0)$, $\forall r \in \{\text{R6, R7, Rres, Growth}\}$, $(l_r, u_r) = (0, 9999)$ and for Oxygen, $(l_r, u_r) = (0, 15.0)$.

The simulation shown in Fig. 2a is composed of 70 metabolic steady states. By applying the binarization β, these 70 metabolic steady states correspond to 5 different binarized metabolic steady states, which are shown in Table 1. These binarized metabolic steady states capture the main features of the simulation.

More precisely, the simulation shows that until $0.5\,h$ only Carbon1 and Oxygen are consumed to produce biomass. This corresponds to a first time period where the behavior of the system is monotone: the binarized metabolic steady states are equal on this time range. The presence of Carbon1 activates the regulatory protein RPcl inhibiting the reaction Tc2 according to the regulatory rules. At $0.5\,h$, Carbon1 is depleted and the current Boolean state $x \in \mathbb{B}^{15}$ is such that $x_{\text{Carbon1}} = 0$, $x_{\text{RPcl}} = 1$, $x_{\text{Tc2}} = 0$ (second qualitative behavior with equal binarization of the metabolic steady states). At $0.51\,h$, as shown in Fig. 2b, the Boolean state x is updated to x' so that the Boolean state of RPcl becomes $x'_{\text{RPcl}} = f_{\text{RPcl}}(x) = x_{\text{Carbon1}} = 0$. The Boolean state of *Tc2* remains unchanged because $x_{\text{RPcl}} = 1$. No biomass is produced at $0.51\,h$. This corresponds to a third qualitative behavior. At $0.52\,h$, the Boolean state x' is updated to x'': all the node states remain unchanged except for $x''_{\text{Tc2}} = f_{\text{Tc2}}(x') = \neg x'_{\text{RPcl}} = 1$. This

Table 1. Binarization of the metabolic steady states of simulation in Fig. 2. It contains the binarized values of the metabolic steady state computed by the rFBA simulation. A timepoint t appears in the table if and only if the binarization of the simulated steady state is different from the binarized metabolic steady state of time $t-1$.

	External metabolites				Regulatory proteins		Reactions								
Time	$\overline{w_{Biomass}}$	$\overline{w_{Carbon1}}$	$\overline{w_{Carbon2}}$	$\overline{w_{Oxygen}}$	$\overline{x_{RPO2}}$	$\overline{x_{RPcl}}$	$\overline{v_{Tc1}}$	$\overline{v_{Tc2}}$	$\overline{v_{To2}}$	$\overline{v_{Td}}$	$\overline{v_{Te}}$	$\overline{v_{Growth}}$	$\overline{v_{Rres}}$	$\overline{v_{R6}}$	$\overline{v_{R7}}$
0	0	1	1	1	0	1	0	0	0	0	0	0	0	0	0
0.1	1	1	1	1	0	1	1	0	1	0	0	1	1	0	0
0.51	1	0	1	1	0	0	0	0	0	0	0	0	0	0	0
0.52	1	0	1	1	0	0	0	1	1	0	0	1	1	0	0
0.59	1	0	0	1	0	0	0	0	0	0	0	0	0	0	0

corresponds to a fourth qualitative behavior. The reaction *Tc2* is not inhibited anymore, and biomass is produced due to the uptake of Carbon2 and Oxygen (through *Tc2*, *Growth* and *Rres*) until Carbon2 depletion at $t = 0.59\,h$ (fifth qualitative behavior).

3 Boolean Abstraction of Dynamic rFBA

In the previous example, we illustrated how the simulation of a regulated metabolic network may generate time-periods for which the qualitative behavior is similar, meaning that the variation of all the metabolic variables is monotone and the Boolean values of the regulatory proteins are constant. In this section, we introduce a discrete definition of steady states to capture the monotone behaviors observed in rFBA simulations. This allows introducing a discretized form of rFBA, which will be used in the next section for the reverse-engineering framework.

3.1 Boolean Metabolic Steady States

Given a metabolic network $\mathcal{N} = (\mathrm{Int}, \mathrm{Ext}, \mathcal{R}, S)$, we derive a logical characterization of the notion of steady state, considering that reactions are either inactive or active, and metabolites either absent or present. This will result in a set of *Boolean* metabolic steady states that form an over-approximation of the continuous steady states.

We associate all reactions with propositional variables $\mathcal{V} = \{\overline{v_r}\}_{r \in \mathcal{R}}$. For each metabolite $m \in \mathrm{Int} \uplus \mathrm{Ext}$, we introduce a variable $\overline{z_m}^+$ as a Boolean abstraction of the production of m and a variable $\overline{z_m}^-$ as a Boolean abstraction of the consumption of m:

$$\forall m \in \mathrm{Int} \uplus \mathrm{Ext}, \quad \overline{z_m}^+ \stackrel{\mathrm{def}}{=} \bigvee_{r \in \mathcal{R}, S_{mr} > 0} \overline{v_r}, \qquad \overline{z_m}^- \stackrel{\mathrm{def}}{=} \bigvee_{r \in \mathcal{R}, S_{mr} < 0} \overline{v_r},$$

(where an empty disjunction is considered to be false).

For each internal metabolite m, we introduce a variable $\widehat{z_m}$ which is equal to 1 iff m is in a logical steady state:

$$\forall m \in \text{Int}, \quad \widehat{z_m} \stackrel{\text{def}}{=} (\overline{z_m}^{+} \Leftrightarrow \overline{z_m}^{-}).$$

For the external metabolites, we introduce propositional variables $\mathcal{V}_{ext} = \{\overline{z_m}\}_{m \in \text{Ext}}$ indicating whether or not m is present in the environment. The formula

$$\widehat{\mathcal{N}}_{\text{Ext}} \stackrel{\text{def}}{=} \bigwedge_{m \in \text{Ext}} (\overline{z_m}^{-} \Rightarrow \overline{z_m})$$

then states that an external metabolite can only be consumed if it is present in the environment.

Definition 6 (Boolean metabolic steady state). *A* Boolean metabolic steady state *of a metabolic network* $\mathcal{N} = (\text{Int}, \text{Ext}, \mathcal{R}, S)$ *is a Boolean vector* $\hat{\nu} \in \mathbb{B}^{|\text{Ext}|+|\mathcal{R}|}$ *which is a satisfying assignment of the following logical steady state formula:*

$$\widehat{\mathcal{N}} \stackrel{def}{=} \widehat{\mathcal{N}}_{\text{Ext}} \wedge \bigwedge_{m \in \text{Int}} \widehat{z_m}$$

We denote by $\text{MSS}^{\mathbb{B}}(\mathcal{N}) \subseteq \mathbb{B}^{|\text{Ext}|+|\mathcal{R}|}$ *the set of all the Boolean metabolic steady states of the metabolic network* $\mathcal{N}$.

As an immediate consequence of this definition, we get the following property:

Property 1. For each metabolic-regulatory steady state (v, w, x) of the regulated metabolic network $(\mathcal{N}, \mathcal{P}, f)$, the binarized value $\beta(w, v)$ of the external metabolite concentrations w and the reaction fluxes v is a Boolean metabolic steady state, i.e., $\beta(w, v) \in \text{MSS}^{\mathbb{B}}(\mathcal{N})$.

Note that the converse is not true: since the logical characterization neglects the stoichiometry, Boolean metabolic steady states may have no real-valued counterpart.

Applied to the example, the internal metabolic constraints are the following:

$$\overline{z_A}^{+} = \overline{v_{Tc1}} \vee \overline{v_{Tc2}}, \quad \overline{z_A}^{-} = \overline{v_{R6}} \vee \overline{v_{R7}} \vee \overline{v_{Growth}}$$
$$\overline{z_D}^{+} = \overline{v_{R6}}, \quad \overline{z_D}^{-} = \overline{v_{Td}}, \quad \overline{z_E}^{+} = \overline{v_{R7}}, \quad \overline{z_E}^{-} = \overline{v_{Te}}$$
$$\overline{z_{O2}}^{+} = \overline{v_{To2}}, \quad \overline{z_{O2}}^{-} = \overline{v_{Rres}}$$
$$\overline{z_{ATP}}^{+} = \overline{v_{R6}} \vee \overline{v_{Rres}}, \quad \overline{z_{ATP}}^{-} = \overline{v_{Growth}}$$
$$\overline{z_{NADH}}^{+} = \overline{v_{Growth}}, \quad \overline{z_{NADH}}^{-} = \overline{v_{R7}} \vee \overline{v_{Rres}}$$

The logical steady state constraints equivalent to $\widehat{\mathcal{N}} = 1$ are obtained by gathering contraints on internal and external metabolites:

$$\overline{v_{Tc1}} \vee \overline{v_{Tc2}} = \overline{v_{R6}} \vee \overline{v_{R7}} \vee \overline{v_{Growth}}$$
$$\overline{v_{R6}} = \overline{v_{Td}} \qquad \overline{v_{R7}} = \overline{v_{Te}} \qquad \overline{v_{To2}} = \overline{v_{Rres}}$$
$$\overline{v_{R6}} \vee \overline{v_{Rres}} = \overline{v_{Growth}} \qquad \overline{v_{R7}} \vee \overline{v_{Rres}} = \overline{v_{Growth}}$$
$$\overline{v_{Tc1}} \Rightarrow \overline{z_{Carbon1}} \qquad \overline{v_{Tc2}} \Rightarrow \overline{z_{Carbon2}} \qquad \overline{v_{To2}} \Rightarrow \overline{z_{Oxygen}}$$

From these equations, we deduce that there are 38 Boolean metabolic steady states for the example shown in Fig. 1. These Boolean metabolic steady states are detailed in Appendix A. Among them, we recover the five binarized metabolic-regulatory steady states (Table 1) appearing in the rFBA simulations of Fig. 2.

3.2 Boolean Dynamics

Using the logical characterization of metabolic steady states, we define a Boolean counterpart of dynamic rFBA (Sect. 2.2). A Boolean state of the regulated metabolic network $(\mathcal{N}, \mathcal{P}, f)$ assigns a Boolean value to external metabolites, reactions, and regulatory proteins, which gives a Boolean vector of dimension $n = k + m + d$. Such a Boolean state $x \in \mathbb{B}^n$ should match with a Boolean metabolic steady state. Denoting by $\mathcal{M} = \text{Ext} \cup \mathcal{R}$ the external metabolites and reactions, $x_{\mathcal{M}}$ should verify the Boolean metabolic steady state constraints described in the previous section ($x_{\mathcal{M}} \in \text{MSS}^{\mathbb{B}}(\mathcal{N})$). The general idea is then to capture the possible successions of such Boolean states, subject to the regulations through the regulatory proteins specified by the Boolean network f.

A key ingredient of dynamic rFBA is the objective function to maximize, typically the fluxes of reactions producing biomass. However, at the Boolean level, it is not possible to directly rank metabolic steady states according to their biomass production, as this will be either absent or present. Thus, a specific *Boolean objective function* has to be provided to score a Boolean metabolic steady state. This takes the form of a function $\hat{o}$ mapping Boolean metabolic steady states to natural numbers: $\hat{o} : \mathbb{B}^{k+m} \to \mathbb{N}$. The Boolean dynamics will only select Boolean metabolic steady states maximizing this supplied objective.

When considering possible next states, it is crucial to account for those where the input metabolites change their value. Hereafter, we consider any possible change.

The Boolean dynamic rFBA is formalized by a function $\text{next}^{\mathbb{B}}_{(\mathcal{N},\mathcal{P},f,\hat{o})}$ which associates any Boolean state of the regulated metabolic network to a set of admissible next states:

Definition 7 (Boolean dynamic rFBA: $\text{next}^{\mathbb{B}}_{(\mathcal{N},\mathcal{P},f,\hat{o})} : \mathbb{B}^n \to 2^{\mathbb{B}^n}$). *For any Boolean states $x, y \in \mathbb{B}^n$, $y \in \text{next}^{\mathbb{B}}_{(\mathcal{N},\mathcal{P},f,\hat{o})}(x)$ if and only if for $x' = (y_{\text{Inp}}, x_{\mathcal{R}\cup\mathcal{P}}) \in \mathbb{B}^n$,*

1. *the values of the regulatory proteins are computed synchronously from x' according to f: $y_{\mathcal{P}} = f_{\mathcal{P}}(x')$,*
2. *y matches with a Boolean metabolic steady state: $y_{\mathcal{M}} \in Z(x')$, and*
3. *the matching Boolean metabolic steady state maximizes the supplied objective function: $\forall y'_{\mathcal{M}} \in Z(x'), \hat{o}(y_{\mathcal{M}}) \geq \hat{o}(y'_{\mathcal{M}})$.*

Here $Z(x') = \{z \in \text{MSS}^{\mathbb{B}}(\mathcal{N}) \mid z_{\text{Inp}} = x'_{\text{Inp}}, z_{\mathcal{R}} \preceq f_{\mathcal{R}}(x')\}$ is the set of Boolean metabolic steady states that match with the value of external metabolites and with the regulations from x'.

Let us consider the regulated metabolic network from Fig. 1. It appears that the steady states maximizing the growth maximize the input fluxes. Thus, we set the Boolean objective function $\hat{o}$ as the sum of input reactions:

$$\hat{o}(x) = x_{\mathrm{Tc1}} + x_{\mathrm{Tc2}} + x_{\mathrm{To2}} \ .$$

Consider the Boolean state from Table 1 at time 0, which we name x, and the next Boolean state at time 0.51, which we name y, with the same input metabolite values ($x_{\mathrm{Inp}} = y_{\mathrm{Inp}}$). Using the notation from the above definition, we set $x' = x$. Imagine the case where no reactions is regulated, i.e., the regulatory BN is of the form $f'_r(x) = 1$ for every $r \in \mathcal{R}$. Among the Boolean metabolic steady states z matching the input values ($z_{\mathrm{Inp}} = x'_{\mathrm{Inp}}$), the ones that maximize $\hat{o}$ always verify $z_{\mathrm{Tc2}} = 1$ (Boolean metabolic steady states 26, 29, 32, 38 in the Table 3 in Appendix A), which does not match with y. Thus y would not be an admissible next state.

Considering now the regulatory BN f of Fig. 1, we obtain $f_{\mathrm{Tc2}}(x') = \neg x'_{\mathrm{RPcl}} = 0$ and for each other reaction $r \in \mathcal{R} \setminus \{\mathrm{Tc2}\}$, $f_r(x') = 1$. The set $Z(x')$ contains 4 matching optimal Boolean steady states (rows 25, 28, 31, 37 of Table A.3), among them the one matching with y. Thus $y \in \mathrm{next}^{\mathbb{B}}_{(\mathcal{N},\mathcal{P},f,\hat{o})}(x)$.

Let x be now the Boolean state at time 0.1, and y the next Boolean state at time 0.51, where the input metabolites have a different state (Carbon1 switched to 0). Let x' be equal to x except for the input metabolites, which are equal to y_{Inp}. We obtain that $f_{\mathrm{RPO2,RPcl}}(x') = (\neg x'_{\mathrm{Oxygen}}, x'_{\mathrm{Carbon1}}) = (0, 0) = y_{\mathrm{RPO2,RPcl}}$. Moreover, $f_{\mathrm{Tc2}}(x') = \neg x'_{\mathrm{RPcl}} = 0$ and for each other reaction $r \in \mathcal{R}$, $r \neq \mathrm{Tc2}$, $f_r(x') = 1$. In this case, there is only one Boolean metabolic steady state z such that $z_{\mathrm{Inp}} = x'_{\mathrm{Inp}}$ and $z_{\mathcal{R}} \preceq f_{\mathcal{R}}(x')$. It appears that it matches with y, i.e., $z = y_{\mathcal{M}}$; thus $y \in \mathrm{next}^{\mathbb{B}}_{(\mathcal{N},\mathcal{P},f,\hat{o})}(x)$.

4 Inference of Regulations from rFBA Time Series

Given sequences of metabolic-regulatory steady states obtained by dynamic rFBA from a ground-truth regulated metabolic network under different conditions, our objective is to infer all the regulatory Boolean networks that can reproduce the observed behaviors. Besides the ground-truth model, the inference may suggest alternative regulatory logics.

Definition 8 (Search domain for BNs). *The search domain for BNs, denoted by* $\mathbb{F}$*, is constrained by an influence graph* $\mathcal{G}$*: any candidate* $f \in \mathbb{F}$ *should satisfy* $G(f) \subseteq \mathcal{G}$, i.e. *uses at most the influences allowed in* $\mathcal{G}$*. Moreover, we assume that* f *is locally monotone.*

Typically, $\mathcal{G}$ contains the putative influences from and to regulatory proteins. In our case study, $\mathcal{G}$ is obtained from the ground-truth regulatory model f° by "forgetting" the sign of influences (for each $(i, s, j) \in G(f^{\circ})$, $\{(i, +, j), (i, -, j)\} \subseteq \mathcal{G}$), and adding putative influences.

Our inference problem mixes both linear constraints for characterizing the optimal steady states of the metabolic network with Boolean constraints for characterizing the value changes of regulatory proteins. To express the inference problem, we rely on the Boolean abstraction of dynamic rFBA presented in the previous section .

4.1 Approximation as a Boolean Satisfiability Problem

We propose a relaxation of the inference problem by the means of the Boolean dynamic rFBA interpretation given in Sect. 3.

Inputs of the Relaxed Inference Problem. The inputs of the problem are **(i)** a metabolic network $\mathcal{N}$ and a set of regulatory proteins $\mathcal{P}$, **(ii)** sequences of metabolic-regulatory steady states, represented by sets of pairs (s^t, s^{t+1}), with $s^t = (v^t, w^t, x^t)$ and $s^{t+1} = (v^{t+1}, w^{t+1}, x^{t+1})$ following the notation from Definition 5: the observed changes of metabolic-regulatory steady states are given as $T \subseteq \mathbb{S} \times \mathbb{S}$ with $\mathbb{S} = \mathbb{R}^{|\mathrm{Inp}|+|\mathcal{R}|} \times \mathbb{B}^{|RPs|}$, **(iii)** a domain of putative regulatory BNs $\mathbb{F}$ of dimension $n = |\mathrm{Inp}| + |\mathcal{R}| + |\mathcal{P}|$, **(iv)** a Boolean state objective score $\hat{o} : \mathbb{B}^n \to \mathbb{N}$.

Relaxed Inference Problem. The relaxed inference problem consists then in identifying the $f \in \mathbb{F}$ such that for each $(s, s') \in T$,

$$\beta(s') \in \mathrm{next}^{\mathbb{B}}_{(\mathcal{N},\mathcal{P},f,\hat{o})}(\beta(s)).$$

Formulation as a Satisfiability Problem. Relying on the Boolean dynamic rFBA abstraction, the inference problem boils down to a satisfiability problem in propositional Boolean logic using two levels of quantifiers (2-QBF):

$$\exists f \in \mathbb{F}, \forall (s, s') \in T, \exists y \in \mathrm{MSS}^{\mathbb{B}}(\mathcal{N}), y_{\mathrm{Inp}} = x'_{\mathrm{Inp}}, y_{\mathcal{P}} = f_{\mathcal{P}}(x'), y_{\mathcal{R}} \preceq f_{\mathcal{R}}(x'),$$
$$\forall z \in \mathrm{MSS}^{\mathbb{B}}(\mathcal{N}), (z_{\mathrm{Inp}} \neq x'_{\mathrm{Inp}} \vee z_{\mathcal{P}} \neq f_{\mathcal{P}}(x') \vee z_{\mathcal{R}} \npreceq f_{\mathcal{R}}(x') \vee \hat{o}(z) \leq \hat{o}(y))$$

with $x' \in \mathbb{B}^n$ defined as $x'_{\mathrm{Inp}} = \beta(s')_{\mathrm{Inp}}$ and $x'_{\mathcal{R}\cup\mathcal{P}} = \beta(s)_{\mathcal{R}\cup\mathcal{P}}$.

Note that without the Boolean optimization criteria $\hat{o}$ (equivalently $\hat{o}(z) = c$), the problem reduces to a SAT problem where the only constraints relate to the local functions of the regulatory proteins:

$$\exists f \in \mathbb{F}, \exists y \in \mathrm{MSS}^{\mathbb{B}}(\mathcal{N}), y_{\mathrm{Inp}} = x'_{\mathrm{Inp}}, y_{\mathcal{P}} = f_{\mathcal{P}}(x')$$

Indeed, $y_{\mathcal{R}} \preceq f_{\mathcal{R}}(x')$ is always verified whenever $f_r(x) = 1$ for each $r \in \mathcal{R}$.

Since the Boolean dynamic rFBA gives an over-approximation of metabolic steady states, and even assuming that the Boolean objective function $\hat{o}$ matches with the optimal metabolic steady states, our formulation leads to an approximation of admissible regulatory BN f: it may happen that a spurious Boolean metabolic steady state (having no real counter part) has a strictly higher value with $\hat{o}$ than non-spurious ones.

4.2 Implementation in Answer-Set Programming

Answer-Set Programming (ASP) [1,12] is a declarative framework allowing solving combinatorial satisfaction problems. It relies on the stable model semantics [10]. The basic idea of ASP is to express a problem in a logical format so that the (logic) models of its representation provide the solutions to the original problem. Problems are expressed as logic programs (first order logic predicates expressed with rules with the shape `<head> :- <body> .`). Stable models of the logic programs are referred to as *answer sets*. Although determining whether a program has an answer set is the fundamental decision problem in ASP, modern ASP solvers like clingo [13] support various combinations of reasoning modes, among them, regular and projective enumeration, intersection and union, multi-criteria optimization and subset minimal and maximal model enumeration [15].

The stable model semantics of ASP combined with disjunctive programming are the key ingredients that enable expressing two quantification levels Boolean formulas (2-QBF problem), *i.e.* $\exists x, \forall y, \phi(x,y)$ where $\phi(x,y)$ is a quantifier-free propositional formula (Σ_2^P-complete) [10]. The encoding of 2-QBF relies on the so-called *saturation technique* [11,14]. Essentially, for fixed x and y, the encoding ensures that a maximal (saturated) answer-set is returned if and only if $\phi(x,y)$. Thus, whenever there exists y such that $\phi(x,y)$ does not hold (counter-example), a smaller answer-set is returned. Following the subset-minimal stable semantics, the 2 QBF problem is satisfiable if and only if only saturated answer-set are subset-minimal.

5 Case Study

As a proof of concept, we apply our inference framework to the simplified core carbon metabolism described in Fig. 1. First, from this ground-truth model, we generate sample dynamic rFBA simulations for different input conditions, reproducing existing biological observations [9]. Next we take these simulations as input for our method, together with an influence graph extending the one from the ground truth model with additional putative regulations. Using our inference method, we then enumerate BNs that are compatible with both the simulations and the influence graph. The results show that the ground truth model is well recovered, together with some alternative BNs. In particular, a simpler BN matching the data is identified, which uses fewer regulations. It turns out that the missing regulation is not needed to reproduce the expected biological behaviour. Our implementation relying on the ASP solver CLINGO [13] together with the case study is available at https://github.com/bioasp/boolean-caspo-flux. They can be reproduced using the notebooks and docker image at https://doi.org/10.5281/zenodo.5060984.

Input Simulations. We designed six dynamic r-FBA simulations of the BN of Fig. 1(b) to mimic the studies of the core carbon metabolism in [9]. They correspond to different sets of initially available input metabolites and regulatory proteins (Table 3a, and Fig. 4 in Appendix B). For instance, Experiment 1 assumes that all input metabolites (Carbon1, Carbon2, Oxygen) are available. Experiment 2 assumes that Carbon1, Carbon2 are present at initialization but not Oxygen.

For each case, we use FLEXFLUX with an initial biomass value of 0.1 and a time step of 0.01 to simulate the system. Each of the 6 simulations involves 200 metabolic steady states. For initial external metabolite values ($\overline{z_{\text{Carbon1}}}$, $\overline{z_{\text{Carbon2}}}$, $\overline{z_{\text{Oxygen}}}$), the regulatory proteins are initialized such that $x_{\text{RPcl}} = \overline{z_{\text{Carbon1}}}$ and $x_{\text{RPO2}} = \neg\overline{z_{\text{Oxygen}}}$ (Table 3a). Each simulation $S = \{(v, w, x)_0, ..., (v, w, x)_{200}\}$ includes 201 continuous metabolic-regulatory steady states (1 for the initialization and 200 for the simulation). The simulations are then binarized with $S^{\mathbb{B}} = \{(\overline{v_t}, \overline{z_t}) = \beta((v_t, w_t)) \mid \forall v_t \in S\}$, and consecutive identical Boolean states are removed. Table 1 shows the binarized metabolic-regulatory steady states from the simulation of the first experiment. From the 201 continuous metabolic steady states, 5 Boolean metabolic-regulatory steady states remain, corresponding to the time steps {0, 1, 51, 52, 59} (see Table 4 in Appendix B for the resulting states in each simulation).

Candidate Models. The search domain $\mathbb{F}$ for the candidate BNs is delimited by the influence graph $\mathcal{G}$ of Fig. 3b, which extends the influence graph from the ground-truth model by additional putative regulations, and by relaxing the sign constraints. Since the influence graph $G(f)$ of the ground-truth BN f is included in $\mathcal{G}$, we have $f \in \mathbb{F}$. In addition, $\mathbb{F}$ contains all the BNs such that $f_i(x) = 1$, for all $i \in \text{Inp} \cup \mathcal{R} \setminus \{\text{Tc1, Tc2, Res}\}$. Furthermore, f_{RPcl} can depend on Carbon1, Carbon2, Tc1, and Tc2, f_{RPO2} can depend on Oxygen, Rres, f_{Tc1} and f_{Tc2} can depend on RPcl, and f_{Rres} can depend on Rres. Overall, $\mathbb{F}$ contains 1 944 320 BNs.

	Input Metabolite			Regulatory Protein	
Experiment	$\overline{z_{\text{Carbon1}}}$	$\overline{z_{\text{Carbon2}}}$	$\overline{z_{\text{Oxygen}}}$	x_{RPcl}	x_{RPO2}
1	1	1	1	1	0
2	1	1	0	1	1
3	0	1	0	0	1
4	1	0	0	1	1
5	1	0	1	1	0
6	0	1	1	0	0

(a) Initial states of the six rFBA simulations used to create the dataset for the case study.

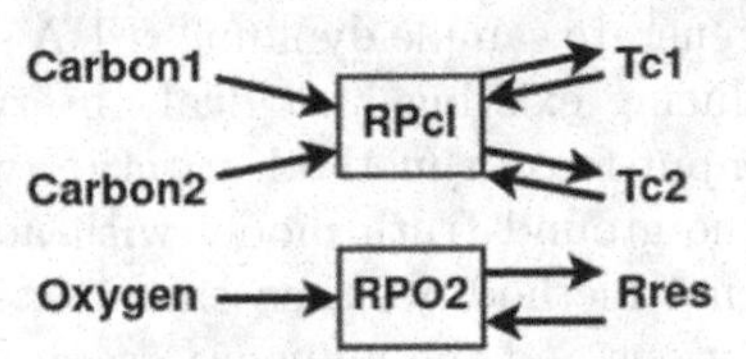

(b) Influence graph $\mathcal{G}$ delimiting the domain of putative regulatory BNs $\mathbb{F}$. Nodes without in-going or out-going edges are not represented. Black regular tipping arrows are unsigned edges, *i.e.* both positive and negative edges.

Fig. 3. Input data for the case study. Table (a) summarizes the experimental conditions used to generate the input simulations. Figure (b) shows the influence graph delimiting the search domain for the inference problem.

Table 2. Inferred models having subset minimal local functions. The not shown local functions $f_{\text{Carbon1}}(x)$, $f_{\text{Carbon2}}(x)$, $f_{\text{Oxygen}}(x)$, $f_{\text{To2}}(x)$, $f_{\text{Td}}(x)$, $f_{\text{Te}}(x)$, $f_{\text{Growth}}(x)$, $f_{\text{R6}}(x)$, $f_{\text{R7}}(x)$ are set to 1.

	$f_{RPO2}(x)$	$f_{RPcl}(x)$	$f_{Tc1}(x)$	$f_{Tc2}(x)$	$f_{Rres}(x)$	Subset minimal	Ground truth
Model 1	$\neg x_{\text{Oxygen}}$	x_{Carbon1}	1	$\neg x_{\text{RPcl}}$	1	✓	
Model 2	$\neg x_{\text{Oxygen}}$	x_{Carbon1}	1	$\neg x_{\text{RPcl}}$	$\neg x_{\text{RPO2}}$		✓
Model 3	$\neg x_{\text{Oxygen}}$	x_{Carbon1}	x_{RPcl}	$\neg x_{\text{RPcl}}$	1		
Model 4	$\neg x_{\text{Oxygen}}$	x_{Carbon1}	x_{RPcl}	$\neg x_{\text{RPcl}}$	$\neg x_{\text{RPO2}}$		

Boolean Objective Function. Our inference framework requires defining an objective function $\hat{o}$ over the Boolean metabolic steady states. Given the set of input metabolites Inp = {Carbon1, Carbon2, Oxygen}, the objective function is defined as $\hat{o}(x) = \sum_{e \in \text{Inp}} x_e, \forall x \in \text{MSS}^{\mathbb{B}}(\mathcal{N})$. This is motivated by the observation that maximizing biomass production often corresponds to maximizing the uptake of inputs according to the QSS constraints. Therefore, if an available input metabolite is not used in the observed Boolean metabolic network, then this must be explained by at least one regulation. This objective function allows capturing more refined behaviors at the discrete level than a standard biomass optimization function, which may be too rough when considering discretized values.

Results. Applying the constraints from above allows inferring 40 models. All these models share 3 local functions whose value is not constantly 1 ($f_{\text{RPO2}}(x), f_{\text{RPcl}}(x), f_{\text{Tc2}}(x)$). They also share 9 local functions equal to 1 ($f_{\text{Carbon1}}(x)$, $f_{\text{Carbon2}}(x)$, $f_{\text{Oxygen}}(x)$, $f_{\text{To2}}(x)$, $f_{\text{Td}}(x)$, $f_{\text{Te}}(x)$, $f_{\text{Growth}}(x)$, $f_{\text{R6}}(x)$, $f_{\text{R7}}(x)$). Finally, 2 functions can be set both to 1 or different from 1 according to the model. The 4 *smallest* inferred models are described in Table 2. They can be considered as the smallest because each local function f_i of these 4 models is contained in the local function f_i of the 36 other models. Note that the ground truth, *i.e.* the model used to generate the input data, is correctly inferred (Model 2).

As we represent the local Boolean functions using their disjunctive normal form (DNF), we can focus on the *simplest* models by looking at the *subset-minimal* ones: a Boolean function f_i is smaller than a Boolean function g_i if each of the clauses of f_i is a subset of a clause of g_i. In this case study, there is a single subset-minimal model: the BN 1 of Table 2. The two functions $f_{\text{Rres}}(x)$, $f_{\text{Tc1}}(x)$ are set to 1 due to the subset-minimal constraint. The inferred model is thus $f_{\text{RPO2}}(x) = \neg x_{\text{Oxygen}}$, $f_{\text{RPcl}}(x) = x_{\text{Carbon1}}$, $f_{\text{Tc2}}(x) = \neg x_{\text{RPcl}}$ and all the others local functions are set to 1. Note that only $f_{\text{Rres}}(x)$ differs between the inferred subset-minimal model and the ground truth model.

In order to check whether this subset-minimal model could be considered as an alternative to the ground truth one, we performed dynamic rFBA simulations with the six experimental conditions described in Table 3a. We observe that the resulting time series are strictly identical to the simulations of the ground truth model used to generate the dataset. This suggests that the regulation on *Rres* is not necessary to reproduce the observed behaviours. The proposed subset-minimal model allows inferring all the needed regulations and can be considered as the simplest regulated metabolic model matching the experimental conditions of Table 3a. Already in [9], the authors recognize that unlike others regulations, *Rres* "regulation is not necessary for the solution". Biologically, this regulation is only present to ensure that unnecessary enzymes decay. However, since enzyme amounts are not explicitly represented in the rFBA framework, the dataset does not reflect this biologic behavior, making it impossible to infer properly the regulation. Taking into account enzymatic resources using methods such as r-deFBA [17], should allow solving this issue. However, the inference approach will also have to be adapted to this kind of extended metabolic modeling.

6 Discussion

We proposed a formal framework to infer Boolean rules for the regulation of a metabolic network. The formulation of dynamic rFBA as sequence of steady states of the regulated metabolic network enables inferring the Boolean rules from time series under multiple conditions. A proof of concept was performed on the simulation of the diauxic shift in carbon metabolism on a small model.

Our method builds on a Boolean abstraction of the dynamic rFBA framework. It enables a formulation of the inference problem as a pure Boolean satisfiability problem using two levels of quantifiers, which can be efficiently solved using Answer Set Programming. One important parameter is the Boolean objective function, which aims at identifying Boolean metabolic steady states that match the optimal real-valued ones. This function is currently specified manually, based on biological expertise. Future work may explore how to derive an objective function automatically. An alternative direction is to solve directly the inference problem by mixing linear programming and Boolean constraints. Future work will investigate the scalability of solving these different inference problems.

Several other perspectives are to be explored. First, all regulations were considered as synchronous, which may not be the case *in vivo*, where regulations can have different time scales. This choice was actually imposed by the use of the FlexFlux implementation. Nevertheless, our method can be easily adapted to support fully-asynchronous and asynchronous updating modes, enabling potential alternative solutions. Second, the production and degradation times of regulatory proteins and enzymes were not taken into account. Moreover, the regulations were considered to be binary. However, we know that metabolism proceeds by finer regulations than the abstraction proposed here, as captured for instance by regulatory dynamic enzyme-cost FBA [17].

Acknowledgments. Work of LC and CB is supported by the French Laboratory of Excellence project "TULIP" (grant number ANR-10-LABX-41; ANR-11-IDEX-0002-02). Work of LP is supported by the French Agence Nationale pour la Recherche (ANR) in the scope of the project "BNeDiction" (grant number ANR-20-CE45-0001).

A Binarized Metabolic Steady State

Table 3. All the Boolean metabolic steady states admissible for the metabolic network $\mathcal{N}$ show Fig. 1a. The external metabolite *Biomass* is not shown since its value can be both 0 and 1 for each Boolean metabolic steady state. The experimentation column indicates the numbers of the experiments where the Boolean metabolic steady states occurs.

	External metabolites			Reactions									
	$\overline{z_{Carbon1}}$	$\overline{z_{Carbon2}}$	$\overline{z_{Oxygen}}$	$\overline{v_{Tc1}}$	$\overline{v_{Tc2}}$	$\overline{v_{To2}}$	$\overline{v_{Td}}$	$\overline{v_{Te}}$	$\overline{v_{Growth}}$	$\overline{v_{Rres}}$	$\overline{v_{R0}}$	$\overline{v_{R7}}$	Experimentation
1	0	0	0	0	0	0	0	0	0	0	0	0	2, 3, 4
2	0	0	1	0	0	0	0	0	0	0	0	0	1, 5, 6
3	0	1	0	0	0	0	0	0	0	0	0	0	2, 3
4	0	1	0	0	1	0	1	1	1	0	1	1	2, 3
5	0	1	1	0	0	0	0	0	0	0	0	0	1, 6
6	0	1	1	0	1	1	0	0	1	1	0	0	1, 6
7	0	1	1	0	1	1	0	1	1	1	0	1	
8	0	1	1	0	1	1	1	0	1	1	1	0	
9	0	1	1	0	1	0	1	1	1	0	1	1	
10	0	1	1	0	1	1	1	1	1	1	1	1	
11	1	0	0	0	0	0	0	0	0	0	0	0	4
12	1	0	0	1	0	0	1	1	1	0	1	1	4
13	1	0	1	0	0	0	0	0	0	0	0	0	5
14	1	0	1	1	0	1	0	0	1	1	0	0	5
15	1	0	1	1	0	1	0	1	1	1	0	1	
16	1	0	1	1	0	1	1	0	1	1	1	0	
17	1	0	1	1	0	0	1	1	1	0	1	1	
18	1	0	1	1	0	1	1	1	1	1	1	1	
19	1	1	0	0	0	0	0	0	0	0	0	0	2
20	1	1	0	0	1	0	1	1	1	0	1	1	
21	1	1	0	1	0	0	1	1	1	0	1	1	2
22	1	1	0	1	1	0	1	1	1	0	1	1	
23	1	1	1	0	0	0	0	0	0	0	0	0	1
24	1	1	1	0	1	1	0	0	1	1	0	0	
25	1	1	1	1	0	1	0	0	1	1	0	0	1
26	1	1	1	1	1	1	0	0	1	1	0	0	
27	1	1	1	0	1	1	0	1	1	1	0	1	
28	1	1	1	1	0	1	0	1	1	1	0	1	
29	1	1	1	1	1	1	0	1	1	1	0	1	
30	1	1	1	0	1	1	1	0	1	1	1	0	
31	1	1	1	1	0	1	1	0	1	1	1	0	
32	1	1	1	1	1	1	1	0	1	1	1	0	
33	1	1	1	0	1	0	1	1	1	0	1	1	
34	1	1	1	1	0	0	1	1	1	0	1	1	
35	1	1	1	1	1	0	1	1	1	0	1	1	
36	1	1	1	0	1	1	1	1	1	1	1	1	
37	1	1	1	1	0	1	1	1	1	1	1	1	
38	1	1	1	1	1	1	1	1	1	1	1	1	

B Experiments and Simulations

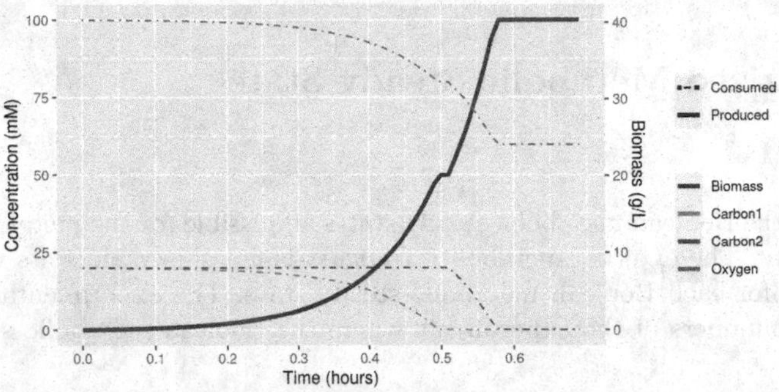

(a) Simulation of experiment 1.

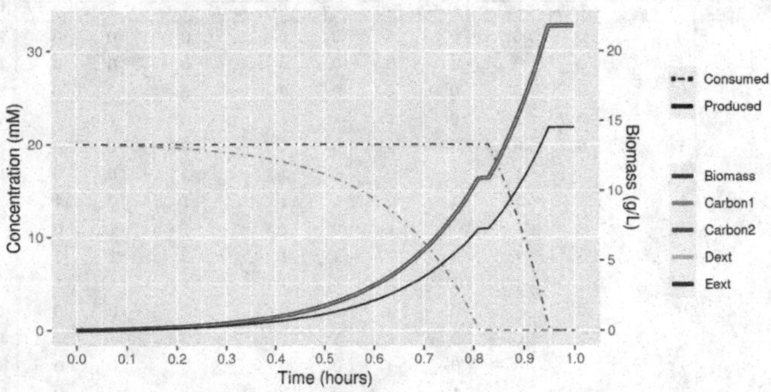

(b) Simulation of experiment 2.

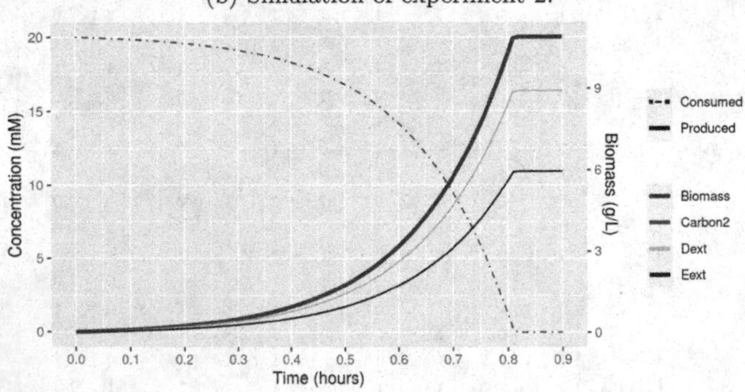

(c) Simulation of experiment 3.

Fig. 4. Simulation made with FLEXFLUX of the regulated metabolic network in Fig. 1 for each experiment (Table 3a). Time step is set to 0.01. Reaction domains are $\forall r \in$ {Tc1, Tc2}, $(l_r, u_r) = (0, 10.5)$, $\forall r \in$ {Td, Te}, $(l_r, u_r) = (0, 12.0)$, $\forall r \in$ {R6, R7, Rres, Growth}, $(l_r, u_r) = (0, 9999)$ and for Oxygen, $(l_r, u_r) = (0, 15.0)$. The same simulation graphs are obtained using the local function $f_{\text{Rres}} = \neg x_{\text{RPO2}}$ and $f_{\text{Rres}} = 1$.

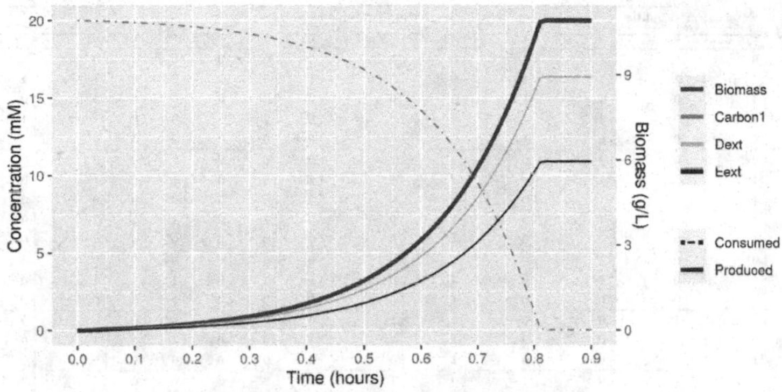

(d) Simulation of experiment 4.

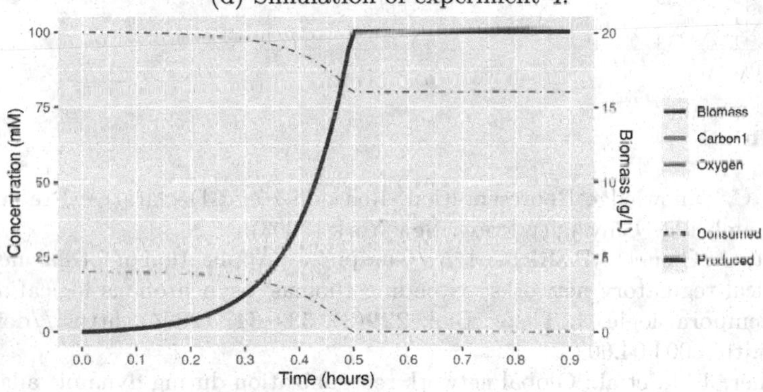

(e) Simulation of experiment 5.

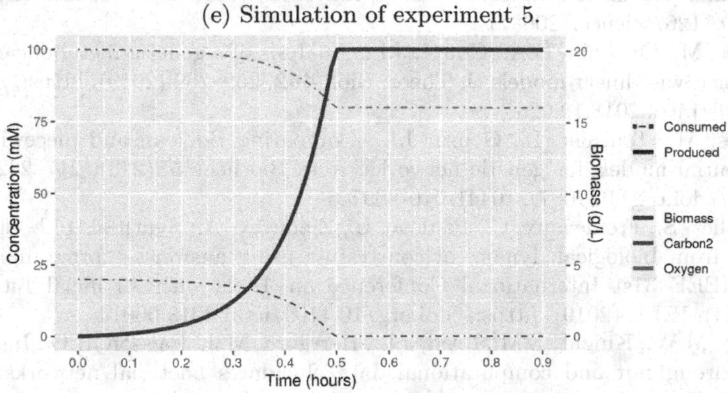

(f) Simulation of experiment 6.

Fig. 4. (*continued*)

Table 4. All the different binarized metabolic steady states of each experiment. They are the input data used to solve the inference problem.

		External metabolites				Regulatory proteins		Reactions								
Experiment	Time	$\overline{z_{Biomass}}$	$\overline{z_{Carbon1}}$	$\overline{z_{Carbon2}}$	$\overline{z_{Oxygen}}$	$\overline{x_{RPO2}}$	$\overline{x_{RPcl}}$	$\overline{v_{Tc1}}$	$\overline{v_{Tc2}}$	$\overline{v_{To2}}$	$\overline{v_{Td}}$	$\overline{v_{Te}}$	$\overline{v_{Growth}}$	$\overline{v_{Rres}}$	$\overline{v_{R6}}$	$\overline{v_{R7}}$
1	0	0	1	1	1	0	1	0	0	0	0	0	0	0	0	0
	1	1	1	1	1	0	1	1	0	1	0	0	1	1	0	0
	51	1	0	1	1	0	0	0	0	0	0	0	0	0	0	0
	52	1	0	1	1	0	0	0	1	1	0	0	1	1	0	0
	59	1	0	0	1	0	0	0	0	0	0	0	0	0	0	0
2	0	0	1	1	0	1	1	0	0	0	0	0	0	0	0	0
	1	1	1	1	0	1	1	1	0	0	1	1	1	0	1	1
	83	1	0	1	0	1	0	0	0	0	0	0	0	0	0	0
	84	1	0	1	0	1	0	0	1	0	1	1	1	0	1	1
	97	1	0	0	0	1	0	0	0	0	0	0	0	0	0	0
3	0	0	0	1	0	1	0	0	0	0	0	0	0	0	0	0
	1	1	0	1	0	1	0	0	1	0	1	1	1	0	1	1
	83	1	0	0	0	1	0	0	0	0	0	0	0	0	0	0
4	0	0	1	0	0	1	1	0	0	0	0	0	0	0	0	0
	1	1	1	0	0	1	1	1	0	0	1	1	1	0	1	1
	83	1	0	0	0	1	0	0	0	0	0	0	0	0	0	0
5	0	0	1	0	1	0	1	0	0	0	0	0	0	0	0	0
	1	1	1	0	1	0	1	1	0	1	0	0	1	1	0	0
	51	1	0	0	1	0	0	0	0	0	0	0	0	0	0	0
6	0	0	0	1	1	0	0	0	0	0	0	0	0	0	0	0
	1	1	0	1	1	0	0	0	1	1	0	0	1	1	0	0
	51	1	0	0	1	0	0	0	0	0	0	0	0	0	0	0

References

1. Baral, C.: Knowledge Representation. Reasoning and Declarative Problem Solving., Cambridge University Press, New York (2003)
2. Bernot, G., Comet, J.P., Richard, A., Guespin, J.: Application of formal methods to biological regulatory networks: extending thomas' asynchronous logical approach with temporal logic. J. Theor. Biol. **229**(3), 339–347 (2004). https://doi.org/10.1016/j.jtbi.2004.04.003
3. Buescher, J.M., et al.: Global network reorganization during dynamic adaptations of bacillus subtilis metabolism. Science **335**(6072), 1099–1103 (2012). https://doi.org/10.1126/science.1206871
4. Chaves, M., Oyarzún, D.A., Gouzé, J.L.: Analysis of a genetic-metabolic oscillator with piecewise linear models. J. Theor. Biol. **462**, 259–269 (2019). https://doi.org/10.1016/j.jtbi.2018.10.026
5. Chaves, M., Tournier, L., Gouzé, J.L.: Comparing Boolean and piecewise affine differential models for genetic networks. Acta Biotheor **58**(2–3), 217–232 (2010). https://doi.org/10.1007/s10441-010-9097-6
6. Chevalier, S., Froidevaux, C., Pauleve, L., Zinovyev, A.: Synthesis of boolean networks from biological dynamical constraints using answer-set programming. In: 2019 IEEE 31st International Conference on Tools with Artificial Intelligence (ICTAI). IEEE (2019). https://doi.org/10.1109/ictai.2019.00014
7. Covert, M.W., Knight, E.M., Reed, J.L., Herrgard, M.J., Palsson, B.O.: Integrating high-throughput and computational data elucidates bacterial networks. Nature **429**(6987), 92–96 (2004). https://doi.org/10.1038/nature02456
8. Covert, M.W., Palsson, B.Ø.: Transcriptional regulation in constraints-based metabolic models of Escherichia coli. J. Biol. Chem. **277**(31), 28058–28064 (2002). https://doi.org/10.1046/j.1462-2920.2002.00282.x
9. Covert, M.W., Schilling, C., Palsson, B.: Regulation of gene expression in flux balance models of metabolism. J. Theor. Biol. **213**(1), 73–88 (2001). https://doi.org/10.1006/jtbi.2001.2405

10. Eiter, T., Gottlob, G.: On the computational cost of disjunctive logic programming: propositional case. Ann. Math. Artif. Intell. **15**(3–4), 289–323 (1995). https://doi.org/10.1007/bf01536399
11. Eiter, T., Ianni, G., Krennwallner, T.: Answer Set Programming: A Primer, pp. 40–110. Springer, Berlin (2009). https://doi.org/10.1007/978-3-642-03754-2_2
12. Gebser, M., Kaminski, R., Kaufmann, B., Schaub, T.: Answer Set Solving in Practice. Synthesis Lectures on Artificial Intelligence and Machine Learning. Morgan and Claypool Publishers (2012)
13. Gebser, M., Kaminski, R., Kaufmann, B., Schaub, T.: Clingo = ASP + control. Preliminary report. CoRR abs/1405.3694 (2014)
14. Gebser, M., Kaminski, R., Schaub, T.: Complex optimization in answer set programming. Theor. Pract. Logic Prog. **11**(4–5), 821–839 (2011). https://doi.org/10.1017/s1471068411000329
15. Gebser, M., Kaufmann, B., Romero, J., Otero, R., Schaub, T., Wanko, P.: Domain-specific heuristics in answer set programming. In: Proceedings of the AAAI Conference on Artificial Intelligence, vol. 27, no. 1 (2013). https://ojs.aaai.org/index.php/AAAI/article/view/8585
16. de Jong, H.: Modeling and simulation of genetic regulatory systems: a literature review. J. Comput. Biol. **9**, 67–103 (2002). https://doi.org/10.1089/10665270252833208
17. Liu, L., Bockmayr, A.: Regulatory dynamic enzyme-cost flux balance analysis: a unifying framework for constraint-based modeling. J. Theor. Biol. **501**, 110317 (2020). https://doi.org/10.1016/j.jtbi.2020.110317
18. Marmiesse, L., Peyraud, R., Cottret, L.: FlexFlux: combining metabolic flux and regulatory network analyses. BMC Syst. Biol. **9**(1), 1–13 (2015). https://doi.org/10.1186/s12918-015-0238-z
19. Orth, J.D., Thiele, I., Palsson, B.Ø.: What is flux balance analysis? Nat. Biotechnol. **28**(3), 245–248 (2010). https://doi.org/10.1038/nbt.1614
20. Ostrowski, M., Paulevé, L., Schaub, T., Siegel, A., Guziolowski, C.: Boolean network identification from perturbation time series data combining dynamics abstraction and logic programming. Biosystems **149**, 139–153 (2016). https://doi.org/10.1016/j.biosystems.2016.07.009
21. Oyarzún, D.A., Chaves, M., Hoff-Hoffmeyer-Zlotnik, M.: Multistability and oscillations in genetic control of metabolism. J. Theor. Biol. **295**, 139–153 (2012). https://doi.org/10.1016/j.jtbi.2011.11.017
22. Razzaq, M., Paulevé, L., Siegel, A., Saez-Rodriguez, J., Bourdon, J., Guziolowski, C.: Computational discovery of dynamic cell line specific boolean networks from multiplex time-course data. PLOS Comput. Biol. **14**(10), e1006538 (2018). https://doi.org/10.1371/journal.pcbi.1006538
23. Saez-Rodriguez, J., et al.: Discrete logic modelling as a means to link protein signalling networks with functional analysis of mammalian signal transduction. Mol. Syst. Biol. **5**(1), 331 (2009). https://doi.org/10.1038/msb.2009.87
24. Tournier, L., Goelzer, A., Fromion, V.: Optimal resource allocation enables mathematical exploration of microbial metabolic configurations. J. Math. Biol. **75**(6–7), 1349–1380 (2017). https://doi.org/10.1007/s00285-017-1118-5
25. Tsiantis, N., Balsa-Canto, E., Banga, J.R.: Optimality and identification of dynamic models in systems biology: an inverse optimal control framework. Bioinformatics **34**(14), 2433–2440 (2018). https://doi.org/10.1093/bioinformatics/bty139

26. Videla, S., Saez-Rodriguez, J., Guziolowski, C., Siegel, A.: Caspo: a toolbox for automated reasoning on the response of logical signaling networks families. Bioinformatics p. btw738 (2017). https://doi.org/10.1093/bioinformatics/btw738
27. Zañudo, J.G.T., Yang, G., Albert, R.: Structure-based control of complex networks with nonlinear dynamics. Proc. Natl. Acad. Sci. U.S.A. **114**(28), 7234–7239 (2017). https://doi.org/10.1073/pnas.1617387114

Population Design for Synthetic Gene Circuits

Baptiste Turpin, Eline Y. Bijman, Hans-Michael Kaltenbach, and Jörg Stelling(✉)

Department of Biosystems Science and Engineering (D-BSSE) and SIB Swiss Institute of Bioinformatics, ETH Zurich, 4058 Basel, Switzerland
joerg.stelling@bsse.ethz.ch

Abstract. Synthetic biologists use and combine diverse biological parts to build systems such as genetic circuits that perform desirable functions in, for example, biomedical or industrial applications. Computer-aided design methods have been developed to help choose appropriate network structures and biological parts for a given design objective. However, they almost always model the behavior of the network in an average cell, despite pervasive cell-to-cell variability. Here, we present a computational framework to guide the design of synthetic biological circuits while accounting for cell-to-cell variability explicitly. Our design method integrates a NonLinear Mixed-Effect (NLME) framework into an existing algorithm for design based on ordinary differential equation (ODE) models. The analysis of a recently developed transcriptional controller demonstrates first insights into design guidelines when trying to achieve reliable performance under cell-to-cell variability. We anticipate that our method not only facilitates the rational design of synthetic networks under cell-to-cell variability, but also enables novel applications by supporting design objectives that specify the desired behavior of cell populations.

Keywords: Cell-to-cell variability · Synthetic biology · Computer-aided design

1 Introduction

Synthetic biology aims at establishing novel functions in biological systems, or to re-engineer existing ones, in many areas such as new materials or cell-based therapies that are starting to see real-world applications [21]. The conceptual core of the field's rational engineering approach to establish, for example, the corresponding synthetic gene circuits are a systematic design-build-test cycle and the use of predictive mathematical models throughout this cycle to design, analyze, and tune the circuits [14].

Computer-aided design helps identifying suitable network structures (topologies) as well as biological parts for their implementation to reach a given design

E. Cinquemani and L. Paulevé (Eds.): CMSB 2021, LNBI 12881, pp. 181–197, 2021.
https://doi.org/10.1007/978-3-030-85633-5_11

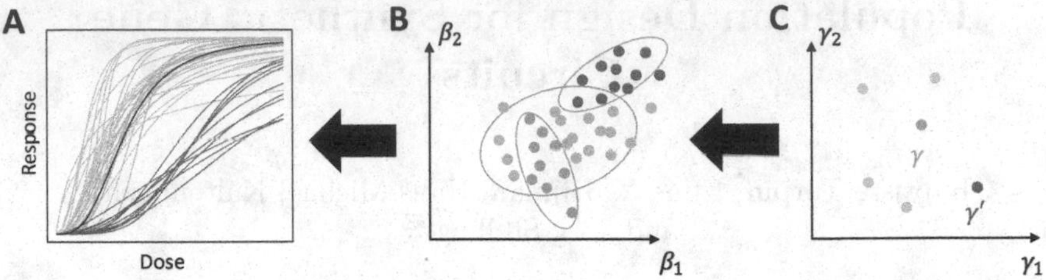

Fig. 1. Cell behaviors relate to parameters at the individual and population level. (A) Dose-response relationships for single cells (lines) drawn from two distinct populations (red and orange) as well as other cells (gray). The design objective for individual cells is represented by an ideal reference curve (black). **(B)** Space of individual parameters β, the set of possible parameter values for a single cell. Dots show parametrizations yielding the behaviors in **(A)** of the corresponding color. The blue ellipse encloses the individual viable space where an individual cost measuring consistency of the single-cell behavior with the design objective for individual cells is below a threshold ε. Red and orange dots encircled by ellipses represent individual cells drawn from the two distinct cell populations. **(C)** Space of population parameters γ, where each parameter vector (dot) describes a full distribution of individual parameters in a population, typically via mean vector and covariance matrix. The orange (γ) and red (γ') dots represent the population parameter vectors that generate the corresponding populations in **(A,B)**. (Color figure online)

objective. For the commonly applied models in the form of ordinary differential equations (ODEs), both design problems can be addressed by investigating the space of model parameters to assess (predicted) circuit behaviors in relation to design objectives encoded by a reference for the desired behavior. With sampling-based methods such as (approximate) Bayesian computation, this defines a 'viable' subspace of the parameter space where the behavior is consistent with the design objective (Fig. 1**A**,**B**) [2,10,17].

The ODE-based approach captures the behavior of an 'average' cell and thus only allows design with respect to such an assumed cell. Yet, for the biological implementation it is critical that a circuit functions under conditions of uncertainty (e.g., in changing environmental conditions or because the models do not capture relevant interactions between parts or with the cellular context [7]) as well as cell-to-cell variability that is present even in isogenic populations (e.g., due to extrinsic or intrinsic stochastic noise, or different cell cycle phases and ages of cells in a population [4]). One can account for uncertainty in ODE-based design, for example, via measures of robustness that quantify parameter uncertainty [10]. It is also possible to tackle cell-to-cell variability with stochastic models, where temporal logic specifications are written as Continuous Stochastic Logic (CSL) [23]. However, the pure ODE and CSL frameworks are limited in two main aspects: First, they cannot account for all aspects of cell-to-cell variability directly; stochastic models do not represent extrinsic variability resulting, for example, from variable cell sizes. This is particularly important when an 'average' cell poorly represents the population dynamics, for example, when subpopulations of cells show different qualitative behaviors. Second, and related, it is not possible to define design objectives for the population, such as requiring a certain fraction of the cells to have a coherent behavior.

To address these limitations, here we propose a framework for robust synthetic circuit design that takes into account cell-to-cell variability, and clearly separates it from experimental noise and impact of variable environmental conditions and interacting parts. For this population design, we extend an existing algorithm for ODE-based design [10] to the NLME (NonLinear Mixed-Effect) models framework [8]. Specifically, this entails augmenting the ODE model with a statistical model at the population level that induces probability distributions over the parameter space at the individual cell level (see Fig. **1B**,**C**). This allows a designer to impose cell-to-cell variability constraints on synthetic networks. We demonstrate the approach with the a posteriori analysis of a recently developed transcriptional controller [1], a class of circuits that is often designed to minimize cell-to-cell variability.

2 Population Design Framework

Individual Cell Model. For any individual cell, the dynamics of the synthetic circuit are governed by the *individual cell model*

$$\Sigma(\beta):\begin{cases}\frac{dx(t)}{dt}=v(x(t),u(t),\alpha)\\ x(0)=x^0\\ y(t)=h(x(t))\,,\end{cases}\tag{1}$$

where x are the system states such as concentrations of chemical species, v is a rate function, and u is an input function. Usually, states cannot be observed directly and the observations y of the system result from a (known) observation function h. We subsume the parameters α and initial conditions x^0 into the parameter $\beta=(\alpha,x^0)\in B$, where B is a bounded set.

Average Cell Design. We first consider the *average cell design problem* of determining the parameter β^* that minimizes the divergence between the circuit's behavior and a desired reference behavior. We model the *behavior of Σ* as an input-output map $D:\mathbb{R}\times B\times\mathcal{U}\to\mathbb{R}$ that provides a (time-dependent) function $D(\tau;\beta,u)$ in τ for each parameter $\beta\in B$ and any input $u\in\mathcal{U}$, where $\mathcal{U}$ is a finite set of relevant inputs. The *reference behavior* $D^{\text{ref}}:\mathbb{R}\times\mathcal{U}\to\mathbb{R}$ is a user-specified (time-dependent) function for each $u\in\mathcal{U}$ that encodes the desired input-output relation; it need not be realizable by Σ. A simple example is a dose-response curve, where a constant input u is mapped to a constant response for the reference, and to the output at steady state for $t\to\infty$ for the circuit. Another example identifies $D(\tau;\beta,u)=y(\tau)$ as the observations of Σ at time τ for a given input and parameter.

We measure the divergence between system and reference behavior by the *individual cost function*

$$s(\beta) = \frac{1}{|\mathcal{U}|} \sum_{u \in \mathcal{U}} \left|\left| D(\tau; \beta, u) - D^{\mathrm{ref}}(\tau; u) \right|\right| \ , \tag{2}$$

which averages some norm $||\cdot||$ between the system and reference behavior over the considered inputs.

In principle, the average cell design problem could be solved directly to identify the optimal average cell parameter $\beta^* = \mathrm{argmin}_\beta s(\beta)$. However, additional uncertainties arise due to unmodelled system components and from combining previously characterized biological parts into a circuit [11]. We account for these uncertainties by defining a threshold $\varepsilon > 0$ on the cost function to encode which solutions are 'good enough', and determine the *viable region* $V^{\mathrm{avg}} = \{\beta \in B \mid s(\beta) \leq \varepsilon\}$ of all parameters that fulfill this criterion. An output of the average cell design problem is then a description of V^{avg} rather than a single parameter.

Population Model. To capture cell-to-cell variability, we postulate a *population model*, where all cells share the same model structure Σ, but each cell i has its own parameter β_i drawn from a common *population distribution*

$$\beta_i \sim P_\gamma \tag{3}$$

with *population parameters* $\gamma \in \Gamma$. This is known as a *nonlinear mixed-effects model* and P_γ is often chosen to be a normal or log-normal distribution, in which case γ are the expected values and (co)variances of the parameters in β_i.

Population Design. The population model allows us to consider the distribution of behaviors of a circuit under cell-to-cell heterogeneity. In particular, each population parameter γ yields a specific distribution P_γ of the individual cell parameters β, and this induces a distribution over the values of the individual cost functions $s(\beta)$. The *population design problem* then consists of finding a population parameter that minimizes a corresponding *population cost function*, given by a functional

$$c : \{P_\gamma \mid \gamma \in \Gamma\} \to \mathbb{R}^+ \ . \tag{4}$$

For example $c(\gamma) = \mathbb{E}_\gamma(s(\beta))$ considers the expected value of the individual costs over the population, and $c(\gamma) = \mathbb{P}_{P_\gamma}(s(\beta) \geq \varepsilon) = \mathbb{P}_{P_\gamma}(\beta \notin V^{\mathrm{avg}})$ considers the percentage of cells whose behavior deviates from the reference by more than a user-defined threshold ε (cf. **Fig. 1B**); this percentage depends on the specific population distribution P_γ, and therefore on the population parameter γ.

Again, the population design problem can in principle be solved directly to yield $\gamma^* = \mathrm{argmin}_\gamma c(\gamma)$. Here, we again relax this problem and seek to identify the *population viable space* $V^{\mathrm{pop}} = \{\gamma \in \Gamma \mid c(\gamma) \leq \delta\}$ to account for additional uncertainties, where δ is again a user-defined parameter. In particular for design objectives such as requiring a minimal fraction of cells with 'acceptable' behavior that will have multiple optima, the population viable space also yields equivalent design alternatives.

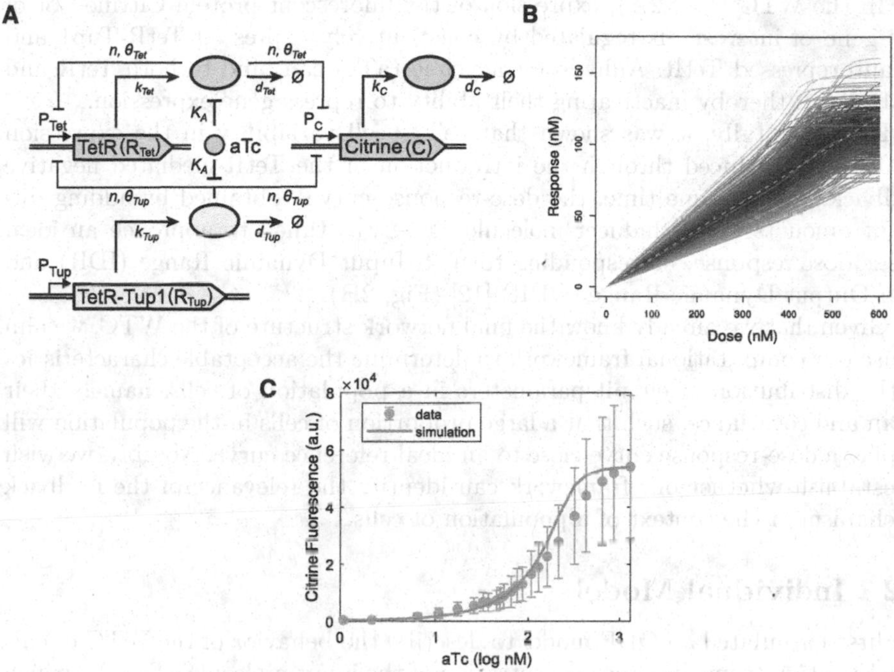

Fig. 2. Well-tempered controller (WTC) circuit. (A) Schematic representation of the circuit structure and its parametrization. Rectangles: genes with associated promoters; ellipses: proteins (corresponding color); bold lines with arrows: molecular reactions; normal lines with bar heads: regulatory interactions for inactivation. **(B)** Simulated dose-response curves of a population of cells for a given population parameter γ with a coefficient of variation $CV \approx 10\%$. Red line: median response; blue to purple lines: responses of individual cells colored by cost: the lower the cost, the darker the color; dashed orange line: reference linear dose-response curve, used to compute the individual cost. **(C)** Experimental and simulated aTc dose-response curve for the WTC. Blue: mean (circles) and standard deviation (error bars) of experimental data obtained by flow cytometry; green line: simulation results for the estimated parameter values in Table 1. Additionally, we used estimated values $d_C = 0.0031\ \text{min}^{-1}$, $d_{Tet} = 0.005\ \text{min}^{-1}$, $\theta_{Tet} = 1.2$ nM, and $\theta_{Tup} = 10^{-4}$ nM. To match the model output (Citrine concentration) to fluorescence (a.u.), we determined a scaling factor as in [10]. (Color figure online)

3 Case Study: Design of a Transcriptional Controller

3.1 Overview

To demonstrate the framework, we use a transcriptional controller termed well-tempered controller (WTC) that was experimentally designed by Azizoglu *et al.*

[1]. In the WTC (Fig. 2**A**), expression of the fluorescent protein Citrine—or of any gene of interest—is regulated by constitutively expressed TetR-Tup1 and by autorepressed TetR. Anhydrotetracycline (aTc) can bind to both TetR and TetR-Tup1, thereby inactivating their ability to repress gene expression.

Experimentally, it was shown that cell-to-cell variability in the expression of Citrine is reduced through the introduction of the TetR-mediated negative feedback. At the same time, the dose-response curve—obtained by adding different amounts of the inducer molecule aTc—was tuned to approach an ideal linear dose-response, corresponding to high Input Dynamic Range (IDR) and high Output Dynamic Range (ODR) [12] (Fig. 2**B**).

Given that we already know the final network structure of the WTC, we aim to use our computational framework to determine the acceptable characteristics of the distribution of circuit parameters in a population of cells, namely their mean and covariance, such that a large proportion of cells in the population will display a dose-response curve close to an ideal reference curve. Notably, we wish to establish whether our framework can identify the relevance of the feedback mechanism in the context of a population of cells.

3.2 Individual Model

We first formulated an ODE model to describe the behavior of the WTC circuit (see Fig. 2**A**). It involves the concentration of the input molecule aTc (a)—which can be added to the cell culture—and three states for the total concentrations of the repressor TetR (R_{Tet}), the repressor TetR-Tup1 (R_{Tup}) and the fluorescent protein Citrine (C):

$$\frac{\mathrm{d}R_{Tet}}{\mathrm{d}t} = \frac{k_{Tet}}{1+\left(\frac{f\cdot R_{Tet}}{\theta_{Tet}}\right)^n+\left(\frac{f\cdot R_{Tup}}{\theta_{Tup}}\right)^n} - d_{Tet}\cdot R_{Tet} \tag{5}$$

$$\frac{\mathrm{d}R_{Tup}}{\mathrm{d}t} = k_{Tup} - d_{Tup}\cdot R_{Tup} \tag{6}$$

$$\frac{\mathrm{d}C}{\mathrm{d}t} = \frac{k_C}{1+\left(\frac{f\cdot R_{Tet}}{\theta_{Tet}}\right)^n+\left(\frac{f\cdot R_{Tup}}{\theta_{Tup}}\right)^n} - d_C\cdot C. \tag{7}$$

Parameters k_{Tet}, k_{Tup} and k_C are maximal expression constants that capture both transcription and translation to keep the model simple. Parameters d_{Tet}, d_{Tup} and d_C are the degradation constants.

For TetR and Citrine production we added a control term representing a Hill function that depends on the active concentrations of the repressors TetR and TetR-Tup1. Active TetR and TetR-Tup1 molecules are those that are not bound to the inducer aTc. Assuming rapid equilibrium for the binding of aTc to TetR and TetR-Tup1 (as in Lormeau *et al.* [10]), the fraction of active TetR and TetR-Tup1 (f) is given by:

$$f = \frac{1}{2} - \frac{1+K_a a - \sqrt{(1+K_a(R_{Tet}+R_{Tup}-a))^2+4K_a a}}{2K_a(R_{Tet}+R_{Tup})}. \tag{8}$$

Experimental data showed that TetR and TetR-Tup1 have different repression efficiencies [1], represented by θ in the model. We therefore decided to model the action of the two repressors on their controlled genes as an 'OR'-gate. This means that we are not taking into account that the repressors might bind to the same DNA sequences. In contrast, we do not expect a difference in Hill coefficient (n) or affinity (K_a) to aTc between TetR and TetR-Tup1.

3.3 Population Model

To simplify computations, we fixed the means of 6 out of 10 parameters of the ODE model (see Table 1). To obtain these values, we estimated the model parameters using data from Azizoglu *et al.* [1] and additional data on the WTC's biological parts. As shown in Fig. **2C**, the parametrized WTC model captures the experimental dose-response curve.

The four remaining parameters ($d_{Tet}, d_C, \theta_{Tet}$, and θ_{Tup}) are the protein degradation constants, and the effective concentrations relative to the repression (including feedback) mechanisms. We fixed the mean value of d_{Tup} because this parameter is not identifiable together with θ_{Tup} using only steady-state information. If d_{Tup} were to be sampled along with θ_{Tup}, the strong negative correlation of these two parameters would not have any biological meaning. For the same reason, we fixed production constants and only allowed degradation constants to vary.

Regarding variances, only the production constants (k_{Tet}, k_{Tup}, k_C) and degradation constants (d_{Tet}, d_{Tup}, d_C) were assumed to display cell-to-cell variability. Without data on the variance of these parameters, we assumed that they all follow a log-normal distribution (to ensure positivity) based on the same variance σ^2 of the underlying Normal distribution. This implies that all the parameter distributions have the same coefficient of variation $CV = \sqrt{e^{\sigma^2} - 1}$. Since $\sigma \leq 0.1$ for our data, we use the approximation $CV \approx \sigma$ to simplify our analysis slightly.

3.4 Design Problem

Reference Dose-Response Curve. Our objective for the behavior of individual cells endowed with the WTC is a linear dose-response curve over an IDR of [0 nM, 600 nM] for aTc with a desired ODR of [0 nM, 120 nM]. We encode a dose-response curve as a reference behavior. It takes the aTc concentration a as a constant input $u(t) \equiv a$, and yields a constant response $D^{\text{ref}}(a) \equiv D^{\text{ref}}(\tau; a)$ for all τ. We encode the high-IDR, high-ODR objective by defining $(a, D^{\text{ref}}(a))$ to be the straight line between (0 nM, 0 nM) and (600 nM, 120 nM).

Individual Cost. To quantify the deviation between an individual cell's behavior and the reference curve, we use the individual cost from Eq. 2 based on the dose-response curve $(a, D(\tau; \beta, a))$, where cell i has individual parameter set $\beta_i = (k_{Tet}^{(i)}, k_{Tup}^{(i)}, k_C^{(i)}, d_{Tup}^{(i)}, n, K_a, d_{Tet}^{(i)}, d_C^{(i)}, \theta_{Tet}, \theta_{Tup})$, and $D(\tau; \beta, a) \equiv D(\beta, a)$ is the steady-state ($t \to \infty$) response to aTc concentration a. In our

Table 1. Parameter specifications for the WTC model. Parameters k_{Tet}, k_{Tup}, k_C, and d_{Tup} are cell-to-cell variable but their mean is fixed to the indicated value.

Fixed parameters					
Name	Description	Units	Fixed value		Cell-specific
k_{Tet}	Max production rate of TetR	$nM \cdot min^{-1}$	1.115		Yes
k_{Tup}	Max production rate of TetR-Tup1	$nM \cdot min^{-1}$	0.7919		Yes
k_C	Max production rate of Citrine	$nM \cdot min^{-1}$	0.8395		Yes
d_{Tup}	Degradation constant of TetR-Tup1	min^{-1}	1.2745		Yes
n	Hill coefficient for promoter repression by TetR and TetR-Tup1	(-)	1.5656		No
K_a	Association constant for TetR and TetR-Tup1 binding to aTc	nM^{-1}	144.37		No
Sampled parameters					
Name	**Description**	**Units**	**Bounds**	**Explored in log space**	**Cell-specific**
d_{Tet}	Degradation constant of TetR	min^{-1}	$[10^{-5}\ 10]$	Yes	Yes
d_C	Degradation constant of Citrine	min^{-1}	$[10^{-10}\ 10]$	Yes	Yes
θ_{Tet}	Repression coefficient TetR	nM	$[10^{-6}\ 10^{6}]$	Yes	No
θ_{Tup}	Repression coefficient TetR-Tup1	nM	$[10^{-6}\ 20]$	Yes	No

implementation, the individual cost function is calculated via a discrete version of the L_2-norm based on N aTc input doses $\mathcal{U} = \{a_1, \ldots, a_N\}$, regularly spaced between 0 and 600 nM:

$$s(\beta) = \sqrt{\frac{1}{N}\sum_{k=1}^{N}\left(D(\beta, a_k) - D^{\text{ref}}(a_k)\right)^2}. \tag{9}$$

We consider an individual cell's dose-response acceptable if $s(\beta) \leq \varepsilon$; the corresponding parameters β constitute the viable space. For our analysis, we use $\varepsilon = 5$ nM and $\varepsilon = 2$ nM, which represent approximately 5% and 2% of the ODR we wish to achieve, respectively.

Population Cost. For our population design, we consider the percentage of individual cells in a population with parameter γ that fulfill the criterion Eq. 9 as our population cost function:

$$c(\gamma) = \mathbb{P}_{P_\gamma}(s(\beta) \geq \varepsilon) \,. \tag{10}$$

We define the population viable space as those γ that yield at least 80% individual cells with behavior sufficiently close to the reference and $c(\gamma) \leq 20\%$.

We estimate the population cost by drawing individual parameter sets β_i from the distribution P_γ and by determining the proportion of sampled parameter sets that yield acceptable individual costs.

Sampling in Parameter Spaces. We sampled from both the individual parameter space and the population parameter space, according to the individual cost s and the population cost c, respectively. We used an adaptive version of the Metropolis-Hastings algorithm [6] in both cases, implemented in the R [15] package 'fmcmc' [20], with pseudo-likelihoods based on individual cost and population cost. The package 'deSolve' [19] was used to solve the ODE model, with derivatives computed in C code. We defined the pseudo-likelihood for the individual parameter space as:

$$l(\beta) = \mathbb{1}(s(\beta) \leq \epsilon) \tag{11}$$

with $\epsilon \in \{5\,\text{nM}, 2\,\text{nM}\}$, therefore sampling uniformly the viable region $V^{\text{avg}} = \{\beta \in B \mid s(\beta) \leq \epsilon\}$. The pseudo-likelihood for the population parameter space was:

$$L(\gamma) = \mathbb{1}(c(\gamma) \leq \delta) \tag{12}$$

with $\delta = 0.2$. We then obtain uniformly distributed samples from the population viable space $V^{\text{pop}} = \{\gamma \in \Gamma \mid c(\gamma) \leq \delta\}$. Note that, as $c(\gamma)$ depends on the value of ε (Eq. 10), $L(\gamma)$ and the associated population viable space will also depend on its value.

To compute this population pseudo-likelihood, however, we need to approximate $c(\gamma)$, as it is the functional of a distribution (in this case study, a probability). For each value of γ, 300 individual parameters were drawn randomly from the underlying log-normal distribution P_γ. For each individual parameter vector, we computed the individual cost s and approximated $c(\gamma)$ as the fraction of samples with individual costs above the corresponding threshold $\epsilon \in \{5\,\text{nM}, 2\,\text{nM}\}$. Note that we are interested in the resulting distribution of the individual costs and not in describing P_γ. Thus, even though we consider 6 cell-to-cell variable parameters, a sample size of 300 proved sufficient to reliably represent this distribution of individual costs as the underlying distance measure between a constant reference and the output of an ODE model is sufficiently smooth. An illustration is given in Fig. 2**B**, where 300 individual dose-response curves from a population distribution with high coefficient of variation cover the graph sufficiently.

The log-normal population distribution for our example allows us to reduce the required amount of random sampling and to provide more consistent results for the approximation of the population pseudo-likelihood. Note that we can reconstruct the mean vector $\mu \in \mathbb{R}^6$ and the 6×6 covariance matrix C of the underlying multivariate Normal distribution from the population parameter γ. We therefore once generated 300 samples S_i from the standard multivariate Normal distribution $N(0, I)$ in $\mathbb{R}^6$. For each value of γ, we constructed the corresponding samples of the individual parameters as $\beta_i = \mu + C^{1/2} \cdot S_i$, where $C^{1/2}$ is the lower triangular matrix from a Cholesky decomposition of C. This ensures that repeated calls to our approximation of the population cost function with the same population parameter γ yields the same cost and requires only a single sample of size 300. On a standard laptop with Intel i7 processor, we obtained $\approx$900 samples from the population space per hour, corresponding to $\approx 2.7 \cdot 10^5$ samples from the individual parameter space.

Example. To illustrate the interplay between the individual and the population level in our design problem, Fig. **2B** shows an example of the dose-response relationship of the WTC model for a population of cells. The NLME formulation takes into account the variance in parameters, that is, cell-to-cell variability. Here, although the median response is close enough to the ideal response, approximately 83% of the response curves are not within the acceptable range due to variance in the individual parameters. This leads to a population cost of $\approx$0.83, given an individual cost threshold of $\varepsilon = 5$ nM (corresponding to approximately 5% deviation from the reference curve). The example illustrates a key difference between traditional design assuming an 'average' cell and population design. If the design objective were to achieve a median response close to the reference, the example would be a valid solution, although the vast majority of individual cells would not comply with the design objective.

3.5 Sampling the Individual Parameters

Figure 3 shows the results of sampling the individual parameter space according to the value of the individual cost $s(\cdot)$. We first note that the protein degradation constant of Citrine, d_C, displays a substantially narrower marginal distribution than all other parameters. Citrine is the system response, and therefore this distribution shape is not surprising: with all other parameters kept identical, a change in d_C will directly impact the shape of the dose-response curve.

In the two-dimensional projections of the joint distribution over the individual viable space V^{avg}, the two parameters for protein degradation, d_{Tet} and d_C, are correlated, but mainly in the high-viability region. This indicates that either of the two parameters could be used to fine-tune the circuit.

Most importantly, the pattern of the projection across $(\theta_{Tet}, \theta_{Tup})$, which capture the strength of transcriptional repression, reveals insights into the relevance of negative feedback for WTC performance. Specifically, θ_{Tet} is the parameter for auto-repression, whereas θ_{Tup} is the parameter for constitutive repression. A smaller value of θ_{Tet} (resp. θ_{Tup}) means a stronger auto-repression (resp. constitutive repression). For the viability threshold of 5 nM used to define the viable space for the data shown in Fig. 3, most values for both θs are allowed, including high values that would effectively nullify the corresponding repressive

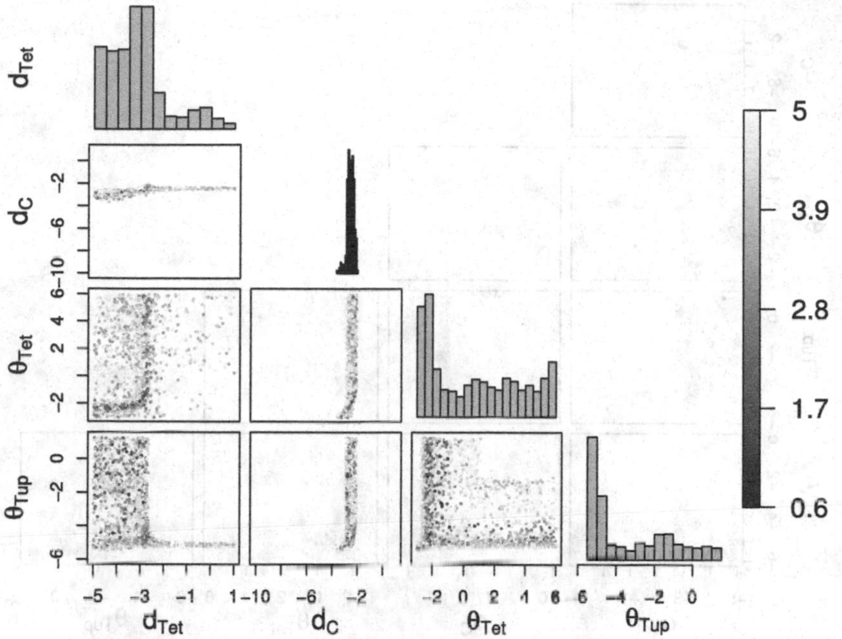

Fig. 3. Viable samples in the individual parameter space. Histograms show marginal distributions, and scatter plots samples in all two-dimensional projections of the parameter space. In the projections, samples are colored according to their individual cost from light blue to purple: a darker blue indicates a lower cost, and thus a higher consistency of the WTC dose-response with the reference curve for a given point. Only the parameters present in the plot were allowed to vary, all others were fixed to values specified in Table 1. Additionally, all parameters were sampled in $\log_{10}$-scale, and are displayed as such. (Color figure online)

effect. However, the upper right quadrant does not contain viable samples, indicating that at least one type of repression is needed for the circuit to achieve the desired behavior. Importantly, samples with lower values of the individual cost are located in the region of low θ_{Tet} (notice the color gradient). If we wish to achieve even closer correspondence of the WTC's dose-response with the reference curve for an individual cell (e.g., with an individual threshold $\varepsilon = 2\,\text{nM}$), auto-repression becomes mandatory. Note as well that θ_{Tet} becomes strongly correlated with both degradation constants d_{Tet} and d_C, whenever auto-repression is strong. This is logical because auto-repression reduces the mean expression of TetR and Citrine, and should thus be compensated for by lower degradation constants to keep mean expressions in the desired range.

3.6 Sampling the Population Parameters

We next applied the population design framework described in Sect. 2 to the WTC model. Our aim is to obtain design guidelines for a reasonably good tran-

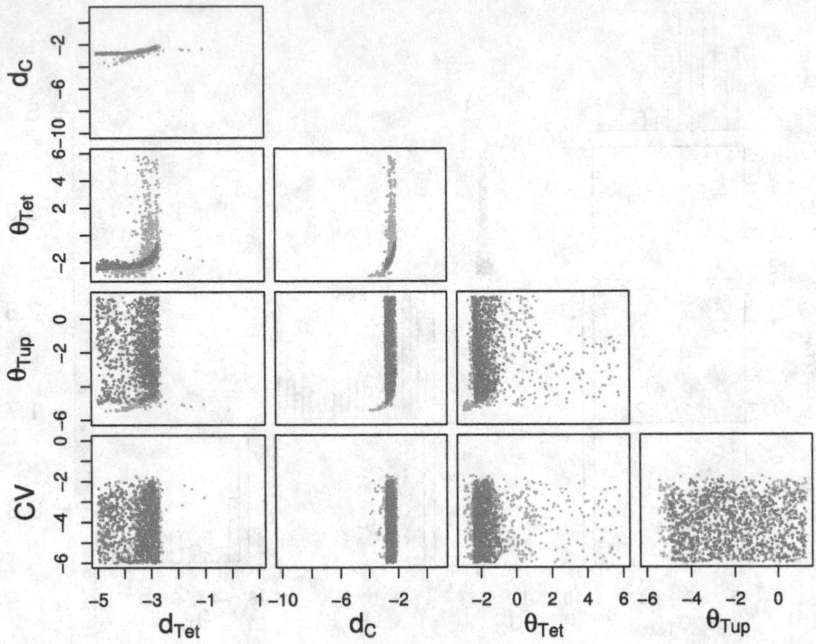

Fig. 4. Viable samples in the population parameter space. Samples in all two-dimensional projections of the parameter space; note that CV is the common coefficient of variation for all cell-to-cell variable population parameters. Orange dots: viable samples for the threshold on the individual cost $\varepsilon = 5\,\text{nM}$; red dots: viable samples for $\varepsilon = 2\,\text{nM}$. All parameters are in $\log_{10}$-scale.(Color figure online)

scriptional controller with low cell-to-cell variability in the steady-state dose-response, which we encode via the population cost Eq. 10 with a threshold of $\delta = 20\%$. The resulting samples according to Eq. 10 are shown in Fig. 4 for the two individual cost thresholds $\varepsilon = 5\,\text{nM}$ (orange) and $\varepsilon = 2\,\text{nM}$ (red). For both values of ε, we ran the Markov Chain Monte-Carlo (MCMC) chain twice from two different starting points. This explains the apparent density differences between regions, particularly visible in the planes (d_{Tet}, θ_{Tup}) and (d_{Tet}, CV).

Compared to the individual parameter samples (Fig. 3), we observe a clear upper bound of about $10^{-2.5}\text{min}^{-1}$ for the population mean of d_{Tet}, which needs to be considered in the population design of the transcriptional controller.

Moreover, we find two distinct 'modes' of parameter combinations that lead to the desired population behavior, clearly visible in the (d_{Tet}, θ_{Tet}) panel of Fig. 4: if the average degradation constant d_{Tet} is large enough, this process alone ensures a level of $TetR$ compatible with the desired output and the auto-repression with θ_{Tet} can be chosen almost arbitrarily. Conversely, a low degradation constant requires strong auto-repression to achieve the population behavior, and thus low values for θ_{Tet}. These two modes are connected via a region with strong correlation between these parameters, indicating that both parameters need to be tuned simultaneously to achieve the population behavior in this region. In contrast, a strong auto-repression cannot compensate for a low degradation of Citrine (parameter d_C), while sufficiently high degradation of Citrine does not require tuning the auto-repression constants, as seen in panels (d_C, θ_{Tet}) and (d_C, θ_{Tup}).

To generate the data in Fig. 4, we allowed the coefficient of variation (CV, which is multiplicative in linear space) to vary up to a value of one. However, viable samples are essentially all below 0.02, pointing to this value as a possible maximum for the admissible cell-to-cell variability for reaching the design objective under the model's assumptions. For future studies, it is, hence, of interest to experimentally quantify the cell-to-cell variability of the parameters, and check the results against our inferred value. Note, however, that higher coefficients of variation would be allowed in the presence of negative correlations between parameters. In the plane (θ_{Tet}, CV) exists also a slightly decreasing slope for the case $\varepsilon = 5\,\text{nM}$: when the value of θ_{Tet} increases, leading to weaker auto-repression, the maximum admissible value for the coefficient of variation decreases. Indeed, the maximum CV for all samples has a value of ≈1.8%, whereas the maximum CV for the samples fulfilling the condition $\theta_{Tet} > 10^4\,\text{nM}$ is only ≈0.45%. This indicates that auto-repression can help compensate for cell-to-cell variability.

Regarding the repression parameters, θ_{Tet} and θ_{Tup}, we observe what could be expected from the individual samples: for $\varepsilon = 5\,\text{nM}$, the pattern of the projection of the samples over the plane $(\theta_{Tet}, \theta_{Tup})$ is very similar, if not exactly identical, to the one observed in Fig. 3. When the individual threshold is decreased to $\varepsilon = 2\,\text{nM}$, the viable region is reduced to low values of θ_{Tet}, indicating that auto-repression becomes necessary to achieve the design objective. Just as in the individual parameter space, θ_{Tet} strongly correlates with both degradation constants for low θ_{Tet}, *i.e.* for strong auto-repression. This applies particularly for $\varepsilon = 2\,\text{nM}$. If auto-repression is mandatory in a circuit, as here, particular attention should be given to tuning repression constants and degradation constants together.

Finally, we assumed that neither of the two repression parameters θ_{Tet} and θ_{Tup} displays cell-to-cell variability because the corresponding (microscopic) binding affinities are related to protein and DNA sequences that should be identical in each cell of an isogenic population. To assess the impact of this assumption, we performed an analogous sampling where the two parameters were assumed to vary from cell to cell just as the production and degradation constants; this yielded results very similar to Fig. 4 (data not shown).

4 Discussion

Nearly all current methods for synthetic circuit design assume an 'average' cell that needs to be optimized to fulfill the design objectives, potentially by considering parameter variations to achieve robustness of the biological implementation [10]. Stochastic design frameworks that account for cell-to-cell variability due to intrinsic noise with low molecule copy numbers are beginning to emerge, but computational complexity currently limits them to small networks, steady-state, and homogeneous model parameters in a cell population [18]. Here, we therefore proposed population design via NLMEs as an alternative to both approaches. We argue that it has the potential to bring information about cell-to-cell variability to synthetic biological design in realistic settings, and to help infer the impact of said variability on the system of interest.

Our case study considers a problem synthetic circuit designers often face, namely to tune their system in order to reduce cell-to-cell variability [1]. For the WTC, the population sampling highlighted the importance of fine-tuning jointly the degradation constant of TetR and its auto-repression constant to achieve low cell-to-cell variability—the parameters could assume a wider range of values to achieve mere individual cell viability. Feedback mechanisms were necessary in both cases, at least under our assumption of a common variance parameter. This indicates that constitutive repression, and even more so auto-repression, are useful to linearize dose-response curves of individual cells. While constitutive repression had no impact on cell-to-cell variability, auto-repression could increase the admissible CV from $\approx$0.45% to $\approx$1.8%. However, we could not achieve higher values of the CV, most likely because variability reduction is directly linked to repression strength: increasing repression would decrease cell-to-cell variability as well as mean expression of the repressed component. To weaken or eliminate this link between mean and variability, one may need to consider more complex topologies [3]. Note also that we limited our analysis to a small number of dimensions. Future studies could include more parameters or allow all variance parameters to be sampled independently. With independently sampled variances, it would be particularly interesting to see how autorepression affects (presumably relaxes) variance constraints across the network.

One limitation of our study (and an impediment to the extended analysis of the WTC) is the sampling technique we used. MCMC sampling does not scale well with dimensions, but one could use dedicated methods for sampling in higher dimensional spaces [22] instead. We also noted a tendency of the MCMC chain to get stuck in some parameter regions for population sampling, thus requiring multiple starting points to explore the whole space; this was not needed for the individual parameter space. However, keeping in mind that the number of variance parameters (including correlations) grows quadratically with the number of individual parameters, it is likely that one will not be able to tune the variance of each parameter individually. As a possible strategy, one could fix the covariance matrix to an experimentally determined one, for example, by using well-established NLME inference approaches [4,8] to obtain a parametrization of the cell-to-cell variability of biological parts. Other (not mutually exclusive) alternatives include the use of approximation methods for the individual cost [16] and of small sets sampling techniques such as the sigma-point approximation [9]. A different approach could be to replace exact MCMC sampling by approximate methods. For example, variational inference can be much faster than MCMC and still provide accurate results, provided that the correlation structure of the likelihood is properly accounted for [5].

For the present case study, we explored the population parameter space of a network topology we knew should work for some parameter values. In the broader context of synthetic biology, a working, simple topology that has the potential to achieve the design objective is not necessarily known. In many cases, one may want to explore different topologies and select the one that performs best while still being simple enough. To achieve this goal while taking into account cell-to-cell variability, we propose to apply the method described by Lormeau *et al.* [10] to the objective function defined at the population level. Briefly, the algorithm will explore a number of possible topologies by simplifying an initial (complex) starting network, removing its edges. The viability (existence of parameters making the network viable) of each network is assessed. One can then choose robust networks according to the size of the viable region, for instance. The case study presented here only aimed at providing a first insight into the relevance of sampling from the population viable space, but we did not sample the population viable space for multiple topologies. However, our findings for the WTC on the importance of feedback mechanisms (to achieve the design objective, without impacting cell-to-cell variability) and of fine-tuning TetR degradation (to reduce cell-to-cell variability) indicate that the concept is promising.

Overall, the population design framework could then be used to recommend network structures, together with their parameter values, that are best suited to fulfill a design objective incorporating cell-to-cell variability. Such an approach could also help exploring situations where cell-to-cell variability and a given distribution over behaviors of cells in a population is desirable. One example is bet-hedging in bacterial populations, where non-genetic variability across a population increases the chances of survival in the face of antibiotics [13].

5 Conclusion

We propose a general framework we call population design that aims to help biologists interested in synthetic circuit design to account for cell-to-cell variability via ODE-based NLMEs. We implemented a simple version of the concept and demonstrated its usefulness for a transcriptional controller in an a posteriori case study. The current implementation is restricted to small models with few parameters. We hope to augment it with advanced numerical methods and extend it to the problem of topology design. In perspective, this could enable the rational design of synthetic gene circuits that induce prescribed (distributions of) behaviors at the population level, and thereby allow to exploit cell-to-cell variability for novel applications.

Acknowlegements. We thank Asli Azizoglu and Claude Lormeau for discussions. This work was supported in part by the Swiss National Science Foundation via the NCCR Molecular Systems Engineering (grant 182895).

References

1. Azizoğlu, A., Brent, R., Rudolf, F.: A precisely adjustable, variation-suppressed eukaryotic transcriptional controller to enable genetic discovery. bioRxiv p. 2019.12.12.874461 (2020). https://doi.org/10.7554/eLife.69549
2. Barnes, C.P., Silk, D., Sheng, X., Stumpf, M.P.H.: Bayesian design of synthetic biological systems. Proc. Nat. Acad. Sci. **108**(37), 15190–15195 (2011). https://doi.org/10.1073/pnas.1017972108
3. Bonny, A.R., Fonseca, J.P., Park, J.E., El-Samad, H.: Orthogonal control of mean and variability of endogenous genes in a human cell line. Nature Commun. **12**(1), 1–9 (2021). https://doi.org/10.1038/s41467-020-20467-8
4. Dharmarajan, L., Kaltenbach, H.M., Rudolf, F., Stelling, J.: A simple and flexible computational framework for inferring sources of heterogeneity from single-cell dynamics. Cell Syst. **8**(1), 15–26.e11 (2019). https://doi.org/10.1016/j.cels.2018.12.007
5. Ghosh, S., Birrell, P., De Angelis, D.: Variational inference for nonlinear ordinary differential equations. In: Banerjee, A., Fukumizu, K. (eds.) Proceedings of The 24th International Conference on Artificial Intelligence and Statistics. Proceedings of Machine Learning Research, vol. 130, pp. 2719–2727. PMLR (2021). http://proceedings.mlr.press/v130/ghosh21b.html
6. Haario, H., Saksman, E., Tamminen, J.: An adaptive Metropolis algorithm. Bernoulli **7**(2), 223–242 (2001). https://doi.org/10.2307/3318737
7. Karamasioti, E., Lormeau, C., Stelling, J.: Computational design of biological circuits: putting parts into context. Mol. Syst. Des. Eng. **2**(4), 410–421 (2017). https://doi.org/10.1039/C7ME00032D
8. Lavielle, M.: Mixed effects models for the population approach: models, tasks, methods, and tools. CPT: Pharmacometrics Syst. Pharmacol. **4**(1), (2015). https://doi.org/10.1002/psp4.10
9. Loos, C., Moeller, K., Fröhlich, F., Hucho, T., Hasenauer, J.: A hierarchical, data-driven approach to modeling single-cell populations predicts latent causes of cell-to-cell variability. Cell Syst. **6**(5), 593–603.e13 (2018). https://doi.org/10.1016/j.cels.2018.04.008
10. Lormeau, C., Rudolf, F., Stelling, J.: A rationally engineered decoder of transient intracellular signals. Nature Commun. **12**(1), 1886 (2021). https://doi.org/10.1038/s41467-021-22190-4
11. Lormeau, C., Rybiński, M., Stelling, J.: Multi-objective design of synthetic biological circuits. IFAC-PapersOnLine **50**(1), 9871–9876 (2017). https://doi.org/10.1016/j.ifacol.2017.08.1601
12. Mannan, A.A., Liu, D., Zhang, F., Oyarzún, D.A.: Fundamental design principles for transcription-factor-based metabolite biosensors. ACS Synth. Biol. **6**(10), 1851–1859 (2017). https://doi.org/10.1021/acssynbio.7b00172
13. Martín, P.V., Muñoz, M.A., Pigolotti, S.: Bet-hedging strategies in expanding populations. PLOS Comput. Biol. **15**(4), e1006529 (2019). https://doi.org/10.1371/journal.pcbi.1006529
14. Nielsen, A.A.K., et al.: Genetic circuit design automation. Science **352**, 6281 (2016). https://doi.org/10.1126/science.aac7341
15. R Core Team: R: A Language and Environment for Statistical Computing. R Foundation for Statistical Computing, Vienna, Austria (2021). https://www.R-project.org/
16. Rasmussen, C.E., Williams, C.K.I.: Gaussian Processes for Machine Learning. MIT Press, Cambridge, Mass, Adaptive Computation and Machine Learning (2006)

17. Ryan, E.G., Drovandi, C.C., McGree, J.M., Pettitt, A.N.: A review of modern computational algorithms for Bayesian optimal design. Int. Stat. Rev. **84**(1), 128–154 (2016). https://doi.org/10.1111/insr.12107
18. Sakurai, Y., Hori, Y.: Optimization-based synthesis of stochastic biocircuits with statistical specifications. J. Royal Soc. Interface **15**,(2018). https://doi.org/10.1098/rsif.2017.0709
19. Soetaert, K., Petzoldt, T., Setzer, R.W.: Solving differential equations in R: Package deSolve. J. Stat. Soft. **33**(9), 1–25 (2010). https://doi.org/10.18637/jss.v033.i09
20. Vega Yon, G., Marjoram, P.: fmcmc: A friendly MCMC framework. J. Open Source Softw. **4**(39), (2019). https://doi.org/10.21105/joss.01427
21. Voigt, C.A.: Synthetic biology 2020–2030: six commercially-available products that are changing our world. Nature Commun. **11**, 6379 (2020). https://doi.org/10.1038/s41467-020-20122-2
22. Zamora-Sillero, E., Hafner, M., Ibig, A., Stelling, J., Wagner, A.: Efficient characterization of high-dimensional parameter spaces for systems biology. BMC Syst. Biol. **5**(1), 142 (2011). https://doi.org/10.1186/1752-0509-5-142
23. Češka, M., Dannenberg, F., Paoletti, N., Kwiatkowska, M., Brim, L.: Precise parameter synthesis for stochastic biochemical systems. Acta Informatica **54**(6), 589–623 (2017). https://doi.org/10.1007/s00236-016-0265-2

Nonlinear Pattern Matching in Rule-Based Modeling Languages

Tom Warnke[(✉)] and Adelinde M. Uhrmacher

Institute for Visual and Analytic Computing, University of Rostock, Rostock, Germany
{tom.warnke,adelinde.uhrmacher}@uni-rostock.de

Abstract. Rule-based modeling is an established paradigm for specifying simulation models of biochemical reaction networks. The expressiveness of rule-based modeling languages depends heavily on the expressiveness of the patterns on the left side of rules. Nonlinear patterns allow variables to occur multiple times. Combined with variables used in expressions, they provide great expressive power, in particular to express dynamics in discrete space. This has been exploited in some of the rule-based languages that were proposed in the last years. We focus on precisely defining the operational semantics of matching nonlinear patterns. We first adopt the usual approach to match nonlinear patterns by translating them to a linear pattern. We then introduce an alternative semantics that propagates values from one occurrence of a variable to other ones, and show that this novel approach permits a more efficient pattern matching algorithm. We confirm this theoretical result by benchmarking proof-of-concept implementations of both approaches.

Keywords: Rule-based modeling · Pattern matching · Formal semantics

1 Introduction

Rule-based modeling is an established paradigm for specifying (simulation) models of biochemical reaction networks [9]. By using patterns on the left side of rules, a single rule can express a whole class of reactions, which can additionally be parametrized with the variables matched in the pattern. Simulation algorithms find the reactions possible in a given state by matching the patterns to the current model state. The expressiveness of a given rule-based modeling language and the efficiency of the corresponding simulation algorithm depends on what kind of patterns are allowed. In this paper, we study a particular kind of patterns: patterns in which variables are allowed to occur multiple times, so called *nonlinear patterns.*

Rule-based languages employ the patterns on the left rule side to constrain the reactants participating in a reaction. Thus, nonlinear patterns allow expressing the relation between reactants through common attribute values. For example, a pattern like $A(x) + A(x) \rightarrow \dots$ can be used to express that two A entities can only react if they share an attribute value. The attribute could encode

E. Cinquemani and L. Paulevé (Eds.): CMSB 2021, LNBI 12881, pp. 198–214, 2021.
https://doi.org/10.1007/978-3-030-85633-5_12

the enclosing compartment, effectively requiring that reactants are located in the same compartment [15]. More generally, nonlinear patterns facilitate expressive left rule sides, in particular if patterns are also allowed to include expressions. Then neighboring locations can be described with a simple pattern like $C(x) + C(x + 1)$. However, more complex patterns require more sophisticated pattern matching algorithms that are able to find all pattern matches efficiently.

So far, nonlinear patterns have not been studied explicitly in the context of (simulation) modeling languages. There is some work on nonlinear patterns in the context of term rewriting, for example for computer algebra systems and other kinds of symbolic computing [16]. Algorithmically, nonlinear patterns are typically matched in a linearized form and the matches are subsequently refined [2]. Oury and Plotkin [20] have proposed a modeling language rooted in term rewriting, but require patterns to be linear. The Kappa calculus [6] and the BioNetGen language [5] both employ nonlinear patterns to denote existing graph edges (without calling them such).

In this paper, we study nonlinear patterns as an important ingredient for rule-based modeling languages. We show that the expressiveness of modeling languages benefits from including nonlinear patterns, and discuss algorithms to match nonlinear patterns in simulation algorithms. Our contributions are as follows:

- We illustrate the usefulness of nonlinear patterns for modeling in computational biology by surveying examples from existing modeling languages.
- Based on an introduction to rule-based modeling with linear patterns, we give two definitions of operational semantics for nonlinear pattern matching. The first approach is the one most commonly used in literature and based on translating nonlinear patterns to linear ones (*linearization*). As an alternative, we introduce an *inline substitution* approach that propagates matches from one occurrence of a variable to other occurrences.
- We present two algorithms implementing both semantics definitions and characterize their performance by benchmarking proof-of-concept implementations. The results suggest that the inline substitution approach allows more efficient pattern matching than the traditional linearization approach.

A proof-of-concept Scala implementation of the algorithms discussed in this paper is available at https://git.informatik.uni-rostock.de/mosi/pattern-matching.

2 Rule-Based Modeling

We start with a short informal introduction of rule-based modeling. Fundamentally, two types of rule-based modeling languages can be distinguished. Some languages, most prominently Kappa [6] and BNGL [5], are based on the idea of graph rewriting. In those languages, entities and their bindings form attributed graphs, and reaction rules use graph patterns to describe the reactants. Mapping such reaction rules to stochastic graph rewriting rules gives a straightforward interpretation to reaction rules. Textual description of graph patterns in those languages employ variables to denote edges. Variables always occur once or twice to denote one end or both ends of a graph edge, which maps directly to graph patterns and requires no special treatment.

Whereas graph rewriting languages like Kappa and BNGL use pattern variables to denote edges, multiset rewriting languages can use variables with more complex domains. Here, the reaction rules operate on multisets of attributed entities (often called a *solution*). For example, attributes might denote a position in continuous space [4] or, with an entity denoting a compartment, the contained entities [19,21]. In this paper, we focus on languages based on multiset rewriting with simple integer-valued attributes. Examples for multiset rewriting languages with formal definitions are CSMMR [20], *React*(C) [15], ML-Rules [24], or Chromar [14]. The foundation of this approach is equational (associative and commutative) term rewriting, augmented with stochastics [1,21]. Thus, multisets are typically denoted as terms of entities, where entities are written as function symbols with attribute values as their arguments. For example, given a species A with two integer attributes, a multiset $\{\!\!\{A(1,1), A(1,2), 2A(2,2)\}\!\!\}$ with four entities can be written as $A(1,1) + A(1,2) + 2A(2,2)$. A pattern on the left side of a reaction rule may then use variables for attributes, and these variables can then be used on the right rule side. For example, a rule for the solution above could be $A(x,1) + A(x,y) \rightarrow A(x,y+1)$. The pattern on the left rule side is nonlinear, as the variable x occurs twice. It matches the solution above with the variable substitution $x = 1, y = 2$, leading to the successor state $\{\!\!\{A(1,3), 2A(2,2)\}\!\!\}$. In addition, patterns can contain expressions. For example, a pattern $A(x, x+1)$ is nonlinear, as x occurs twice, but the second occurrence of x is wrapped in an expression.

Syntactically, the different rule-based modeling languages based on multiset rewriting differ. For example, some languages use a comma instead of a plus to denote multiset patterns. Another difference is that some languages use nominal pattern matching (attributes are addressed by name), whereas other languages use structural pattern matching (attributes are addressed by position). Semantically, the difference between these approaches is minimal [14]. In this paper, we use the notation as above, with a plus symbol for multiset addition and structural pattern matching.

Rule-based modeling languages use reaction rules to denote reactions, that is state transitions of the model. Typically, reaction rules are equipped with a rate expression, which might depend on the variables matched on the left rule side, associating each resulting state transition with a propensity. A continuous-time Markov chain (CTMC) is then defined by interpreting the propensities as the rate of an exponential distribution of the waiting time until the corresponding reaction occurs [13]. In addition, mass-action kinetics are often applied to factor the multiplicity of reactants into the propensity. Simulation runs of a rule-based model can be executed by sampling a path through the CTMC with a stochastic simulation algorithm (SSA) [11]. It is important to note that the CTMC is only well-defined when state transitions with the same source and target state are conflated by adding their rate [7]. SSAs, however, operate correctly also if parallel transitions are not conflated. Therefore, we will not consider this caveat any further in this paper.

3 Nonlinear Patterns in the Wild

The advantages of rule-based modeling are most evident if a single rule can denote a large number of reactions. Common examples of such situations are discrete spaces such as 1D or 2D grids, where one rule describes similar reactions in all locations. For example, the publications introducing Chromar [14] and CSMMR [20] apply the respective languages to model a sequence of locations in which reactions occur and between which entities diffuse. The presentation of ML-Rules [19] and the protocol of a modeling language "contest" at a Dagstuhl seminar [10, Sect. 4.4.6] feature entities reacting and diffusing in a two-dimensional grid.

There are two ways to express such models [20], and both are related to nonlinear patterns. The first variant equips each entity with attributes representing its location. Then diffusion can be very easily modeled as the change of those attributes. For example, an entity in a one-dimensional sequence of locations diffusing from one location to the neighboring one can be expressed as $A(x) \rightarrow A(x+1)$. However, describing a reaction that occurs between entities in the same location requires a nonlinear pattern: $A(x) + A(x) \rightarrow B(x)$. Alternatively, the locations can be modeled explicitly as compartments containing entities, where the compartments now carry their location as attributes. Then, a reaction between colocated entities is easily expressed as $C(x, A+A+\mathit{rest}) \rightarrow C(x, B+\mathit{rest})$. But in this approach, diffusion requires a nonlinear pattern to capture neighboring compartments: $C(x, A+\mathit{rest}_1)+C(x+1, \mathit{rest}_2) \rightarrow C(x, \mathit{rest}_1)+C(x+1, A+\mathit{rest}_2)$. Note that the location attribute of the second compartment C is matched by the expression $x+1$.

In modeling languages that do not support nonlinear patterns, these situations are typically resolved by linearizing the left rule side. For example, the pattern $A(x) + A(x)$ is translated to $A(x) + A(y)$ with the additional constraint that $x = y$. Compared to the formulation as a nonlinear pattern, this has two disadvantages. First, the rule is more complex—it has two variables instead of one as well as an additional constraint. Second, matching the pattern is less efficient, as all combinations of values for x and y need to be checked for whether they satisfy the constraint [20].

Although the significance of nonlinear patterns is evident from the examples above, so far the research on rule-based modeling has not focused on them. In particular, their semantics and their impact on pattern matching algorithms have not been studied. In the next sections, we formalize the intuitions above and precisely define linear and nonlinear pattern matching for rule-based modeling languages. We also relate these definitions to the handling of nonlinear patterns in existing languages.

4 Linear Pattern Matching

We define the syntax and semantics of a "canonical" rule-based modeling language based on multiset rewriting. This is explicitly not intended as the definition of a novel multiset rewriting language, but rather as a unifying formalization that captures this notion of rule-based modeling. We start with a simple version of the language, with linear patterns and no expressions on the left rule side. In the next section, we add both these features to illustrate the difference in syntax and semantics.

4.1 Abstract Syntax

First, we assume a number of attributed (chemical) *species* with a non-negative arity. Let the set of species be $\mathcal{S}$ and $ar : S \to \mathbb{N}_0$ be the function mapping each species to its arity. Thus, each *entity* belonging to species $S \in \mathcal{S}$ has $ar(S)$ attributes. The values of the entity's attributes come from a set $\mathcal{V}$—we do not type attributes here to keep the definition simple. Let $\{\!\!\{\}\!\!\}$ denote the empty multiset, and $\uplus$ and $\ominus$ denote multiset addition and subtraction, respectively. Further, for a multiset $\mathcal{A}$ and an element a, let $n_{\mathcal{A}}(a)$ be the multiplicity of a in $\mathcal{A}$ (how often a is contained in $\mathcal{A}$), and $a \in \mathcal{A}$ iff $n_{\mathcal{A}}(a) \geq 1$.

Now we can define a state of our model, a *solution*, as a multiset of entities:

$$
\begin{aligned}
ent &:= S(v_1, \ldots, v_n) && S \in \mathcal{S}, n = ar(S), v_1, \ldots, v_n \in \mathcal{V} \\
sol &:= \{\!\!\{ent\}\!\!\} \uplus sol \\
&\mid\ \ \{\!\!\{\}\!\!\}
\end{aligned}
$$

A rule is then a triple of a pattern on the left rule side, a rate expression, and products on the right rule side. The pattern can contain variables and plain values for attributes, the products can also contain expressions for attributes. To define this, we additionally need a set $\mathcal{X}$ of pattern *variables*. We also assume some set $\mathcal{E}$ of expressions over $\mathcal{V}$ and $\mathcal{X}$. The syntax also defines a dedicated symbol $\emptyset$ to explicitly denote empty left or right sides of rules. We will usually omit the $\emptyset$ when writing nonempty left or right rule sides.

$$
\begin{aligned}
a &:= v \mid x && v \in \mathcal{V}, x \in \mathcal{X} \\
pat &:= S(a_1, \ldots, a_n) + pat && S \in \mathcal{S}, n = ar(S) \\
&\mid\ \ \emptyset \\
prod &:= S(e_1, \ldots, e_n) + prod && S \in \mathcal{S}, n = ar(S), e_1, \ldots, e_n \in \mathcal{E} \\
&\mid\ \ \emptyset \\
rule &:= pat \xrightarrow{e} prod && e \in \mathcal{E}
\end{aligned}
$$

The definition of the rule syntax is not complete without restricting which variables may occur where. Let *var* be a function that maps each term to the set of contained variables:

$$
\begin{array}{ll}
var(v) & := \{\} \\
var(x) & := \{x\} \\
var(S(a_1,\ldots,a_n)+pat) & := var(a_1)\cup\ldots\cup var(a_n)\cup var(pat) \\
var(S(e_1,\ldots,e_n)+prod) & := var(e_1)\cup\ldots\cup var(e_n)\cup var(prod) \\
var(\emptyset) & := \{\} \\
var(e) & := \{x\in\mathcal{X} \mid x \text{ occurs in } e\}
\end{array}
$$

Then we require for a rule $pat \xrightarrow{e} prod$ that $var(prod) \subseteq var(pat)$ and $var(e) \subseteq var(pat)$. For now, we also require that the pattern *pat* is *linear*, that is each variable in *pat* occurs only once.

4.2 Pattern Matching Semantics

To apply a rule to a solution, the pattern on the left rule side is matched to the solution. This results in a set of variable *substitutions* that map the pattern variables to values. Substitutions $\sigma : \mathcal{X} \rightharpoonup \mathcal{V}$ are partial functions that map a finite number of variables to values. The domain of a substitution σ, that is the variables on which it is defined, is $dom(\sigma)$. While being defined on variables, a substitution σ is trivially extended to compound syntactic terms, where it replaces each variable $v \in dom(v)$ with $\sigma(v)$ and leaves everything else unchanged. Sometimes it is useful to write a substitution as a set of mappings $x \mapsto v$, with $x \in \mathcal{X}$ and $v \in \mathcal{V}$.

The big-step semantics of matching a linear pattern to a solution is given in Fig. 1 with an operator $\Downarrow_{\mathbf{s},\sigma}$ that is parametrized with the current model state $\mathbf{s}$ (which is a solution) and the variable substitution σ. The big-step operator takes a rule to a reaction, that is a transition from one solution to another, equipped with a propensity taking mass-action kinetics into account. As mentioned in Sect. 2, this is interpreted as a state transition in a CTMC.

The key point in this definition is the composition of the variable assignment in the rule (PAT), highlighted in red. As the pattern is assumed to be linear, each variable only occurs once in the pattern. Therefore, variable assignments resulting from evaluating subterms (for attributes or recursion to the next pattern) will never overlap. This can be exploited by combining them via set addition. In fact, the set of all valid substitutions is the cartesian product of the valid substitutions for all individual variables. The linearity of the patterns is reflected in the linearity of combining substitutions.

This kind of semantics is used, for example, in Oury's and Plotkin's (C)SMMR [20,21]. Similar as in the semantics defined above, linearity of patterns is exploited there to decompose matching a complex pattern into matching the subpatterns independently and combining the results.

4.3 Algorithm

The rule (PAT) in the operational semantics suggests a simple algorithm for matching a pattern and finding all valid substitutions. For each entity in the pattern and for each entity in the current solution, if both belong to the same

$$\text{(VAL)}\frac{}{v \Downarrow_{\mathbf{s},\emptyset} v} \quad \text{(VAR)}\frac{}{x \Downarrow_{\mathbf{s},\{x\mapsto v\}} v} \quad \text{(EXP)}\frac{\sigma(e) \text{ evaluates to } v}{e \Downarrow_{\mathbf{s},\sigma} v}$$

$$\text{(EMPTY)}\frac{}{\emptyset \Downarrow_{\mathbf{s},\sigma} \{\!\!\{\}\!\!\}}$$

$$\text{(PAT)}\frac{S(v_1,\ldots,v_n) \in \mathbf{s} \quad a_1 \Downarrow_{\mathbf{s},\sigma_1} v_1 \quad \ldots \quad a_n \Downarrow_{\mathbf{s},\sigma_n} v_n \qquad pat \Downarrow_{\mathbf{s},\sigma'} sol \qquad \sigma = \bigcup_{i=1}^{n} \sigma_i \cup \sigma'}{S(a_1,\ldots,a_n) + pat \Downarrow_{\mathbf{s},\sigma} S(v_1,\ldots,v_n) \uplus sol}$$

$$\text{(PROD)}\frac{e_1 \Downarrow_{\mathbf{s},\sigma} v_1 \quad \ldots \quad e_n \Downarrow_{\mathbf{s},\sigma} v_n \quad prod \Downarrow_{\mathbf{s},\sigma} sol}{S(e_1,\ldots,e_n) + prod \Downarrow_{\mathbf{s},\sigma} S(v_1,\ldots,v_n) \uplus sol}$$

$$\text{(RULE)}\frac{pat \Downarrow_{\mathbf{s},\sigma} \mathbf{r} \quad prod \Downarrow_{\mathbf{s},\sigma} \mathbf{p} \quad e \Downarrow_{\mathbf{s},\sigma} v \quad k = v \prod_{ent \in \mathbf{r}} \binom{n_{\mathbf{s}}(ent)}{n_{\mathbf{r}}(ent)}}{pat \xrightarrow{e} prod \Downarrow_{\mathbf{s},\sigma} \mathbf{s} \xRightarrow{k} \mathbf{s} \ominus \mathbf{r} \uplus \mathbf{p}}$$

Fig. 1. Operational semantics of linear pattern matching

species, match all attributes and combine the resulting substitutions. Algorithm 1 shows a recursive formulation of this idea in pseudocode. The algorithm deconstructs a pattern following the abstract syntax and returns all substitutions that are valid according to the formal semantics, given a solution. In a simulation algorithm, each substitution can be used to evaluate the rate expression and products of a rule to instantiate a state transition. Note that the algorithm recurs at most once (in line 4); thus, it does not backtrack. With n distinct entities in the solution and m variables in the pattern, the runtime is bounded by $O(n^m)$.

5 Nonlinear Pattern Matching with Expressions

As demonstrated in Sect. 3, nonlinear patterns as well as expressions in the pattern are useful for modeling discrete space. To allow for nonlinear patterns, the semantics needs to be adapted to allow matching nonlinear patterns containing expressions. In the following, we first adapt the definition of the syntax and then present two approaches to define the semantics of pattern matching.

5.1 Abstract Syntax

The abstract syntax of this extended language now does not distinguish between the left and right rule sides. On both sides, we have lists of entities with expressions for attributes.

Input: A solution *sol* and a pattern *pat*
Output: A set of substitutions

Function *match(sol, pat)*
  switch *pat* **do**
    case $S(a_1, a_2, \ldots, a_n) + pat'$ **do**
      $tailResults = match(sol, pat')$;
      $subs \leftarrow \emptyset$;
      foreach $S(v_1, v_2, \ldots, v_n) \in sol$ **do** `// filters by species name`
        if $\forall a_i \in \mathcal{V} : a_i = v_i$ **then**
          foreach $r \in tailResults$ **do**
            $subs \leftarrow subs \cup \{r \cup \{a_i \mapsto v_i \mid a_i \in \mathcal{X}\}\}$
          end
        end
      end
      return *subs*;
    end
    case $\emptyset$ **do**
      return $\{\emptyset\}$;
    end
  end
end

Algorithm 1: An algorithm for matching linear patterns

$$\begin{aligned} pat &:= S(e_1, \ldots, e_n) + pat && S \in \mathcal{S}, n = ar(S), e_1, \ldots, e_n \in \mathcal{E} \\ &\mid \emptyset \\ rule &:= pat_l \xrightarrow{e} pat_r && e \in \mathcal{E} \end{aligned}$$

We assume $\mathcal{V} \subseteq \mathcal{E}$ and $\mathcal{X} \subseteq \mathcal{E}$, that is the set of expressions includes individual values and variables.

We now allow variables in pat_l to occur multiple times, that is pat_l may be a *nonlinear pattern.* We still require that $var(pat_r) \subseteq var(pat_l)$ and $var(e) \subseteq var(pat_l)$. In addition, however, we must now constrain the pattern on the left side such that it is possible to match it with reasonable effort. For example, we want to prohibit rules like $A(x+y) \xrightarrow{k} A(x)+A(y)$ which, depending on the value range of variables, might yield infinitely many pattern matches. In this paper, we require that every variable appears at least once in place of an attribute directly rather than nested in an expression. We call this a *direct occurrence.* Intuitively, such a direct occurrence allows us to obtain all possible values for that variable by looking at the corresponding attribute values in the solution.

Alternatively, we could require that the *first* occurrence of a variable must be directly in place of an attribute. This would make the order of the reactants significant, and also preclude patterns such as $A(x, y+1)+A(x+1, y)$. This approach is used (although not documented) in the implementation of ML-Rules [12].

A more ambitious requirement could allow simple algebraic transformations. For example, the rule $A(x+1) \xrightarrow{k} A(x)$ could be equivalently expressed as $A(x) \xrightarrow{k} A(x-1)$. In this fashion, patterns on the left side can be simplified automatically. We leave further considerations in this direction to future work.

Finally, note that the extended syntax also facilitates syntactic sugar for reversible rules, provided the above constraints for variable usage are satisfied on both rule sides. For example, a pair of rules $A \xrightarrow{k_1} B$ and $B \xrightarrow{k_2} A$ could be written as $A \underset{k_2}{\overset{k_1}{\rightleftharpoons}} B$.

5.2 Pattern Matching Semantics

We now present two different ways to define the semantics of nonlinear patterns with expressions.

Linearization. The first, simplest, and arguably most common approach to match nonlinear patterns is to "linearize" the pattern. This term refers to the transformation of a nonlinear pattern to a linear one by replacing all but one occurrences of a multiply occurring variable with a fresh variable [2]. Chromar [14] uses the same idea and replaces expressions in which variables occur with fresh variables. That linearized pattern can then be matched according to the semantics of linear patterns, and the resulting substitutions are then filtered to obtain those in which the replaced expressions have the same value as the variable they were replaced with. For example, the nonlinear pattern $A(x)+A(x+1)$ is transformed to the linear pattern $A(x)+A(y)$ with the constraint $y = x+1$, which can be evaluated for any substitution that assigns values to x and y by matching the pattern $A(x)+A(y)$.

To formalize this idea, we assume a relation *lin* that associates a nonlinear pattern pat_l with a linear pattern $pat_{l,lin}$ and the set of resulting constraints *con*. Then the matches of the linear pattern are defined by the operational semantics in Sect. 4.2. The resulting substitution σ will map the original and the newly introduced variables to values, and only if all the constraints hold under σ, it is a valid result also for the nonlinear pattern.

$$\text{(LIN)}\frac{\begin{array}{c} lin(pat_l, pat_{l,lin}, con) \quad pat_{l,lin} \xrightarrow{e} pat_r \Downarrow_{\mathbf{s},\sigma} \mathbf{s} \overset{k}{\Rightarrow} \mathbf{t} \\ \text{for all } c \in con : \sigma(con) \text{ holds} \end{array}}{pat_l \xrightarrow{e} pat_r \Downarrow_{\mathbf{s},\sigma} \mathbf{s} \overset{k}{\Rightarrow} \mathbf{t}}$$

This is also directly translatable into a three-step algorithm. First, the pattern is linearized, saving the replacements made as constraints. Then, the linearized pattern is matched with the linear pattern matching algorithm from Sect. 4.3, resulting in a set of substitutions. Those substitutions are then filtered using the saved constraints. Figure 2 shows a schematic visualization of the matching algorithm.

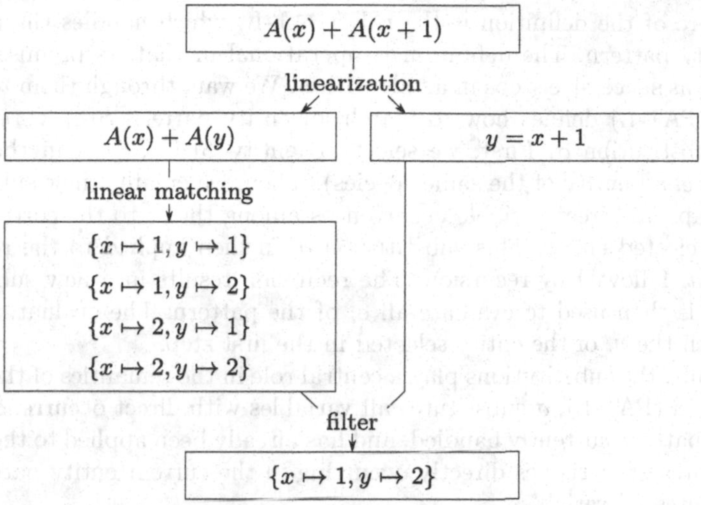

Fig. 2. Visualization of the linearization algorithm for matching the pattern $A(x) + A(x+1)$ in the solution $\{\!|A(1), A(2)|\!\}$. The linearized pattern yields 4 substitutions, of which only one satisfies the constraints obtained during the linearization.

This procedure is problematic if many of the linear pattern matching results have to be filtered out [20]. As the constraints are only evaluated in the end, the intermediate results are produced in any case and only then discarded. For example, the pattern $A(x) + A(x+1)$ matched to a solution $\{\!|A(1), A(2), \ldots, A(n)|\!\}$ has $n-1$ matches, but the linearized version will produce n^2 intermediate results. In general, the algorithm's runtime is bounded by $O(n^m)$ with n distinct entities in the solution and m the number of attributes in the pattern that contain at least one variable occurrence.

In the following, we introduce an algorithm that exploits the links between different occurrences of the same variable in the pattern to avoid unnecessary work.

Inline Substitution. So far, we have defined the semantics with an operator *term* $\Downarrow_{\mathbf{s},\sigma}$ *value* that includes the variable substitution as a parameter of the relation. This emphasizes that the variable substitution is created as a by-product of pattern matching, but has no significant influence on the actual pattern matching process. In the following, we additionally use a relation *term* $\mid \sigma \Downarrow_{\mathbf{s}}$ *value* $\mid \sigma'$ for defining the matching of the pattern on the left rule side (Fig. 3). It reads as follows: given a substitution σ, matching *term* to $\mathbf{s}$ results in *value* and the modified substitution σ'. This allows us to define how the variable substitution is created and transformed during pattern matching, similar to how mutable references are represented in formal semantics of programming languages [22, Sect. 13]. This alternative evaluation relation is "invoked" in the premises of (RULE) and resolved in (PAT-L) and (EMPTY-L).

The core of the definition is the rule (PAT-L) which handles the matching of an entity pattern. The definition is operational in that its premises can be thought of as successive steps in an algorithm. We walk through them to explain the rule. (PAT-L) defines how to match an entity pattern $S(e_1 \ldots, e_n) + pat$, given a substitution σ. First, we select an entity $S(v_1 \ldots, v_n)$ in the current solution (i.e. an entity of the same species). Then, we create a new substitution σ' that maps all direct variable occurrences among the e_i to the corresponding v_i of the selected entity. This substitution σ' is then applied to the remaining pattern *pat*, followed by recursion. The recursion results in a new substitution σ'', which is then used to evaluate all e_i of the pattern. The evaluation results must match the v_i of the entity selected in the first step.

Naturally, the substitutions play a central role in the semantics of the pattern matching. In (PAT-L), σ substitutes all variables with direct occurrences left of the entity pattern currently handled, and has already been applied to the pattern. σ' substitutes all variables directly occurring in the current entity pattern, and σ'' substitutes all variables.

$$\text{(PAT-L)}\frac{S(v_1 \ldots, v_n) \in \mathbf{s} \qquad \sigma' = \{e_i \mapsto v_i \mid e_i \in \mathcal{X}\} \\ \sigma'(pat) \mid (\sigma \cup \sigma') \Downarrow_{\mathbf{s}} sol \mid \sigma'' \quad e_1 \Downarrow_{\mathbf{s},\sigma''} v_1 \quad \ldots \quad e_n \Downarrow_{\mathbf{s},\sigma''} v_n}{S(e_1 \ldots, e_n) + pat \mid \sigma \Downarrow_{\mathbf{s}} S(v_1 \ldots, v_n) \uplus sol \mid \sigma''}$$

$$\text{(EMPTY-L)}\frac{}{\emptyset \mid \sigma \Downarrow_{\mathbf{s}} \{\!\!\{\}\!\!\} \mid \sigma}$$

$$\text{(PAT-R)}\frac{e_1 \Downarrow_{\mathbf{s},\sigma} v_1 \quad \ldots \quad e_n \Downarrow_{\mathbf{s},\sigma} v_n \quad pat \Downarrow_{\mathbf{s},\sigma} sol}{S(e_1 \ldots, e_n) + pat \Downarrow_{\mathbf{s},\sigma} S(v_1 \ldots, v_n) \uplus sol}$$

$$\text{(EXP)}\frac{\sigma(e) \text{ evaluates to } v}{e \Downarrow_{\mathbf{s},\sigma} v} \qquad \text{(EMPTY-R)}\frac{}{\emptyset \Downarrow_{\mathbf{s},\sigma} \{\!\!\{\}\!\!\}}$$

$$\text{(RULE)}\frac{pat_l \mid \emptyset \Downarrow_{\mathbf{s}} \mathbf{r} \mid \sigma \quad pat_r \Downarrow_{\mathbf{s},\sigma} \mathbf{p} \quad e \Downarrow_{\mathbf{s},\sigma} v \quad k = v \prod_{ent \in \mathbf{r}} \binom{n_{\mathbf{s}}(ent)}{n_{\mathbf{r}}(ent)}}{pat_l \xrightarrow{e} pat_r \Downarrow_{\mathbf{s},\sigma} \mathbf{s} \overset{k}{\Rightarrow} \mathbf{s} \ominus \mathbf{r} \uplus \mathbf{p}}$$

Fig. 3. Operational semantics of inline substitution

Inline Substitution Algorithm. Algorithm 2 shows an algorithm that implements the operational semantics of inline substitution. Similarly as the semantics, the algorithm operates in two phases. In the first phase, all direct occurrences of variables are found from left to right and all possible values for these variables are iterated (line 8). When the end of the pattern is reached, the accumulated substitution contains values for all variables in the pattern (line 27). In the second phase now, the recursive calls retreat from right to left through all entity patterns with the complete substitution. This allows evaluating all expressions

Input: A solution *sol*, a pattern *pat*, and a substitution σ (initially $\emptyset$)
Output: A set of substitutions

```
1  Function match(sol, pat, σ)
2      switch pat do
3          case S(e1 ..., en) + pat' do
4              if one of ei is a direct variable occurrence then
5                  subs ← ∅;
6                  x ← variable with the first direct occurrence in ei;
7                  j ← index of the first occurrence of x in ei;
8                  foreach S(v1 ..., vn) ∈ sol do        // filters by species name
9                      σ' ← {x ↦ vj};
10                     pat'' ← σ'(S(e1 ..., en) + pat');
11                     σ'' ← match(sol, pat'', σ ∪ σ');
12                     subs ← subs ∪ σ'';
13                 end
14                 return subs;
15             else
16                 tailResults ← match(sol, pat', σ);
17                 foreach σ'' ⊂ tailResults do
18                     ent ← σ''(S(e1 ..., en));
19                     if ent ∈ sol then
20                         subs ← subs ∪ σ'';
21                     end
22                 end
23                 return subs;
24             end
25         end
26         case ∅ do
27             return {σ};
28         end
29     end
30 end
```

Algorithm 2: An algorithm for matching nonlinear patterns via inline substitution

to values, obtaining an entity from each entity pattern (line 18). Then the algorithm checks whether that entity exists in the solution (line 19). This last check can be sped up by using appropriate data structures for representing the solution. Figure 4 shows a schematic visualization of the matching algorithm.

Note that the algorithm recurs once for every variable in the pattern (in line 11) and backtracks to cover all possible values for that variable. The algorithm also recurs once for every entity pattern in line 16. Thus, with n distinct entities in the solution and m variables in the pattern the algorithm's runtime is bounded by $O(n^m)$ (assuming that the number of entity patterns is small w.r.t. the number of variables and the lookup in line 19 happens in $O(1)$).

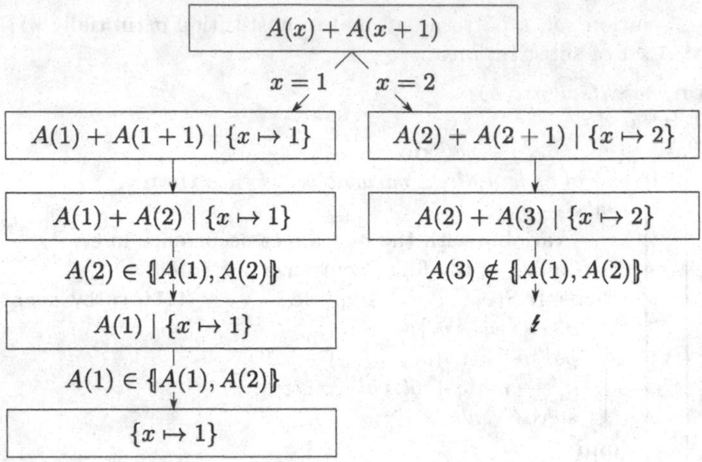

Fig. 4. Visualization of the inline substitution algorithm for matching the pattern $A(x)+A(x+1)$ in the solution $\{\!\!\{A(1), A(2)\}\!\!\}$. The branches denote the iteration in line 8 of Algorithm 2. When substituting x with 2, the pattern $A(x+1)$ is evaluated to $A(3)$, which does not exist in the solution; thus, this branch fails. When substituting x with 1, however, the matching succeeds.

6 Benchmarks

To illustrate the difference in the runtime of the different algorithms, we benchmarked our proof-of-concept implementations of both the linearization and the inline substitution approach. We matched the pattern $A(x) + A(x + 1)$, which expresses neighborhood in a linear chain of entities, in chains $A(1) + \ldots + A(n)$ of increasing length. Figure 5 shows the number of completed pattern matches per second[1]. The benchmarks were executed on a standard laptop with an AMD Ryzen 7 PRO 4750U processor and 32 GB of memory. Note that this benchmark is not meant to be representative of real-world uses of pattern matching in simulation algorithms, but to highlight the impact of the different complexity classes.

The benchmark results indicate that the linearization approach scales quadratically with the length of the chain. This is consistent with our classification of the algorithm's runtime in $O(n^m)$ (Sect. 5.2), where $m = 2$. The pattern is translated to $A(x) + A(y)$ (with the constraint $y = x + 1$). Therefore, doubling the length of the chain doubles the number of values for x and y each, leading to a fourfold increase of substitutions to check for whether they satisfy the constraint.

In contrast, the inline substitution approach scales linearly. Doubling the length of the chain means that the number of values for x doubles, and that for each x it has to be checked whether an $A(x + 1)$ exists in the solution. As long

[1] The source code repository at https://git.informatik.uni-rostock.de/mosi/pattern-matching contains instructions on how to reproduce this plot.

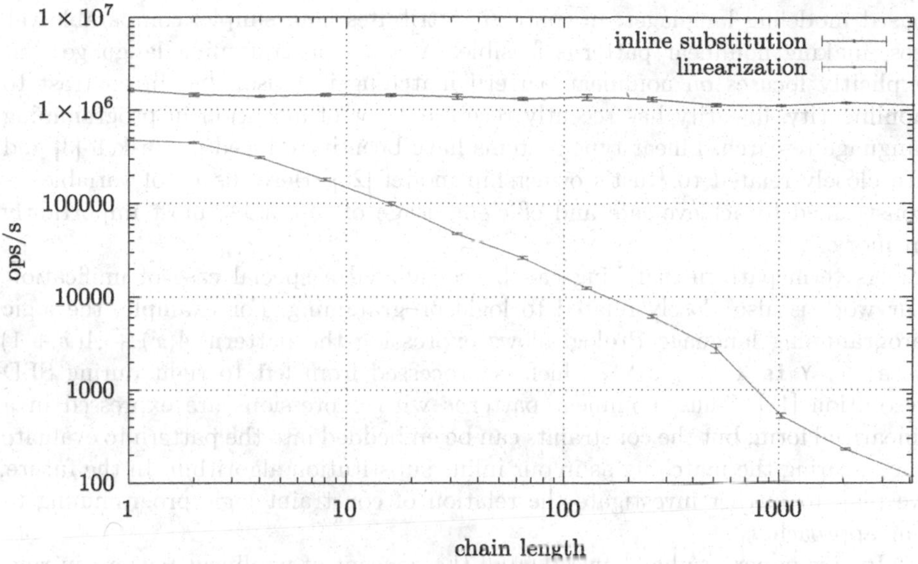

Fig. 5. Number of completed matches per second for the pattern $A(x) + A(x+1)$ in a solution of the form $A(1) + \cdots + A(n)$ with increasing n.

as this check scales well, the overall approach does as well. In our implementation, this check is implemented as a lookup in a hash map and is, thus, largely independent from the length of the chain.

Our performance results have to be interpreted carefully. In fact, simulation algorithms usually achieve efficiency by avoiding pattern matching as much as possible. For example, network-based algorithms perform the pattern matching as a preprocessing step to "unfold" reaction rules to all their instances [23]. But even network-free algorithms can minimize pattern matching by maintaining a dependency graph between reaction instances [17]. Benchmarks with a complete simulation algorithm (and real-world models) would help to characterize the actual impact of improving the pattern matching's performance. In addition, both implementations could be optimized further.

7 Discussion and Conclusion

While nonlinear pattern matching has not been studied in the context of rule-based modeling languages, it has also not received much attention in other domains (with a few exceptions, e.g. [2,16]). For example, Haskell and Scala, which are functional programming languages with good support for pattern matching, do not allow nonlinear patterns. The reason is that these languages allow attribute values that are not easily comparable (e.g., functions)[2]. In rule-

[2] See, for example, this discussion on the haskell-cafe mailing list.

based modeling languages, in contrast, attributes hold simple, comparable values, making nonlinear patterns feasible. A novel programming language that explicitly focuses on nonlinear pattern matching is Egison [8]. In contrast to nonlinearity, linearity has recently received a lot of attention in programming language research. Linear type systems have been introduced in Haskell [3] and are closely related to Rust's ownership model [25]. Here, usage of variables is constrained to achieve safe and efficient usage of resources, most importantly memory.

As term pattern matching can be considered a special case of unification, our work is also closely related to logic programming. For example, the logic programming language Prolog allows expressing the pattern $A(x) + A(x + 1)$ as `a(X), Y is X + 1, a(Y)`, which is processed from left to right during SLD resolution [18]. Thus, nonlinear patterns with expressions are expressed in a linearized form, but the constraints can be embedded into the pattern to evaluate them during the matching as in our inline substitution algorithm. In the future, we plan to further investigate the relation of constraint logic programming to our approach.

In this paper, we have investigated the concept of nonlinear pattern in rule-based modeling languages for biochemical reaction networks. Although some rule-based languages including nonlinear patterns have been proposed, the nature of these patterns was, so far, not studied in detail. We have explored two aspects of nonlinear patterns.

First, we have shown two ways to formally define the operational semantics of matching nonlinear patterns to a multiset of attributed entities. This contributes to defining more expressive rule-based languages formally. As the expressiveness of rule-based languages depends largely on the allowed patterns on the left rule side, including nonlinear patterns increases the expressiveness of the language (as illustrated in Sect. 3).

Second, we have devised and implemented algorithms that implement both formal definitions. Whereas the linearization approach translates the nonlinear pattern to a linear one, the inline substitution approach checks the pattern directly. Based on the algorithms, we can conclude that the inline substitution approach has a higher potential for computational efficiency. Benchmarks of our proof-of-concept implementations confirmed this.

As a continuation of this work, the semantics and algorithmics of nonlinear pattern matching should be studied further. In particular, the impact on real-world modeling problems should be investigated, referring to expressiveness (i.e., how does the model description benefit from allowing nonlinear patterns) as well as performance (i.e., how much can the simulation efficiency be improved by matching nonlinear patterns efficiently). This paper should also contribute to helping to integrate nonlinear pattern matching into existing and future modeling language specifications and implementations.

References

1. Baader, F., Nipkow, T., Franz, B.: Term Rewriting and All That. Cambridge University Press, Cambridge (2006)
2. Bachmair, L., Chen, T., Ramakrishnan, I.V.: Associative-commutative discrimination nets. In: Gaudel, M.-C., Jouannaud, J.-P. (eds.) CAAP 1993. LNCS, vol. 668, pp. 61–74. Springer, Heidelberg (1993). https://doi.org/10.1007/3-540-56610-4_56
3. Bernardy, J.-P., Boespflug, M., Newton, R.R., Jones, S.P., Spiwack, A.: Linear haskell: practical linearity in a higher-order polymorphic language. In: Proceedings of the ACM on Programming Languages, vol. 2, no. POPL, pp. 1–29 (2018)
4. Bittig, A.T., Uhrmacher, A.M.: ML-space: hybrid spatial gillespie and particle simulation of multi-level rule-based models in cell biology. IEEE/ACM Trans. Comput. Biol. Bioinf. **14**(6), 1339–1349 (2017)
5. Blinov, M.L., Faeder, J.R., Goldstein, B., Hlavacek, W.S.: Bionetgen: software for rule-based modeling of signal transduction based on the interactions of molecular domains. Bioinformatics **20**(17), 3289–3291 (2004)
6. Danos, V., Laneve, C.: Formal molecular biology. Theor. Comput. Sci. **325**(1), 69–110 (2004)
7. De Nicola, R., Latella, D., Loreti, M., Massink, M.: A uniform definition of stochastic process calculi. ACM Comput. Surv. **46**(1), 1–35 (2013)
8. Egi, S.: Egison: non-linear pattern-matching against non-free data types (2015)
9. Faeder, J.R., Blinov, M.L., Hlavacek, W.S.: Rule-based modeling of biochemical systems with bionetgen. In: Maly, I.V. (ed.) Methods in Molecular Biology, pp. 113–167. Humana Press (2009)
10. Gilbert, D., Heiner, M., Takahashi, K., Uhrmacher, A.M.: Multiscale spatial computational systems biology (dagstuhl seminar 14481) (2015)
11. Gillespie, D.T.: Exact stochastic simulation of coupled chemical reactions. J. Phys. Chem. **81**(25), 2340–2361 (1977)
12. Helms, T., Warnke, T., Maus, C., Uhrmacher, A.M.: Semantics and efficient simulation algorithms of an expressive multi-level modeling language. ACM Trans. Model. Comput. Simul. 27(2), 8:1–8:25 (2017)
13. Henzinger, T.A., Jobstmann, B., Wolf, V.: Formalisms for specifying markovian population models. Int. J. Found. Comput. Sci. **22**(4), 823–841 (2011)
14. Honorato-Zimmer, R., Millar, A.J., Plotkin, G.D., Zardilis, A.: Chromar, a language of parameterised agents. Theor. Comput. Sci. **765**, 97–119 (2019)
15. John, M., Lhoussaine, C., Niehren, J., Versari, C.: Biochemical reaction rules with constraints. In: Barthe, G. (ed.) ESOP 2011. LNCS, vol. 6602, pp. 338–357. Springer, Heidelberg (2011). https://doi.org/10.1007/978-3-642-19718-5_18
16. Krebber, M., Barthels, H., Bientinesi, P.: Efficient pattern matching in python. ACM Press (2017)
17. Köster, T., Warnke, T., Uhrmacher, A.M.: Partial evaluation via code generation for static stochastic reaction network models. In: Proceedings of the 2020 ACM SIGSIM Conference on Principles of Advanced Discrete Simulation, Miami FL, Spain, pp. 159–170. ACM (2020)
18. Lloyd, J.W.: Foundations of Logic Programming, (2nd Extended. Springer, Heidelberg (1987). https://doi.org/10.1007/978-3-642-83189-8
19. Maus, C., Rybacki, S., Uhrmacher, A.M.: Rule-based multi-level modeling of cell biological systems. BMC Syst. Biol. **5**(1), 166 (2011)

20. Oury, N., Plotkin, G.D.: Coloured stochastic multilevel multiset rewriting. ACM Press (2011)
21. Oury, N., Plotkin, G.D.: Multi-level modelling via stochastic multi-level multiset rewriting. Math. Struct. Comput. Sci. **23**(2), 471–503 (2013)
22. Pierce, B.C.: Types and Programming Languages. MIT Press, Cambridge (2002)
23. Suderman, R., Mitra, E.D., Lin, Y.T., Erickson, K.E., Feng, S., Hlavacek, W.S.: Generalizing gillespie's direct method to enable network-free simulations. Bull. Math. Biol. **81**(8), 2822–2848 (2018)
24. Warnke, T., Helms, T., Uhrmacher, A.M.: Syntax and semantics of a multi-level modeling language. In: Proceedings of the 2015 ACM SIGSIM Conference on Principles of Advanced Discrete Simulation, pp. 133–144. ACM, New York (2015)
25. Weiss, A., Gierczak, O., Patterson, D., Matsakis, N.D., Ahmed, A.: The essence of rust. Oxide (2020)

Protein Noise and Distribution in a Two-Stage Gene-Expression Model Extended by an mRNA Inactivation Loop

Candan Çelik[1](✉), Pavol Bokes[1,2], and Abhyudai Singh[3]

[1] Department of Applied Mathematics and Statistics, Comenius University, 84248 Bratislava, Slovakia
{candan.celik,pavol.bokes}@fmph.uniba.sk

[2] Mathematical Institute, Slovak Academy of Sciences, 81473 Bratislava, Slovakia

[3] Department of Electrical and Computer Engineering, University of Delaware, Newark, DE 19716, USA
absingh@udel.edu

Abstract. Chemical reaction networks involving molecular species at low copy numbers lead to stochasticity in protein levels in gene expression at the single-cell level. Mathematical modelling of this stochastic phenomenon enables us to elucidate the underlying molecular mechanisms quantitatively. Here we present a two-stage stochastic gene expression model that extends the standard model by an mRNA inactivation loop. The extended model exhibits smaller protein noise than the original two-stage model. Interestingly, the fractional reduction of noise is a non-monotonous function of protein stability, and can be substantial especially if the inactivated mRNA is stable. We complement the noise study by an extensive mathematical analysis of the joint steady-state distribution of active and inactive mRNA and protein species. We determine its generating function and derive a recursive formula for the protein distribution. The results of the analytical formula are cross-validated by kinetic Monte-Carlo simulation.

Keywords: Stochastic gene expression · Master equation · Analytical distribution · Generating function · Stochastic simulation

1 Introduction

As many other biochemical mechanisms, gene expression in which protein synthesis occurs is inherently stochastic due to random fluctuations in the copy number of gene products, e.g. proteins [7]. From the viewpoint of biochemical

CÇ is supported by the Comenius University grant for doctoral students Nos. UK/106/2020 and UK/100/2021. PB is supported by the Slovak Research and Development Agency under the contract No. APVV-18-0308 and by the VEGA grant 1/0339/21, and the EraCoSysMed project 4D-Healing. AS acknowledges support by ARO W911NF-19-1-0243 and NIH grants R01GM124446 and R01GM126557.

E. Cinquemani and L. Paulevé (Eds.): CMSB 2021, LNBI 12881, pp. 215–229, 2021.
https://doi.org/10.1007/978-3-030-85633-5_13

reactions, in simplest formulations, gene expression consists of two main steps: transcription and translation. While RNA polymerase enzymes produce mRNA molecules in the former, protein synthesis takes place by ribosomes in the latter, each reaction corresponding to the production and decay of relevant species. Additionally, the two-stage model can be extended by the regulation of transcription factors, which affect gene expression by modulating the binding rate of RNA polymerase [3].

Over the last decades, the two-stage model of gene expression has been extensively studied to understand how the stochastic phenomenon in cellular processes takes place [14,15,19,20]. Specifically, quantifying the number of species in terms of probability distributions has become an interesting and challenging endeavour due to the subtleties involved in finding a solution to the underlying problem. On the other hand, the fluctuations in mRNA and protein levels are considered as a major source of noise, leading to cell-to-cell variability in gene regulatory networks [12,16,18]. The noise emerges from different sources, namely *intrinsic* and *extrinsic* noise [25,27]; yet, structural elements such as stem-loops can also contribute to noise by binding to an untranslated region of mRNA [6]. The untranslated regions of mRNAs often contain these stem-loops that can reversibly change configurations making individual mRNAs translationally active/inactive.

Numerous modelling approaches have been proposed that are based on deterministic and stochastic frameworks, and recently also hybrid ones as a combination of the preceding two [5,10,23]. Only a few of those provide an explicit solution to the two-stage gene-expression model [4,20]; most of the studies are based on Monte Carlo simulations, which are usually computationally expensive.

As a generalisation of the two-stage model, some studies in the literature consider a set of multiple gene states and investigate the dynamics of stochastic transitions among these states [11,29]. Nevertheless, to the best of our knowledge, none of these studies takes an mRNA inactivation into account. Here we extend the two-stage model by an MRNA inactivation loop, by which we mean that after transcription species can switch between active and inactive states. In other words, there exists a pair of reversible chemical reactions occurring at constant rates by turning active mRNA species into inactive ones, and vice versa. Subsequently, the active mRNA is translated, while the inactive mRNA stays dormant. The schematic of reactions describing the model is given in (1). Here we thereafter refer to the aforementioned model as *the extended model.* A possible biological scenario that can implement this extended model is by a regulatory RNA that temporarily blocks mRNA function [17].

This paper is organised as follows. In Sect. 2, the stationary means of active mRNA, inactive mRNA, and protein are obtained from a deterministic formulation the model; the master equation of the stochastic model is formulated, and transformed into a partial differential equation for the generating function. In Sect. 3, the partial differential equation is transformed into one for the factorial cumulant generation function and a power series solution is found; recursive expressions for the coefficients—the factorial cumulants of the three molecular species—are thereby provided. In Sect. 4, the protein Fano factor is expressed in terms of the first two factorial cumulants, and the noise-reduction effect of the mRNA inactivation loop is analysed. The generating function of the stationary distribution of active mRNA, inactive mRNA and protein amounts is represented in the special-function form in Sect. 5. The marginal protein and active and inactive mRNA distributions are derived in Sect. 6. The paper is concluded in Sect. 7.

2 Model Formulation

The extended model involves three species, mRNA, inactive mRNA (imRNA for short), and protein, and consists of the reactions

$$\begin{aligned} &\emptyset \underset{\gamma_1}{\overset{\lambda_1}{\rightleftharpoons}} \text{mRNA}, \quad \text{mRNA} \underset{\beta}{\overset{\alpha}{\rightleftharpoons}} \text{imRNA}, \quad \text{imRNA} \xrightarrow{\tilde{\gamma}_1} \emptyset, \\ &\text{mRNA} \xrightarrow{\lambda_2} \text{mRNA} + \text{protein}, \quad \text{protein} \xrightarrow{\gamma_2} \emptyset. \end{aligned} \tag{1}$$

The reactions in (1) correspond to mRNA transcription and decay, mRNA activation and inactivation, inactive mRNA decay, protein translation, and protein decay, respectively.

Due to the linearity of kinetics in (1), the mean levels of the mRNA (m), inactive mRNA ($\tilde{m}$) and protein (n) exactly satisfy the system of deterministic rate equations

$$\begin{aligned} \frac{\mathrm{d}\langle m\rangle}{\mathrm{d}t} &= \lambda_1 - (\gamma_1+\alpha)\langle m\rangle + \beta\langle\tilde{m}\rangle, \\ \frac{\mathrm{d}\langle \tilde{m}\rangle}{\mathrm{d}t} &= \alpha\langle m\rangle - (\tilde{\gamma}_1+\beta)\langle\tilde{m}\rangle, \\ \frac{\mathrm{d}\langle n\rangle}{\mathrm{d}t} &= \lambda_2\langle m\rangle - \gamma_2\langle n\rangle. \end{aligned} \tag{2}$$

Setting time derivatives in (2) to zero, and solving the resulting algebraic system, the stationary means are obtained as

$$\langle m\rangle = \frac{\lambda_1}{\gamma_1^{\text{eff}}}, \quad \langle\tilde{m}\rangle = \frac{\alpha}{\tilde{\gamma}_1+\beta}\langle m\rangle, \quad \langle n\rangle = \frac{\lambda_2}{\gamma_2}\langle m\rangle, \tag{3}$$

for the mRNA, inactive mRNA, and protein respectively, where

$$\gamma_1^{\text{eff}} = \gamma_1 + \frac{\alpha\tilde{\gamma}_1}{\tilde{\gamma}_1+\beta} \tag{4}$$

denotes the effective rate of mRNA decay. Owing to the linearity of reaction rates, one can find a closed system of differential equations not only for means, but also for higher-order moments [21,24]; however these equations are typically less revealing than the mean dynamics. Here we take a different approach and quantify the protein noise as a by-product of a generating-function analysis in Sect. 4.

The probability $p_{m,\tilde{m},n}(t)$ of having m mRNA, $\tilde{m}$ inactive mRNA, and n protein molecules at time t satisfies the chemical master equation

$$\begin{aligned}\frac{\mathrm{d}p_{m,\tilde{m},n}}{\mathrm{d}t} &= \lambda_1(p_{m-1,\tilde{m},n} - p_{m,\tilde{m},n}) + \alpha((m+1)p_{m+1,\tilde{m}-1,n} - mp_{m,\tilde{m},n}) \\ &+ \tilde{\gamma}_1((\tilde{m}+1)p_{m,\tilde{m}+1,n} - \tilde{m}p_{m,\tilde{m},n}) + \lambda_2 m(p_{m,\tilde{m},n-1} - p_{m,\tilde{m},n}) \\ &+ \gamma_2((n+1)p_{m,\tilde{m},n+1} - np_{m,\tilde{m},n}) + \gamma_1((m+1)p_{m+1,\tilde{m},n} - mp_{m,\tilde{m},n}) \\ &+ \beta((\tilde{m}+1)p_{m-1,\tilde{m}+1,n} - \tilde{m}p_{m,\tilde{m},n}). \end{aligned} \tag{5}$$

Equating the left-hand side of (5) to zero yields the steady-state master equation

$$\begin{aligned} 0 &= \lambda_1(p_{m-1,\tilde{m},n} - p_{m,\tilde{m},n}) + \alpha((m+1)p_{m+1,\tilde{m}-1,n} - mp_{m,\tilde{m},n}) \\ &+ \tilde{\gamma}_1((\tilde{m}+1)p_{m,\tilde{m}+1,n} - \tilde{m}p_{m,\tilde{m},n}) + \lambda_2 m(p_{m,\tilde{m},n-1} - p_{m,\tilde{m},n}) \\ &+ \gamma_2((n+1)p_{m,\tilde{m},n+1} - np_{m,\tilde{m},n}) + \gamma_1((m+1)p_{m+1,\tilde{m},n} - mp_{m,\tilde{m},n}) \\ &+ \beta((\tilde{m}+1)p_{m-1,\tilde{m}+1,n} - \tilde{m}p_{m,\tilde{m},n}), \end{aligned} \tag{6}$$

We additionally require that the normalising condition

$$\sum_{m,\tilde{m},n} p_{m,\tilde{m},n} = 1 \tag{7}$$

hold.

We aim to find the moments of the probability distribution $p_{m,\tilde{m},n}$ by using the generating function approach [8]. In order to solve (6)–(7), we employ the probability generating function

$$G(x,y,z) = \sum_{m,\tilde{m},n} x^m y^{\tilde{m}} z^n p_{m,\tilde{m},n} \tag{8}$$

for the probability distribution $p_{m,\tilde{m},n}$. Multiplying (6) by the factor $x^m y^{\tilde{m}} z^n$ and summing over $m, \tilde{m}$ and n yields

$$\begin{aligned}\lambda_1(1-x)G &= (\lambda_2 x(z-1) + \gamma_1(1-x) + \alpha(y-x))\frac{\partial G}{\partial x} \\ &+ (\tilde{\gamma}_1(1-y) + \beta(x-y))\frac{\partial G}{\partial y} + \gamma_2(1-z)\frac{\partial G}{\partial z}. \end{aligned} \tag{9}$$

Equation (9) is subject to

$$G(1,1,1) = 1, \tag{10}$$

which is implied by the normalisation condition (7).

3 Factorial Cumulant Generating Function

In order to find a particular solution to (9)–(10), we change the variables according to

$$x = 1+u, \quad y = 1+v, \quad z = 1+w, \quad G = \exp(\varphi), \tag{11}$$

and obtain that the factorial cumulant generating function [9] $\varphi = \varphi(u, v, w)$ is a solution of the inhomogeneous linear partial differential equation (PDE),

$$\lambda_1 u = (-\lambda_2(1+u)w + \gamma_1 u + \alpha(u-v))\frac{\partial \varphi}{\partial u} + (\tilde{\gamma}_1 v + \beta(v-u))\frac{\partial \varphi}{\partial v} + \gamma_2 w \frac{\partial \varphi}{\partial w} \quad (12)$$

subject to

$$\varphi(0,0,0) = 0. \quad (13)$$

In order to solve (12)–(13) we shall employ the ansatz

$$\varphi(u, v, w) = \varphi_{00}(w) + u\varphi_{10}(w) + v\varphi_{01}(w). \quad (14)$$

We immediately obtain the partial derivatives

$$\frac{\partial \varphi}{\partial u} = \varphi_{10}(w), \quad \frac{\partial \varphi}{\partial v} = \varphi_{01}(w), \quad \frac{\partial \varphi}{\partial w} = \varphi_{00}'(w) + u\varphi_{10}'(w) + v\varphi_{01}'(w). \quad (15)$$

Inserting (15) into (12) and rearranging the terms yields an inhomogeneous system of ODEs

$$\begin{aligned} &\gamma_2 w \varphi_{00}' - \lambda_2 w \varphi_{10} = 0, \\ &\gamma_2 w \varphi_{10}' + (\gamma_1 + \alpha - \lambda_2 w)\varphi_{10} - \beta\varphi_{01} = \lambda_1, \\ &\gamma_2 w \varphi_{01}' + (\tilde{\gamma}_1 + \beta)\varphi_{01} - \alpha\varphi_{10} = 0. \end{aligned} \quad (16)$$

Let us assume that the functions φ_{00}, φ_{10}, and φ_{01} are of the power series form, i.e.,

$$\varphi_{00}(w) = \sum_{k=0}^{\infty} a_k w^k, \quad \varphi_{10}(w) = \sum_{k=0}^{\infty} b_k w^k, \quad \varphi_{01}(w) = \sum_{k=0}^{\infty} c_k w^k. \quad (17)$$

The coefficients a_k, b_k, and c_k give the factorial cumulants of the joint molecular distribution [9]. Note that $a_0 = 0$ follows immediately from the normalisation condition (13). Evaluating the derivatives in (17) and substituting into (16), we obtain the following recurrence equations:

$$a_k = \frac{\lambda_2}{k\gamma_2} b_{k-1}, \quad k \geq 1, \quad (18)$$

$$(\gamma_1 + \alpha)b_0 - \beta c_0 - \lambda_1 + \sum_{k=1}^{\infty} (\gamma_2 k b_k + (\gamma_1 + \alpha)b_k - \lambda_2 b_{k-1} - \beta c_k)w^k = 0, \quad (19)$$

$$(\tilde{\gamma}_1 + \beta)c_0 - \alpha b_0 + \sum_{k=1}^{\infty} (\gamma_2 k c_k + (\tilde{\gamma}_1 + \beta)c_k - \alpha b_k)w^k = 0. \quad (20)$$

Since we consider (17) as a solution to (12) then all the coefficients in (19)–(20) must be zero. Thus, we get

$$(\gamma_1 + \alpha + \gamma_2 k)b_k - \lambda_2 b_{k-1} - \beta c_k = 0, \quad (21)$$

$$(\tilde{\gamma}_1 + \beta + \gamma_2 k)c_k - \alpha b_k = 0, \tag{22}$$

for b_k and c_k. Solving the algebraic system (21)–(22) in b_k, $k \geq 1$, yields

$$(\gamma_2^2 k^2 + \gamma_2(\tilde{\gamma}_1 + \gamma_1 + \beta + \alpha)k + \tilde{\gamma}_1\gamma_1 + \gamma_1\beta + \tilde{\gamma}_1\alpha)b_k = \lambda_2(\tilde{\gamma}_1 + \beta + k\gamma_2)b_{k-1},$$

i.e.

$$b_k = \frac{\lambda_2(\tilde{\gamma}_1 + \beta + k\gamma_2)}{\gamma_2^2 k^2 + \gamma_2(\tilde{\gamma}_1 + \gamma_1 + \beta + \alpha)k + \tilde{\gamma}_1\gamma_1 + \gamma_1\beta + \tilde{\gamma}_1\alpha} b_{k-1}, \tag{23}$$

where the zeroth term of the sequence b_k is obtained, by equating the terms out of the sums in (19) and (20) to zero, as

$$b_0 = \frac{\lambda_1(\tilde{\gamma}_1 + \beta)}{(\gamma_1 + \alpha)(\tilde{\gamma}_1 + \beta) - \beta\alpha} = \frac{\lambda_1}{\gamma_1^{\text{eff}}}. \tag{24}$$

Equation (24) thus rederives the stationary mRNA mean (3) by means of factorial cumulant analysis; similarly, c_0 and a_1 can be identified as the stationary imRNA and protein means. Thus, the sequence b_k can be calculated iteratively from (23) starting from the initial condition (24). Having calculated b_k, the sequence a_k and c_k can be evaluated via (18) and (22). In Sect. 5, we will utilise these formulas to obtain a special-function representation of the generating function. Before doing that, we show that the first two terms of these sequences determine protein variability.

4 Protein Variability

As outlined in the previous section, the first-order cumulants b_0, c_0, and a_1 ($a_0 = 0$ by normalisation condition), coincide with the stationary mRNA, imRNA, and protein mean values. In this section, we use the second-order cumulants to describe the stationary noise in our model. The noise in mRNA and imRNA is Poissonian (see Sect. 6 for details) and therefore uninteresting: we focus on the protein noise.

This section is divided into two parts: the first expresses the Fano factor in terms of the first and second order cumulants (and is independent of the specifics of the current model); the second part uses the formula to analyse the noise reduction effect of the inactivation loop.

Expressing the Fano Factor in Terms of the Cumulants. The generating function is expanded by the Taylor formula as

$$G(1,1,z) = G(1,1,1) + \frac{\partial G}{\partial z}(1,1,1)(z-1) + \frac{1}{2}\frac{\partial^2 G}{\partial z^2}(1,1,1)(z-1)^2 + \mathcal{O}(z-1)^3. \tag{25}$$

Differentiating (8) with respect to z and setting $(x, y, z) = (1, 1, 1)$ links the derivatives of the generating function to the factorial moments:

$$\frac{\partial G}{\partial z}(1,1,1) = \langle n \rangle, \quad \frac{\partial^2 G}{\partial z^2}(1,1,1) = \langle n(n-1) \rangle. \tag{26}$$

Inserting (10) and (26) into (25), we have

$$G(1,1,z) = 1 + \langle n \rangle (z-1) + \frac{\langle n(n-1) \rangle}{2}(z-1)^2 + \mathcal{O}(z-1)^3. \quad (27)$$

On the other hand, (11), (14), and (17) imply

$$\begin{aligned} G(1,1,z) &= \exp\left(a_1(z-1) + a_2(z-1)^2 + \mathcal{O}(z-1)^3\right) \\ &= \left(1 + a_1(z-1) + \frac{a_1^2}{2}(z-1)^2\right)\left(1 + a_2(z-1)^2\right) + \mathcal{O}(z-1)^3 \\ &= 1 + a_1(z-1) + \left(a_2 + \frac{a_1^2}{2}\right)(z-1)^2 + \mathcal{O}(z-1)^3. \end{aligned} \quad (28)$$

Comparing (27) and (28) gives

$$\langle n \rangle = a_1, \quad \langle n(n-1) \rangle = 2a_2 + a_1^2.$$

The Fano factor,

$$\mathrm{F} = \frac{\langle n^2 \rangle}{\langle n \rangle} - \langle n \rangle = \frac{\langle n(n-1) \rangle}{\langle n \rangle} + 1 - \langle n \rangle = \frac{2a_2}{a_1} + 1, \quad (29)$$

is thus expressed in terms of the first two factorial cumulants a_1 and a_2.

Noise Reduction by mRNA Inactivation Loop. Substituting (18) and (23) into (29) and simplifying gives

$$\mathrm{F} = 1 + \frac{b_1}{b_0} = 1 + \frac{\lambda_2}{\gamma_2 + \gamma_1 + \frac{\alpha(\gamma_2 + \tilde{\gamma}_1)}{\gamma_2 + \tilde{\gamma}_1 + \beta}}. \quad (30)$$

Formula (30) gives the steady-state protein Fano factor as function of the model parameters (degradation rate constants $\gamma_1, \tilde{\gamma}_1, \gamma_2$ of active/inactive mRNA and protein; inactivation/activation rate constants α, β; translation rate constant λ_2).

In order to compare the protein noise in the current model to that exhibited by the classical two-stage model (without the inactivation–activation loop) we define the baseline Fano factor as

$$\mathrm{F}_0 = 1 + \frac{\lambda_2}{\gamma_2 + \gamma_1^{\mathrm{eff}}} = 1 + \frac{\lambda_2}{\gamma_2 + \gamma_1 + \frac{\alpha \tilde{\gamma}_1}{\tilde{\gamma}_1 + \beta}}, \quad (31)$$

which can be obtained from (30) by first setting $\alpha = 0$ (no inactivation) and then replacing the mRNA decay rate γ_1 by its effective value (4). Adjusting the mRNA decay rate maintains the same species means in the baseline model like in the full model extended by the inactivation loop. Note that a comparison in protein variance between the extended and canonical two-stage model can also be done by the mRNA autocovariance function [28].

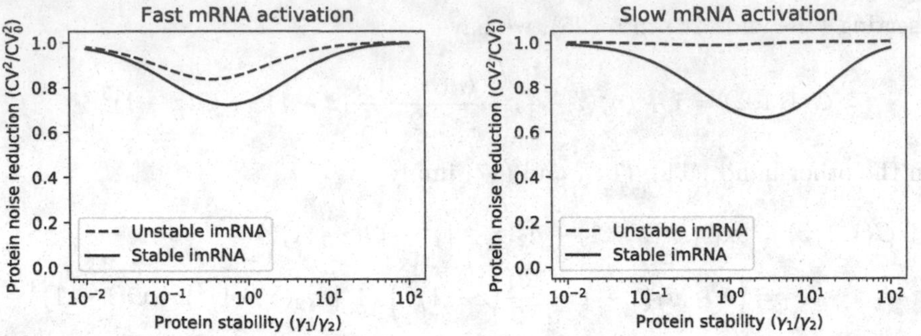

Fig. 1. Fractional protein noise reduction by the mRNA inactivation loop as function of protein stability. The ordinate gives the protein noise (the squared coefficient of variation) in the two-stage model extended by the mRNA inactivation loop relative to the protein noise in a baseline two-stage model without the mRNA inactivation loop (adjusting the mRNA decay rate to obtain the same species means). The protein mean is set to $\langle n \rangle = 500$; the mRNA mean is $\langle m \rangle = 10$; the imRNA decay rate is either the same as that of active mRNA ($\tilde{\gamma}_1 = \gamma_1$; dashed line) or set to zero ($\tilde{\gamma}_1 = 0$; solid line). The inactivation and activation rates are $\alpha = 3$, $\beta = 3$ (left panel) or $\alpha = 1$, $\beta = 0.1$ (right panel); we thereby set $\gamma_1 = 1$ without loss of generality.

The protein variability formulae (30) and (31) can equivalently be expressed in terms of the squared coefficient of variation [13,22] $\mathrm{CV}^2 = \mathrm{F}/\langle n \rangle$ and $\mathrm{CV}_0^2 = \mathrm{F}_0/\langle n \rangle$. Combining (3) and (30)–(31), we find

$$\mathrm{CV}^2 = \frac{1}{\langle n \rangle} + \frac{1}{\langle m \rangle} \frac{\gamma_2}{\gamma_2 + \gamma_1 + \frac{\alpha(\gamma_2 + \tilde{\gamma}_1)}{\gamma_2 + \tilde{\gamma}_1 + \beta}}, \tag{32}$$

$$\mathrm{CV}_0^2 = \frac{1}{\langle n \rangle} + \frac{1}{\langle m \rangle} \frac{\gamma_2}{\gamma_2 + \gamma_1 + \frac{\alpha \tilde{\gamma}_1}{\tilde{\gamma}_1 + \beta}} \tag{33}$$

for the protein coefficient of variation and its baseline value (no activation loop).

Comparing (32) to (33), we see that $\mathrm{CV}^2 < \mathrm{CV}_0^2$, allowing us to conclude that the inclusion of the mRNA inactivation loop decreases protein noise. However, the two coefficients will be very close in many parameter regimes; the necessary conditions for observing a significant difference are given by

$$\tilde{\gamma}_1 \lesssim \min\{\beta, \gamma_2\}, \quad \max\{\gamma_1, \gamma_2\} \lesssim \alpha, \tag{34}$$

where by "$\lesssim$" we mean smaller than or of similar magnitude. Thus, in order to obtain significant reduction of noise, we require that an individual active mRNA molecule be more likely to be inactivated than degraded, and that an individual inactive mRNA molecule be more likely to be activated than degraded. Additionally, we require that inactive mRNA be more stable than protein (which is possible if inactivation protects the mRNA from decay).

One particular consequence of the necessary conditions (34) is that the fraction protein noise reduction, $\mathrm{CV}^2/\mathrm{CV}_0^2$, is a non-monotonous function of protein stability: it tends to one for highly unstable or highly stable proteins, and is less than one for proteins of optimal stability (cf. Fig. 1). The optimal value of protein stability critically depends on the rate constant β of mRNA activation. In case of fast mRNA activation, the optimum noise reduction is achieved by unstable proteins (less stable than mRNA; Fig. 1, left panel. In case of slow mRNA activation, the optimum can be achieved by stable proteins (Fig. 1, right panel). However, slow activation ($\beta \ll 1$) imposes, via (34), a stringent condition on the stability of inactivated mRNA. Indeed, the right panel of Fig. 1 demonstrates that there is hardly any reduction of noise if the inactive mRNA is unstable.

In the next section, we go beyond the mean and noise statistics (the first and second order factorial cumulants), using the higher order cumulants to find a special-function representation of the generating function of the joint distribution of mRNA, imRNA, and protein copy numbers.

5 Special-Function Representation

Factorising the second-order polynomial in k in the denominator of (23) gives

$$b_k = \lambda_2 \frac{\tilde{\gamma_1} + \beta + k\gamma_2}{\gamma_2^2 (k + r_1)(k + r_2)} b_{k-1} \quad \text{for } k \geq 1, \tag{35}$$

where

$$r_{1,2} = \frac{\gamma_1 + \alpha + \tilde{\gamma_1} + \beta \pm \sqrt{(\tilde{\gamma_1} + \beta - \gamma_1 - \alpha)^2 + 4\beta\alpha}}{2\gamma_2}.$$

Note that the sequence b_k in (35) can be rewritten as

$$b_k = b_0 \frac{(1+\tau)_k}{(1+r_1)_k (1+r_2)_k} \left(\frac{\lambda_2}{\gamma_2}\right)^k, \quad k \geq 1, \tag{36}$$

where we set $\tau = (\tilde{\gamma_1} + \beta)/\gamma_2$ for the sake of simplicity and the polynomial

$$(x)_k = x(x+1)(x+2)\dots(x+k-1), \quad (x)_0 = 1$$

represents the rising factorial, also called the Pochammer symbol.

We next find the remaining sequences a_k and c_k. Inserting (36) into (18) gives

$$a_k = \frac{b_0 r_1 r_2}{\tau} \frac{(\tau)_k}{k (r_1)_k (r_2)_k} \left(\frac{\lambda_2}{\gamma_2}\right)^k, \quad k \geq 1. \tag{37}$$

Similarly, substituting (36) into (22) yields

$$c_k = \frac{\alpha b_0}{\tilde{\gamma_1} + \beta} \frac{(\tau)_k}{(1+r_1)_k (1+r_2)_k} \left(\frac{\lambda_2}{\gamma_2}\right)^k, \quad k \geq 1, \tag{38}$$

where $c_0 = \frac{\alpha b_0}{\tilde{\gamma_1}+\beta}$, which can be obtained by combining (20) and (24).

Having found the sequences in (17), we next return to the original variables in (11) to obtain the generating function of the stationary distribution of active mRNA, inactive mRNA, and protein amounts, which is given by

$$G(x,y,z) = \exp\left(\sum_{k\geq 1} a_k(z-1)^k + (x-1)\sum_{k\geq 0} b_k(z-1)^k + (y-1)\sum_{k\geq 0} c_k(z-1)^k\right). \tag{39}$$

Equation (39) can be rewritten as

$$\begin{aligned} G(x,y,z) = \exp\Bigg(&\frac{b_0\lambda_2}{\gamma_2}\int_1^z {}_2F_2\left(\begin{matrix}1, 1+\tau\\ 1+r_1, 1+r_2\end{matrix}; \frac{\lambda_2}{\gamma_2}(s-1)\right) ds \\ &+ b_0(x-1)\,{}_2F_2\left(\begin{matrix}1, 1+\tau\\ 1+r_1, 1+r_2\end{matrix}; \frac{\lambda_2}{\gamma_2}(z-1)\right) \\ &+ \frac{\alpha b_0}{\tilde{\gamma}_1+\beta}(y-1)\,{}_2F_2\left(\begin{matrix}1, \tau\\ 1+r_1, 1+r_2\end{matrix}; \frac{\lambda_2}{\gamma_2}(z-1)\right)\Bigg) \end{aligned} \tag{40}$$

in terms of the generalised hypergeometric functions defined by [2]

$${}_pF_q\left(\begin{matrix}a_1, \ldots, a_p\\ b_1, \ldots, b_q\end{matrix}; \tilde{z}\right) = \sum_{n=0}^{\infty} \frac{(a_1)_n \ldots (a_p)_n}{(b_1)_n \ldots (b_q)_n} \frac{\tilde{z}^n}{n!}. \tag{41}$$

Equation (39) provides the sought-after special function representation of the joint generating function. In the following section, we focus on specific one-dimensional sections of the joint generating function that give the generating functions of the three marginal distributions.

6 Marginal Distributions

In this section, we use the analytic formula (40) for the generating function to determine the marginal active and inactive mRNA, and protein distributions. To do so, we first set $y = z = 1$ in (40) and obtain

$$G(x) = G(x,1,1) = \exp(b_0(x-1))$$

for the marginal active mRNA distribution. Similarly, setting $x = z = 1$ in (40) yields the marginal inactive mRNA distribution

$$G(y) = G(1,y,1) = \exp\left(\frac{\alpha b_0}{\tilde{\gamma}_1+\beta}(y-1)\right).$$

Finally, we set $x = y = 1$ in (40) and get the marginal protein generating function $G(z)$ as

$$G(z) = G(1,1,z) = \exp(\psi(z)),$$

where ψ is given by

$$\psi(z) = \frac{b_0\lambda_2}{\gamma_2}\int_1^z {}_2F_2\left(\begin{matrix}1, 1+\tau\\ 1+r_1, 1+r_2\end{matrix}; \frac{\lambda_2}{\gamma_2}(s-1)\right)ds. \tag{42}$$

In order to obtain the marginal protein distribution, we exploit its generating function

$$p_{\cdot,\cdot,n} = \left.\frac{\mathrm{D}^n(G(z))}{n!}\right|_{z=0}, \tag{43}$$

where D stands for the differential operator d/dz and $p^{st}_{\cdot,\cdot,z}$ gives the probability of having z protein molecules and any number of active and inactive amount of mRNA. The first derivative of the composite function $G(z)$ in (43) is obtained by chain rule as

$$\frac{\mathrm{d}G(z)}{\mathrm{d}z} = G(z)\frac{\mathrm{d}\psi(z)}{\mathrm{d}z}. \tag{44}$$

For the n-th derivative, we evaluate the $(n-1)$th derivative of (44) according to the Leibniz rule, thus we have

$$\mathrm{D}^n(G(z)) = \sum_{i=0}^{n-1}\binom{n-1}{i}\mathrm{D}^i(G(z))\mathrm{D}^{n-i}(\psi(z)). \tag{45}$$

Next, we determine the rth–r is an arbitrary positive integer–derivative of the function $\psi(z)$, which is given by

$$\mathrm{D}^r(\psi(z)) = b_0\left(\frac{\lambda_2}{\gamma_2}\right)^r \frac{(r-1)!(1+\tau)_{r-1}}{(1+r_1)_{r-1}(1+r_2)_{r-1}}\,{}_2F_2\left(\begin{matrix}r, \tau+r\\ r_1+r, r_2+r\end{matrix}; \frac{\lambda_2}{\gamma_2}(z-1)\right), \tag{46}$$

in which we used the formula

$$\frac{\mathrm{d}^s}{\mathrm{d}\tilde{z}^s}\,{}_pF_q\left(\begin{matrix}a_1, \ldots, a_p\\ b_1, \ldots, b_q\end{matrix}; \tilde{z}\right) = \frac{\prod_{i=1}^p (a_i)_s}{\prod_{j=1}^q (b_j)_s}\,{}_pF_q\left(\begin{matrix}a_1+s, \ldots, a_p+s\\ b_1+s, \ldots, b_q+s\end{matrix}; \tilde{z}\right)$$

for the s-th derivative of the generalised hypergeometric function ${}_pF_q$. Inserting the derivatives in (46) into (45), taking $z = 0$, and rearranging the resulting equation according to (43) gives the formula for the marginal protein probabilities

$$\begin{aligned}p_{\cdot,\cdot,n} = {} & \frac{b_0\lambda_2}{n\gamma_2}\sum_{i=0}^{n-1}\left(\frac{\lambda_2}{\gamma_2}\right)^{n-i-1}\frac{(1+\tau)_{n-i-1}}{(1+r_1)_{n-i-1}(1+r_2)_{n-i-1}}\\ & \times {}_2F_2\left(\begin{matrix}n-i, \tau+n-i\\ n-i+r_1, n-i+r_2\end{matrix}; -\frac{\lambda_2}{\gamma_2}\right)p_{\cdot,\cdot,i},\end{aligned} \tag{47}$$

where the first term of the series is given by

$$p_{\cdot,\cdot,0} = G(0) = \exp\left(-\frac{b_0\lambda_2}{\gamma_2}\int_0^1 {}_2F_2\left(\begin{matrix}1, 1+\tau\\ 1+r_1, 1+r_2\end{matrix}; \frac{\lambda_2}{\gamma_2}(s-1)\right)ds\right). \tag{48}$$

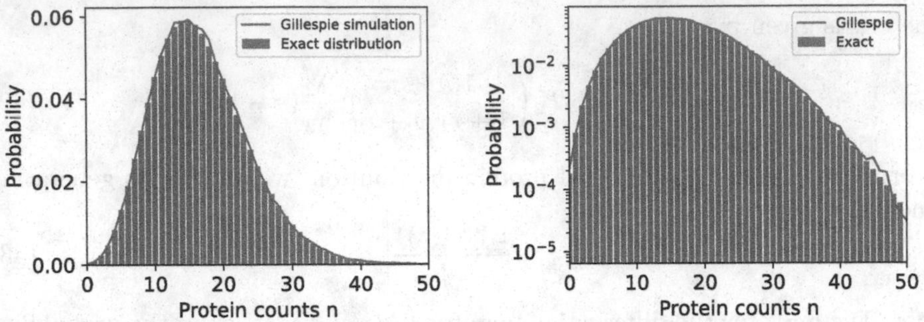

Fig. 2. *Left:* Comparison of the probability mass function (47) of the marginal protein distribution and the probability calculated by Gillespie's stochastic simulation algorithm (the solid line). *Right:* A logarithmic scale plot of the probability, out of 10^5 repeats, obtained by the two approaches. *Parameter values:* The kinetic parameters are: $\lambda_1 = 5$, $\alpha = \gamma_1 = \beta = \tilde{\gamma_1} = \gamma_2 = 1$, $\lambda_2 = 5$.

In order to calculate and compare the marginal protein probabilities (47) with those obtained by stochastic simulations based on Gillespie's algorithm, we implement the recursive formula (47) in a high-level programming language, Python, together with using its numerical computing library NumPy and plotting library Matplotlib. The probabilities in (47) are calculated iteratively starting from its first term given by (48) up to $n = 50$. In Fig. 2, the right panel compares the theoretical probability distribution (47) (blue bars) with the one obtained using stochastic simulations (solid line) at the timepoint $t = 100$, while the left panel shows the same comparison but on a logarithmic scale. The number of Gillespie iterations was set to 10^5 in the Python package GillesPy2 [1]. The initial number of active and inactive mRNA and protein was set to 5. A Python routine `mpmath.hyp2f2` used to calculate the generalised hypergeometric function ${}_2F_2$ in (47)–(48).

7 Conclusion

In this paper, we analysed a formulation of the two-stage model for gene expression that extends the classical version [4,26] by an mRNA inactivation loop. The principal results of our analysis are the characterisation of the mean and noise behaviour, as well as the underlying probability distribution. The principal tool is the factorial cumulant generating function and the factorial cumulant expansion.

The incorporation of the mRNA inactivation loop into the classical two-stage model for gene expression reduces the protein noise. However, in order for the reduction be substantial, several restrictions on the parameter rates have to be in place. In particular, the protein cannot be too stable or unstable, but its stability has to be optimally chosen. The resulting optimal value of protein stability is typically unrealistically low (lower than mRNA stability, in particular). In order to obtain an optimal stability that is greater than mRNA stability, one has to assume that inactivation protects the mRNA from degradation and activation is slow. Thus, our noise analysis points towards a potential role of the mRNA inactivation loop in gene expression noise control; at the same time, it delineates the limits of its application.

In addition to the noise analysis, we provide a comprehensive classification of the underlying probability distributions. Unsurprisingly, the distributions of the active/inactive mRNA are Poissonian. On the other hand, the protein distribution is highly non-trivial, and is characterised in terms of the generalised hypergeometric series. The characterisation is used to derive a recursive expression for the protein probability mass function. The recursive formula is found to be consistent with kinetic Monte-Carlo simulation (by means of the Gillespie direct method).

In summary, the paper provides a systematic mathematical analysis of an mRNA–protein model for gene expression extended by an inactive mRNA species, and hints at possible functional roles of mRNA inactivation loop in the control of low copy number gene-expression noise.

References

1. Abel, J.H., Drawert, B., Hellander, A., Petzold, L.R.: Gillespy: a python package for stochastic model building and simulation. IEEE Life Sci. Lett. **2**, 35–38 (2016). https://doi.org/10.1109/LLS.2017.2652448
2. Abramowitz, M., Stegun, I.A., Romer, R.H.: Handbook of mathematical functions with formulas, graphs, and mathematical tables. Am. J. Phys. **56**(10), 958–958 (1988). https://doi.org/10.1119/1.15378
3. Bartman, C.R., Hamagami, N., Keller, C.A., Giardine, B., Hardison, R.C., Blobel, G.A., Raj, A.: Transcriptional burst initiation and polymerase pause release are key control points of transcriptional regulation. Mol. Cell **73**(3), 519–532 (2019). https://doi.org/10.1016/j.molcel.2018.11.004
4. Bokes, P., King, J.R., Wood, A.T.A., Loose, M.: Exact and approximate distributions of protein and mRNA levels in the low-copy regime of gene expression. J. Math. Biol. **64**(5), 829–854 (2012). https://doi.org/10.1007/s00285-011-0433-5
5. Bokes, P., King, J.R., Wood, A.T.A., Loose, M.: Transcriptional bursting diversifies the behaviour of a toggle switch: hybrid simulation of stochastic gene expression. Bull. Math. Biol. **75**(2), 351–371 (2013). https://doi.org/10.1007/s11538-013-9811-z
6. Dacheux, E., Malys, N., Meng, X., Ramachandran, V., Mendes, P., McCarthy, J.E.G.: Translation initiation events on structured eukaryotic mRNAs generate gene expression noise. Nucleic Acids Res. **45**(11), 6981–6992 (2017). https://doi.org/10.1093/nar/gkx430
7. Elowitz, M.B., Levine, A.J., Siggia, E.D., Swain, P.S.: Stochastic gene expression in a single cell. Science **297**(5584), 1183–1186 (2002). https://doi.org/10.1126/science.1070919
8. Gardiner, C.: Stochastic Methods: A Handbook for the Natural and Social Sciences. Springer Series in Synergetics, iv. Springer, Heidelberg (2009). www.springer.com/gp/book/9783540707127

9. Johnson, N.L., Kemp, A.W., Kotz, S.: Univariate Discrete Distributions. John Wiley & Sons, iii edn., Oct 2005. https://doi.org/10.1002/0471715816
10. Kurasov, P., Mugnolo, D., Wolf, V.: Analytic solutions for stochastic hybrid models of gene regulatory networks. J. Math. Biol. **82**(1), 1–29 (2021). https://doi.org/10.1007/s00285-021-01549-7
11. Li, J., Ge, H., Zhang, Y.: Fluctuating-rate model with multiple gene states. J. Math. Biol. **81**(4), 1099–1141 (2020). https://doi.org/10.1007/s00285-020-01538-2
12. Munsky, B., Neuert, G., van Oudenaarden, A.: Using gene expression noise to understand gene regulation. Science **336**(6078), 183–187 (2012). https://doi.org/10.1126/science.1216379
13. Paulsson, J.: Summing up the noise in gene networks. Nature **427**(6973), 415–418 (2004). https://doi.org/10.1038/nature02257
14. Peccoud, J., Ycart, B.: Markovian modeling of gene-product synthesis. Theor. Popul. Biol. **48**(2), 222–234 (1995). https://doi.org/10.1006/tpbi.1995.1027
15. Pendar, H., Platini, T., Kulkarni, R.V.: Exact protein distributions for stochastic models of gene expression using partitioning of Poisson processes. Phys. Rev. E **87**(4), 042720 (2013). https://doi.org/10.1103/PhysRevE.87.042720
16. Raser, J.M., O'Shea, E.K.: Noise in gene expression: origins, consequences, and control. Science **309**(5743), 2010–2013 (2005). https://doi.org/10.1126/science.1105891
17. Rodríguez Martínez, M., Soriano, J., Tlusty, T., Pilpel, Y., Furman, I.: Messenger RNA fluctuations and regulatory RNAs shape the dynamics of a negative feedback loop. Phys. Rev. E **81**(3), 031924 (2010). https://doi.org/10.1103/PhysRevE.81.031924
18. Sanchez, A., Choubey, S., Kondev, J.: Regulation of noise in gene expression. Annu. Rev. Biophys. **42**, 469–491 (2013). https://doi.org/10.1146/annurev-biophys-083012-130401
19. Schnoerr, D., Sanguinetti, G., Grima, R.: Approximation and inference methods for stochastic biochemical kinetics-a tutorial review. J. Phys. A: Math. Theor. **50**(9), 093001 (2017). https://doi.org/10.1088/1751-8121/aa54d9
20. Shahrezaei, V., Swain, P.S.: Analytical distributions for stochastic gene expression. Presented at the (2008). https://doi.org/10.1073/pnas.0803850105
21. Singh, A., Hespanha, J.P.: Approximate moment dynamics for chemically reacting systems. IEEE Trans. Autom. Control **56**(2), 414–418 (2011). https://doi.org/10.1109/TAC.2010.2088631
22. Singh, A., Bokes, P.: Consequences of mRNA transport on stochastic variability in protein levels. Biophys. J . **103**(5), 1087–1096 (2012). https://doi.org/10.1016/j.bpj.2012.07.015
23. Singh, A., Hespanha, J.P.: Stochastic hybrid systems for studying biochemical processes. Philos. Trans. Roy. Soc. A Math. Phys. Eng. Sci. **368**(1930), 4995–5011 (2010). https://doi.org/10.1098/rsta.2010.0211
24. Soltani, M., Vargas-Garcia, C.A., Singh, A.: Conditional moment closure schemes for studying stochastic dynamics of genetic circuits. IEEE Trans. Biomed. Circuits Syst. **9**(4), 518–526 (2015). https://doi.org/10.1109/tbcas.2015.2453158
25. Swain, P.S., Elowitz, M.B., Siggia, E.D.: Intrinsic and extrinsic contributions to stochasticity in gene expression. Proc. Natl. Acad. Sci. **99**(20), 12795–12800 (2002). https://doi.org/10.1073/pnas.162041399
26. Thattai, M., Oudenaarden, A.v.: Intrinsic noise in gene regulatory networks. Proc. Natl. Acad. Sci. **98**(15), 8614–8619 (2001). https://doi.org/10.1073/pnas.151588598

27. Thomas, P.: Intrinsic and extrinsic noise of gene expression in lineage trees. Sci. Rep. **9**(1), 474 (2019). https://doi.org/10.1038/s41598-018-35927-x
28. Warren, P.B., Tănase-Nicola, S., ten Wolde, P.R.: Exact results for noise power spectra in linear biochemical reaction networks. J. Chem. Phys. **125**(14), 144904 (2006). https://doi.org/10.1063/1.2356472
29. Zhou, T., Liu, T.: Quantitative analysis of gene expression systems. Quantitative Biol. **3**(4), 168–181 (2015). https://doi.org/10.1007/s40484-015-0056-8

Aeon 2021: Bifurcation Decision Trees in Boolean Networks

Nikola Beneš, Luboš Brim, Samuel Pastva(✉), and David Šafránek

Faculty of Informatics, Masaryk University, Brno, Czech Republic
{xbenes3,brim,xpastva,safranek}@fi.muni.cz

Abstract. Aeon is a recent tool which enables efficient analysis of long-term behaviour of asynchronous Boolean networks with unknown parameters. In this tool paper, we present a novel major release of Aeon (Aeon 2021) which introduces substantial new features compared to the original version. These include (i) enhanced static analysis functionality that verifies integrity of the Boolean network with its regulatory graph; (ii) state-space visualisation of individual attractors; (iii) stability analysis of network variables with respect to parameters; and finally, (iv) a novel decision-tree based interactive visualisation module allowing the exploration of complex relationships between parameters and network behaviour. Aeon 2021 is open-source, fully compatible with SBML-qual models, and available as an online application with an independent native compute engine responsible for resource-intensive tasks. The paper artefact is available via https://doi.org/10.5281/zenodo.5008293.

1 Introduction

Boolean networks (BNs) provide an effective mathematical formalism to model regulatory processes in biological systems. A Boolean network consists of Boolean variables which interact together. These interactions are outlined by a set of *regulations*, which (with the variables) form the *regulatory graph* of the network. The state of the variables is governed by associated Boolean *update functions* (one for each variable). We consider that the variables are updated *asynchronously*, i.e. every state transition corresponds to execution of a single update function in the given source state.

The long-term behaviour of a BN, starting from an initial state, has three possible outcomes. Briefly, the first situation is when the network evolves to a single stable state. Such states are the fixed points or *point attractors* (⊙). The second situation is that the network periodically oscillates through a finite sequence of states—an *oscillating attractor* (↻), i.e. the discrete equivalent of a limit cycle in continuous systems. The third case is what we call a *disordered attractor* (⇄), or chaotic oscillation [18] – an attractor that is neither stable not periodically oscillating and in which the system may behave unpredictably, due to the nondeterminism of the asynchronous dynamics of the BN. Together, we refer to the multiplicity of these three attractor types in a system as its *behaviour*

E. Cinquemani and L. Paulevé (Eds.): CMSB 2021, LNBI 12881, pp. 230–237, 2021.
https://doi.org/10.1007/978-3-030-85633-5_14

class. Attractors are particularly relevant in the context of biological modelling as they are used to represent differentiated cellular types or tissues (in the case of fixed points) [2] and biological rhythms or oscillations (in the case of cycles) [10].

A critical problem of BN modelling is to fully determine the update functions. In many cases, BNs with incompletely specified update functions have to be used. We consider *parametrised BNs* where the unknown information is represented in terms of *logical parameters* that appear in logical expressions specifying the update functions [22]. The long-term behaviour of parametrised BNs is affected by the logical parameters. This is observed, e.g., in the count and/or quality of the attractors. A change in the set of attractors (i.e. the behaviour class of the system) is called a *bifurcation.* Since there is no natural notion of a bifurcation point for Boolean networks, we use the so called *bifurcation function* to describe the partition of the parameter space into disjoint regions exhibiting the same behaviour class. Determining the bifurcation function for a network, *attractor bifurcation analysis*, is an important task in the analysis of BNs [4].

Several computational tools have been developed for construction, visualisation and analysis of attractors in non-parametrised BNs. From amongst them, the established tools include ATLANTIS [20], Bio Model Analyzer (BMA) [6], BoolNet [17], PyBoolNet [15], Inet [7], The Cell Collective [13], CellNetAnalyzer [14], and ASSA-PBN [16]. Another group of existing tools targets the parameter synthesis problem for parametrised BNs. The most prominent tools here are GRNMC [12], GINsim [8] (through NuSMV [9]), and TREMPPI [21]. In general, parameter synthesis tools can be used to identify parameters producing a specified long-term behaviour (depending on the logics employed), however, they do not provide a sufficient solution for identification and classification of all attractors in large parametrised BNs. Finally, [1] takes a slightly different approach to bifurcation analysis of logical models – their goal is to study bifurcations in a continuous dynamical system derived from the logical model.

In [3], we introduced the tool Aeon for attractor bifurcation analysis of parametrised BNs. The tool fills the gap in the existing tools by enabling fully automated bifurcation analysis of BNs. However, the first version suffers from several problems that limit its practical applications. First, semi-symbolic algorithms allow it to handle BNs with at most 15–20 variables. Second, the presentation of the results is limited to displaying the classification of attractors without any possibility to explore the structure of individual attractors, or the behaviour of variables in them. Third, the results of the attractor bifurcation analysis do not provide detailed information on the bifurcation function structure.

To address the performance limits of Aeon, we have significantly improved the algorithmics by introducing a novel *fully symbolic* solution utilising a powerful new heuristics [5]. This allows us to analyse parametrised BNs with hundreds of variables in seconds. The presentation of results, notably the bifurcation function structure, has been improved by implementing the automated construction of *bifurcation decision trees* [4]. These allow visualisation of decisions reflecting the effect of bifurcations on achieving particular behaviour classes. This ML-based technology allows the user to comprehend the exhaustive information computed by the tool. All of these features including the added possibility of visual attractor exploration and static consistency checking of the model are included in Aeon 2021 as presented in this paper.

Due to space constraints, we omit the formal definitions of parametrised BNs, their asynchronous semantics, and their attractors in this tool paper. Instead, we refer the interested reader to the Preliminaries section of [3]. In the following, we discuss the novel additions to the Aeon tool and their implementation aspects. We then show a simple case study that demonstrates the usefulness of the implemented techniques.

2 Methods

Here we summarise the improvements made to Aeon since its publication in [3].

Regulatory Graph Consistency. A Boolean network typically corresponds to a *regulatory graph* that is distributed alongside that network and describes high-level properties of interactions between network variables. Such properties typically include the presence of a regulation (essentiality) as well as monotonicity (activation or inhibition). However, especially for large networks, it is easy to accidentally write a function that violates these declared properties.

In Aeon, the update functions are represented symbolically using BDDs. We can then algorithmically verify that the function follows the declared properties by constructing a BDD for each such property, and check that they have a non-empty intersection with the function BDD. In case of functions with parameters, we can even compute the exact number of parametrisations that follow these static constraints. Finally, Aeon can infer which properties are violated, and show them to the user *while* the network is being edited.

Example 1. If the user writes an update function `B∧(B∨A)`, Aeon will notify them that `A` is not essential in this function, i.e. it has no effect on its output. Similarly, if `B` was declared as a negatively monotonous regulation (i.e. an inhibition), Aeon would notify the user that the function is in fact positively monotonous in `B`.

While this may seem like a trivial feature, in our experience, most Boolean network toolboxes do not perform this type of analysis. As a result, we have encountered numerous instances of large models with subtle inconsistencies like this in basically all public Boolean network repositories.

Attractor Visualisation. Originally, Aeon was not able to show the state space of the discovered attractors. In the current version, the interface allows the user to jump to an interactive visualisation of the attractor state space. Here, Aeon highlights the behavioural properties of each attractor (stability, oscillation, disorder), as well as distinguishes the fixed (stable) variables from the ones whose value changes in different attractor states (unstable).

The visualisation uses the `vis.js` JavaScript library and is limited to roughly 5,000 states. If the attractor is larger, Aeon shows a "simplified" view using a two-state loop, with all unstable variables updating synchronously in one step. For Boolean networks with parameters, we always show the attractors for one specific parametrisation only. However the user can specify which behaviour class should be represented. As an example of attractor visualisation, see Fig. 1.

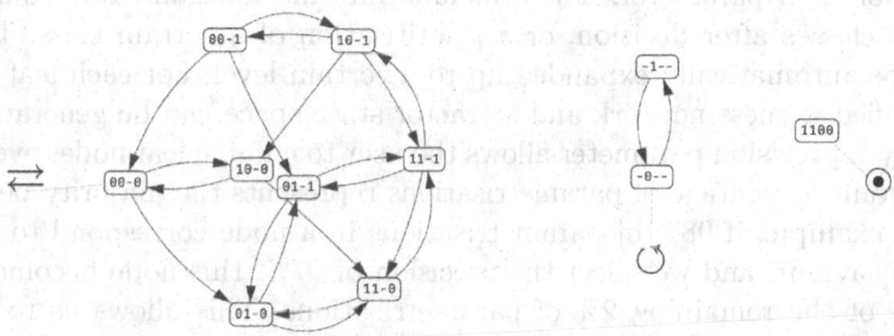

Fig. 1. The state space of three attractors (using variables v_1, v_2, v_3 and v_4, with matching order in the state labels), labelled with behaviour types. The dashes in labels represent stable variables in otherwise unstable attractors. In the ⇄ attractor, v_3 is always *true*, whereas in ↻, v_1 and v_4 are *true* while v_3 is *false* (this information was obtained through other parts of the user interface).

Bifurcation Decision Trees. One of the main challenges when working with parametrised Boolean networks is the fact that the parameter space can be enormous and simultaneously high-dimensional (with tens or even hundreds of parameters). In Aeon, we have long struggled with the visualisation of bifurcation functions for this very reason. In [4], we have proposed a method for taming the multidimensionality of the problem using decision trees, which are commonly used to address this exact problem in machine learning.

The method is now fully implemented in Aeon, with an interactive editor based on the `cytoscape` library [11]. The tree consists of three types of nodes:

- A *leaf* node represents a group of parametrisations that exhibit the same singular behaviour class (for example, bi-stability, i.e. [⊙, ⊙]).
- A *decision* node represents a choice on the value of some parameter. It has an outgoing positive and negative edge, which lead to the subtrees where the parameter is fixed to *true* and *false* respectively.
- Finally, a *mixed* node represents a group of parametrisations that exhibit multiple behaviour classes.

The tree starts as one mixed node associated with a bifurcation function over *all* the admissible parametrisations. Then, the user can expand this mixed node into a decision node by selecting a parameter used as a choice in the newly created node. The bifurcation function is then split into two, based on this parameter. If one of the subtrees contains only parametrisations with a singular behaviour class, it becomes a leaf. Otherwise, a new mixed node is created and the user can repeat this process until all mixed nodes are expanded. An example of such a tree is shown in Fig. 2.

The user can revert each decision and test different combinations of choices. Additionally, the user can make their choice based on various metrics computed by Aeon for each parameter. These include information gain, total number of behaviour classes after decision, or a prioritisation of a certain class. The tree can also be automatically expanded up to a certain level. For each leaf node, a fully specified witness network and attractor state space can be generated.

Finally, a precision parameter allows the user to enforce leaf nodes even when only a certain percentage of parametrisations represents the majority behaviour class. For example, if 98% of parametrisations in a node correspond to a single type of behaviour, and we select the precision of 97%, this node becomes a leaf regardless of the remaining 2% of parametrisations. This allows us to quickly filter out unlikely edge cases and focus on the most prominent behaviour classes.

Stability Analysis. Finally, once the bifurcation decision tree is constructed, the user can perform *stability analysis* for any node in the tree. During stability analysis, Aeon considers all the network variables and all the attractor states associated with the particular tree node (i.e. the attractors appearing for the tree node parametrisations). Then, for each variable, it categorises the parametrisations based on the values appearing in the individual attractors.

Example 2. Consider the attractors shown in Fig. 1. Here:

- v_1 switches between *unstable* (the $\rightleftarrows$ attractor) and *true* (rest).
- v_2 also switches between *unstable* (both the $\rightleftarrows$ and $\circlearrowleft$) and *true* ($\odot$).
- v_3 switches between being always *true* ($\rightleftarrows$) and always *false* ($\circlearrowleft$ and $\odot$).
- v_4 switches between all three cases: *true* in $\circlearrowleft$, *false* in $\odot$ and *unstable* in $\rightleftarrows$.

As a result, aside from the information about attractor behaviour classes, the user can also obtain a more fine-grained categorisation of the parameter space based on whether a particular variable will appear as always *true*, always *false*, *unstable* or some (nondeterministic) combination of these properties in the network attractors. For each of such cases, we can again give a non-parametrised witness network or explore the attractor state space.

3 Case Study

Let us consider the T-LGL model from [19], which we obtained as an SBML file from the Cell Collective database [13]. The model has 60 variables, out of which 6 are constant and are automatically recognised by Aeon as parameters. In this model, a key variable is `Apoptosis`, which when *true* indicates normal cell behaviour, and when *false* indicates a cancerous disorder.

In [19], the authors had to fix these 6 constant values and perform other structural reductions of the network to make their analysis feasible. Aeon can easily analyse the original network with all input combinations at once in roughly 10 s on a basic laptop. This results in 8 distinct behaviour classes.

Upon loading the results in the bifurcation tree editor, a quick stability check uncovers 14 parametrisations with `Apoptosis` always stabilising to *true* and 50 with a switching behaviour, where `Apoptosis` is either *true* or *false* depending on the attractor. Importantly, no parametrisations have `Apoptosis` set always to *false*, or attractors in which it is unstable.

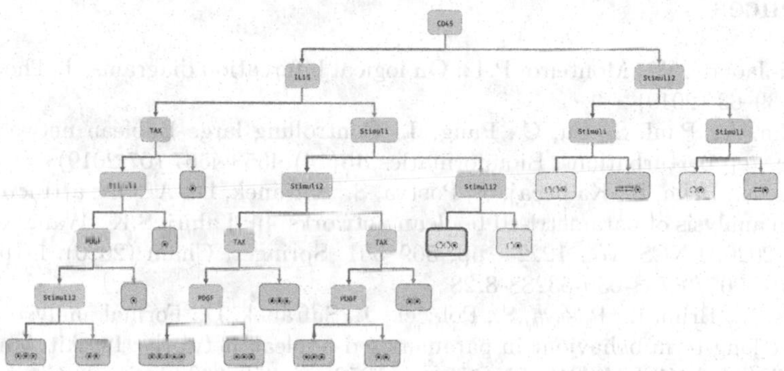

Fig. 2. Bifurcation decision tree of the T-LGL [19] model with 6 inputs (produced by Aeon). The highlighted node contains the parametrisation originally considered in [19].

By expanding the decision tree, we uncover a complete mapping between parametrisations and attractor phenotypes, as seen in Fig. 2. Here, we see a clear transition from stable to oscillating and disordered behaviour, and identify `CD45`, `IL15`, and `Stimuli` as key values leading to this transition. Using stability analysis, we can also easily observe that the parametrisations where `Apoptosis` is always *true* are exactly the ones with a single stable attractor. The tree then shows the conditions guaranteeing the presence of this phenotype class. Such observations could be then used to design experiments and treatments that focus on ensuring these favourable conditions.

Finally, we compare our findings to those in [19]. We confirm that their setting of input values (`Stimuli` = `IL15` = `PDGF` = *true* and `Stimuli2` = `CD45` = `TAX` = *false*) indeed leads to a non-trivial oscillating attractor with `Apoptosis` = *false*, and with `TCR` and `CTLA4` being unstable. Note that our decision tree shows the presence of *two* attractors of these properties. This is due to the structural reductions performed in [19] that simplify the model and consequently merge these two attractors into one. We can also visualise the attractor state space and see that it matches the results presented in [19].

4 Conclusion

We presented a significant update in the functionality of the tool Aeon, which enables completely new workflows for analysis of Boolean networks with partially unknown behaviour. Aeon serves as an interactive editor, simple static analyser, and, thanks to the underlying symbolic algorithms, gracefully handles even very large networks. The visualisation of bifurcation decision trees can be used to uncover interesting dependencies between long-term behaviour and parameters, as well as stability properties of individual variables.

References

1. Abou-Jaoudé, W., Monteiro, P.T.: On logical bifurcation diagrams. J. Theor. Biol. **466**, 39–63 (2019)
2. Baudin, A., Paul, S., Su, C., Pang, J.: Controlling large Boolean networks with single-step perturbations. Bioinformatics **35**(14), i558–i567 (07 2019)
3. Beneš, N., Brim, L., Kadlecaj, J., Pastva, S., Šafránek, D.: AEON: attractor bifurcation analysis of parametrised boolean networks. In: Lahiri, S.K., Wang, C. (eds.) CAV 2020. LNCS, vol. 12224, pp. 569–581. Springer, Cham (2020). https://doi.org/10.1007/978-3-030-53288-8_28
4. Beneš, N., Brim, L., Pastva, S., Poláček, J., Šafránek, D.: Formal analysis of qualitative long-term behaviour in parametrised boolean networks. In: Ait-Ameur, Y., Qin, S. (eds.) ICFEM 2019. LNCS, vol. 11852, pp. 353–369. Springer, Cham (2019). https://doi.org/10.1007/978-3-030-32409-4_22
5. Beneš, N., Brim, L., Pastva, S., Šafránek, D.: Computing bottom SCCs symbolically using transition guided reduction. In: Computer Aided Verification (2021), accepted. Preprint available from authors
6. rg Benque, D., et al.: Bio Model Analyzer: Visual tool for modeling and analysis of biological networks. In: Computer Aided Verification. Lecture Notes in Computer Science, vol. 7358, pp. 686–692. Springer, Heidelberg (2012)
7. Berntenis, N., Ebeling, M.: Detection of attractors of large Boolean networks via exhaustive enumeration of appropriate subspaces of the state space. BMC Bioinf. **14**, 361 (2013)
8. Chaouiya, C., Naldi, A., Thieffry, D.: Logical modelling of gene regulatory networks with GINsim. In: Bacterial Molecular Networks, pp. 463–479. Springer, Heidelberg (2012). https://doi.org/10.1007/978-1-61779-361-5_23
9. Cimatti, A., et al.: NuSMV 2: an opensource tool for symbolic model checking. In: Brinksma, E., Larsen, K.G. (eds.) CAV 2002. LNCS, vol. 2404, pp. 359–364. Springer, Heidelberg (2002). https://doi.org/10.1007/3-540-45657-0_29
10. Feillet, C., et al.: Phase locking and multiple oscillating attractors for the coupled mammalian clock and cell cycle. Proc. Natl. Acad. Sci. **111**(27), 9828–9833 (2014)
11. Franz, M., Lopes, C.T., Huck, G., Dong, Y., Sumer, O., Bader, G.D.: Cytoscape.js: a graph theory library for visualisation and analysis. Bioinformatics **32**(2), 309–311 (2015)
12. Giacobbe, M., Guet, C.C., Gupta, A., Henzinger, T.A., Paixão, T., Petrov, T.: Model checking the evolution of gene regulatory networks. Acta Inf. **54**(8), 765–787 (2017)
13. Helikar, T., et al.: The cell collective: toward an open and collaborative approach to systems biology. BMC Syst. Biol. **6**(96), 1 (2012)

14. Klamt, S., Saez-Rodriguez, J., Gilles, E.D.: Structural and functional analysis of cellular networks with Cell NetAnalyzer. BMC Syst. Biol. **1**(1), 2 (2007)
15. Klarner, H., Streck, A., Siebert, H.: PyBoolNet: a Python package for the generation, analysis and visualization of Boolean networks. Bioinformatics **33**(5), 770–772 (2016)
16. Mizera, A., Pang, J., Su, C., Yuan, Q.: ASSA-PBN: a toolbox for probabilistic Boolean networks. IEEE/ACM Trans. Comput. Biol. Bioinf. (2018)
17. Müssel, C., Hopfensitz, M., Kestler, H.A.: BoolNet-an R package for generation, reconstruction and analysis of Boolean networks. Bioinformatics **26**(10), 1378–1380 (2010)
18. de S. Cavalcante, H.L.D., Gauthier, D.J., Socolar, J.E.S., Zhang, R. : On the origin of chaos in autonomous Boolean networks (2010)
19. Saadatpour, A., et al.: Dynamical and structural analysis of a T-cell survival network identifies novel candidate therapeutic targets for large granular lymphocyte leukemia. PLoS Comput. Biol. **7**(11), e1002267 (2011)
20. Shah, O.S., et al.: ATLANTIS - attractor landscape analysis toolbox for cell fate discovery and reprogramming. Sci. Rep. **8**(1), 3554 (2018)
21. Streck, A., Thobe, K., Siebert, H.: Comparative statistical analysis of qualitative parametrization sets. In: Abate, A., Šafránek, D. (eds.) HSB 2015. LNCS, vol. 9271, pp. 20–34. Springer, Cham (2015). https://doi.org/10.1007/978-3-319-26916-0_2
22. Zou, Y.M.: Boolean networks with multiexpressions and parameters. IEEE/ACM Trans. Comput. Biol. Bioinf. **10**, 584–592 (2013)

LNetReduce: Tool for Reducing Linear Dynamic Networks with Separated Timescales

Marion Buffard[1,2], Aurélien Desoeuvres[1], Aurélien Naldi[3], Clément Requilé[4], Andrei Zinovyev[5,6], and Ovidiu Radulescu[1(✉)]

[1] LPHI UMR CNRS 5235, University of Montpellier, Montpellier, France
ovidiu.radulescu@umontpellier.fr
[2] IRCM, ICM, INSERM, University of Montpellier, Montpellier, France
[3] Lifeware Group, Inria Saclay-Île de France, Palaiseau, France
[4] Department of Mathematics, Uppsala University, Uppsala, Sweden
[5] Institut Curie, PSL Research University, INSERM U900, 75005 Paris, France
[6] Lobachevsky University, 603000 Nizhny Novgorod, Russia

Abstract. We introduce LNetReduce, a tool that simplifies linear dynamic networks. Dynamic networks are represented as digraphs labeled by integer timescale orders. Such models describe deterministic or stochastic monomolecular chemical reaction networks, but also random walks on weighted protein-protein interaction networks, spreading of infectious diseases and opinion in social networks, communication in computer networks. The reduced network is obtained by graph and label rewriting rules and reproduces the full network dynamics with good approximation at all timescales. The tool is implemented in Python with a graphical user interface. We discuss applications of LNetReduce to network design and to the study of the fundamental relation between timescales and topology in complex dynamic networks.

Availability: the code, documentation and application examples are available at https://github.com/oradules/LNetReduce.

1 Introduction

In bioinformatics and systems biology, molecular networks are used as mechanistic models of cell physiology and disease with numerous applications in biology and medicine. Networks are also used by the complex systems community to study social interactions, epidemics, or computer communication. Generally, large scale networks are available as digraphs, in which vertices and edges represent individuals (for instance molecules) and interactions, respectively. Connectivity is supposed essential for the network properties, therefore a large number of tools are dedicated to the analysis of network topology [4]. However, network dynamics and timescales are also very important. The simplest model of dynamic network is obtained by associating to each edge, a number representing the strength of the interaction or its timescale. For molecular networks, this type of information can result from quantitative network analysis approaches such as

E. Cinquemani and L. Paulevé (Eds.): CMSB 2021, LNBI 12881, pp. 238–244, 2021.
https://doi.org/10.1007/978-3-030-85633-5_15

modular response analysis, flux balance analysis, or from direct probing of the interactions by biochemical or biophysical methods. When interaction timescales are not accurate, one can represent their values by integer orders of magnitude instead of real numbers. In many cases, it is important to know that one interaction is much faster than another without having to know by precisely how much. Integer labelled digraphs are thus well suited to study network properties that depend on timescale orders. In this paper we introduce a tool to simplify such networks to an extent that qualitative analysis of their dynamics becomes easy.

2 Model

Dynamic networks are represented as integer edge-labeled digraphs $G = (V, \mathcal{A}, \mathcal{L})$, with V the set of vertices, $\mathcal{A} \subset V \times V$ the set of edges, $\mathcal{L} : \mathcal{A} \to O \subset \mathbb{N}$ the label function. The labels can be obtained from timescales as follows. Fixing a scale basis $0 < \epsilon < 1$, to each edge with kinetic constant k and timescale $\tau = 1/k$ and we associate an integer kinetic order $g = round(\log(k)/\log(\epsilon))$, where *round* stands for round half down. g is the order of magnitude of k, as shown by $\epsilon^{g+0.5} < k \leq \epsilon^{g-0.5}$. The time units are chosen such that the fastest reaction has $k = 1, g = 0$, which results in positive integer orders. When $\epsilon = 1/10$ one recovers the familiar decimal orders of magnitude. Using this power parametrisation we can cope with widely distributed rates. Such networks can be endowed with deterministic or stochastic dynamics.

The deterministic dynamics is defined by a set of ODEs:

$$\dot{c}_i = \sum_{(i,j)\in\mathcal{A}} \epsilon^{g_{ij}} c_j - \Big(\sum_{(j,i)\in\mathcal{A}} \epsilon^{g_{ji}}\Big) c_i,\ i \in V,\ g_{ij} = \mathcal{L}((j,i)). \tag{1}$$

The stochastic dynamics is a random walk on the network, where the probability to jump from i to j is proportional to $\epsilon^{g_{ji}}$. For continuous time random walks, (1) is the backward Kolmogorov equation (master equation) and c_i is the probability to be in i.

3 Reduction Algorithm

We are interested in the reduced model valid in the limit $\epsilon \to 0$. This model can be obtained algorithmically using the following rules [3,5–7]:

1. **Pruning**. For any node with several successors keep the edge with minimum order $g_i = \min\{g_{ji}, j \in Succ(i)\}$ and delete all the other edges. We ask for the **condition 1**: *at a bifurcation the minimum order is attained only once.* The result of this step is the deterministic auxiliary network $Aux(G)$. If the auxiliary network is acyclic, the algorithm stops after this step.
2. **Pooling.** If $Aux(G)$ contains cycles, find a maximal set of disjoint irreducible cycles. Replace these cycles by "glued nodes". As pooling will eventually apply several times, it generates hierarchical glued nodes. Each glued node retains the memory of the cycles it contains (nodes and edges) as follows:

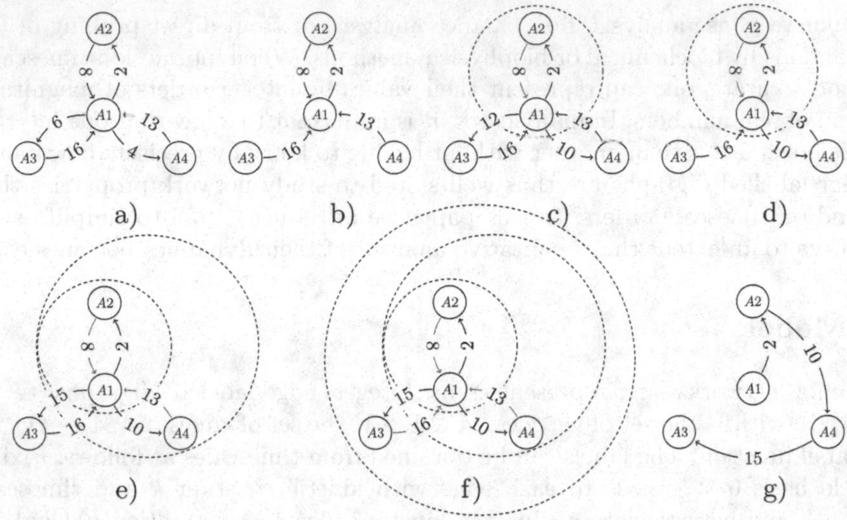

Fig. 1. The successive steps of the reduction algorithm. a) is the initial model; b) is the auxiliary network resulting from pruning; c) is the result of gluing the cycle $\{A_1, A_2\}$ and rewriting the exit edge labels (the labels 6, 4 become $6 + 8 - 2 = 12$, and $4 + 8 - 2 = 10$, respectively); d) is the auxiliary network after one more iteration; e) results from gluing the cycle $\{\{A_1, A_2\}, A_4\}$ and rewriting the exit edge label (12 becomes $12 + 13 - 10 = 15$) ; f) results from gluing the cycle $\{\{\{A_1, A_2\}, A_4\}, A_3\}$; g) restoring the single species without their limiting steps starting with the innermost cycle. Limiting steps of different cycles are represented in red. (Color figure online)

- The glued node inherits all the edges of $Aux(G)$ entering the cycle and also all the edges of G exiting the cycle.
- The labels of edges from $Aux(G)$ are maintained, while the labels of edges of G and not in $Aux(G)$ are recomputed according to the rule:

$$g' = g + g_{lim} - g_c \tag{2}$$

where g', g is the order after and before gluing, respectively, g_{lim} is the largest order edge in the cycle (limiting step) and g_c corresponds to the in-cycle edge sharing the tail with the exit edge. Here we ask for the **condition 2:** *the limiting step is unique in all cycles.*

If the application of pooling results in a non-deterministic graph, apply pruning again. Iterate until there are no more cycles.

3. **Restore** glued vertices through the following steps:
   - Restore all vertices of the glued cycles.
   - Restore all cycle edges except the limiting step.
   - An edge exiting the glued cycle and arriving in an unglued node, is replaced by an edge with the same head and label, but originating from the tail of the limiting step of the glued cycle.
   - An edge exiting the glued cycle and arriving in another glued cycle is replaced as above using its original head within the glued cycle.

The result of restore is path independent: one can start with the most compact or less compact cycles in the hierarchy.

The top level version of the algorithm is given by Algorithm 1. If conditions 1 and 2 are everywhere satisfied, then the reduced graph is acyclic and deterministic. Exceptions lead to stopping the reduction before eliminating all cycles and multiple branching. An example of application of the algorithm to "flower" motifs, consisting of a central hub node and satellite nodes is shown in Fig. 1.

Algorithm 1. Reduce

Require: labeled digraph G
Ensure: reduced labeled digraph
  if condition 1 **then**
    Prune, compute auxiliary deterministic graph;
  else
    break
  end if
  Ncycles:= number of cycles;
  while $Ncycles > 1$ **do**
    Compute disjoint, irreducible cycles;
    if condition 2 **then**
      Glue cycles;
      if condition 1 **then**
        Prune, compute auxiliary deterministic graph;
      else
        break
      end if
    else
      break
    end if
    Ncycles:= number of cycles;
  end while
  Restore glued vertices;

The complexity of the reduction algorithm is $O(n)$ in time where n is the number of nodes, because pruning the full or glued cycles network and computing the disjoint cycles of the auxiliary network ask for $O(n)$ operations. For its implementation we have used Networkx [4].

4 Applications

4.1 Connection Between Topology and Dynamics

For a given topology one can have several reduced models, depending on the kinetic orders. Each reduction corresponds to a particular qualitative dynamics.

In order to illustrate the connection between network topology and its dynamics in the limit of well-separated rate constants, we made a number of experiments on fragments of real-life transcription networks extracted from Dorothea database (https://saezlab.github.io/dorothea/), for the edges of which we assigned random distinct kinetic orders. For example, we extracted all network neighbours of MYCN transcription factor (see Fig. 2A) and randomly assigned kinetic orders from 0 to 9, to network reactions. Application of LNetReduce to 10000 random kinetic order assignments led to 51 topologically distinct (topologically isomorphic with node identity kept) model reductions, where some reductions were much more frequent than others. Of note, in approximately 7% cases, LNetReduce met a conflict in the reduction algorithm. In these cases, two kinetic rates of the same order in the outgoing reaction fork happened at a stage of reduction process. Interestingly, rare reductions (Fig. 2C, D) were characterized by inefficient activation dynamics of MYCN node. This inefficiency was manifested either by leaky dynamics of MYCN, with vanishing probability at long timescales or by existence of reactions with reversed direction in the reduced model. By contrast, the most frequent reductions (Fig. 2A, B) were characterized by activating and efficient (fast) MYCN dynamics. Other computation experiments with fragments of the transcription network are described at the LNetReduce web-site as Python notebooks.

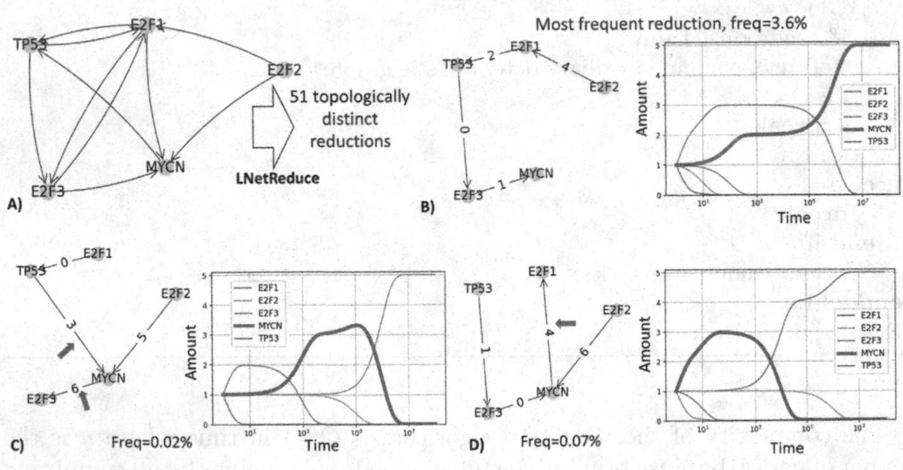

Fig. 2. Example of LNetReduce application for studying the connection between the network topology and dynamics, using a small fragment of experimental transcription regulation network. A) Network fragment. B) Most frequent topologically isomorphic reduction and its dynamics. C, D) Examples of rarely obtained reductions and their dynamics. Red arrows show edges whose direction is reverted with respect to A). (Color figure online)

4.2 Design of Slow Transients

The reduced graph is a forest of inverted (directed towards roots) trees. In this case, relaxation timescales defined as times after which something happens are simply the new labels [3]. However, the relation between initial step timescales g_{ij} and the final labels can be intricate. In particular, relation (2) shows that network timescales tend to be larger than timescales of the initial steps and complex networks tend to have very slow transients. Slow transients are important in biology for a variety of processes from cellular memory to long period circadian oscillations. We can formulate the following design rule for long transients:

Rule 1. *At least at some iteration, the auxiliary network contains cycles such that the in-cycle step at a cycle exit point is not the limiting step, i.e.* $g_{lim} > g_c$. *Then, according to* (2) $g' > g$, *new steps are slower than older.*

Rule 1 is responsible for the slow transient in Fig. 3. In linear networks without separation of rate constants, slow timescales can result from the addition of multiple reaction steps, for instance multiple phosphorylations in signaling networks. This explains why proteins controlling the long 24 h period circadian clock in *Neurospora* have about one hundred phosphorylation sites [1]. In the presence of separation, the main slow-down mechanism is the one described mathematically by Rule 1 and biochemically by *futile cycles with rare output*. In such a mechanism, molecules are processed in a cyclic way and the useful, rare output is generated only after many cycles. Such networks can display the counter-intuitive behavior that a non-uniform increase of all the rate constants can lead to slower transients, as noticed for stochastic networks in [2]. Indeed, it is enough to favor cycling relative to output reactions, thus increasing the number of cycles needed for obtaining the output.

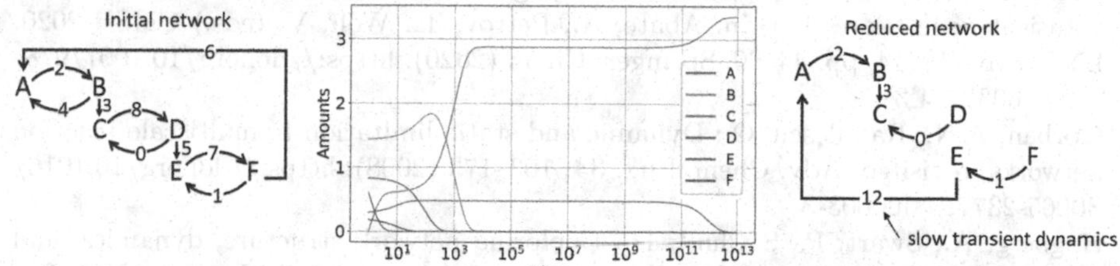

Fig. 3. Application of LNetReduce to identify slow transient dynamics in networks. The cascade shown on the left, after fast initial dynamics achieves a very slowly relaxing state (middle panel). The simulations were done using random initial conditions in the [0;2] range. The timescales of kinetic rates are shown as numbers on reaction arcs (the smaller the faster). On the right the result of application of LNetReduce is shown, explicitly revealing the existence of a very long timescale in the network, four orders of magnitude larger than any reaction in the initial network (12 vs 8). Using random permutations of kinetic orders, we estimate the frequency of emergence of slow transients with this network topology to 4%.

5 Conclusion

We provide a tool allowing to study dynamics of networks with separated rate constants. This tool implements Algorithm 1 that produces a fully reduced network (forest) if the separation Conditions 1,2 are fulfilled. When these conditions are not fulfilled, the reduction is partial and results into graphs with cycles and forks. When the Conditions 1,2 hold, the reduced model provides immediately the relaxation timescales of the network. In this case we can also compute left and right eigenvectors of the reduced model, allowing fast computation of traces [3,5–7]. Although Conditions 1,2 are quite frequent in biology and complex systems applications, in future work we will consider releasing them at least partially when computing the eigenvalues and eigenvectors of the reduced model.

For analysis of protein-protein interaction networks using random walk, our model and tool represent an alternative to uniform jump rate algorithms. For network design it allows to study the interplay between timescales and topology for predicting network dynamics and eventually controllability.

Acknowledgements. This work was supported by Agence Nationale de la Recherche, projects ANR-17-CE40-0036 SYMBIONT and ANR-19-P3IA-0001 (PRAIRIE 3IA Institute), and by the Ministry of Science and Higher Education of the Russian Federation (project No. 14.Y26.31.0022).

References

1. Baker, C.L., Kettenbach, A.N., Loros, J.J., Gerber, S.A., Dunlap, J.C.: Quantitative proteomics reveals a dynamic interactome and phase-specific phosphorylation in the Neurospora circadian clock. Mol. Cell **34**(3), 354–363 (2009)
2. Bokes, P., Klein, J., Petrov, T.: Accelerating reactions at the DNA can slow down transient gene expression. In: Abate, A., Petrov, T., Wolf, V. (eds.) CMSB 2020. LNCS, vol. 12314, pp. 44–60. Springer, Cham (2020). https://doi.org/10.1007/978-3-030-60327-4_3
3. Gorban, A.N., Radulescu, O.: Dynamic and static limitation in multiscale reaction networks, revisited. Adv. Chem. Eng. **34**, 103–173 (2008). https://doi.org/10.1016/S0065-2377(08)00003-3
4. Hagberg, A., Swart, P., S Chult, D.: Exploring network structure, dynamics, and function using networkx. Technical report, Los Alamos National Lab. (LANL), Los Alamos, NM (United States) (2008)
5. Radulescu, O., Gorban, A.N., Zinovyev, A., Noel, V.: Reduction of dynamical biochemical reactions networks in computational biology. Front. Genet. **3**, 131 (2012)
6. Radulescu, O., Gorban, A.N., Zinovyev, A.Y., Lilienbaum, A.: Robust simplifications of multiscale biochemical networks. BMC Syst. Biol. **2**, 86 (2008). https://doi.org/10.1186/1752-0509-2-86
7. Radulescu, O., Swarup Samal, S., Naldi, A., Grigoriev, D., Weber, A.: Symbolic dynamics of biochemical pathways as finite states machines. In: Roux, O., Bourdon, J. (eds.) CMSB 2015. LNCS, vol. 9308, pp. 104–120. Springer, Cham (2015). https://doi.org/10.1007/978-3-319-23401-4_10

Ppsim: A Software Package for Efficiently Simulating and Visualizing Population Protocols

David Doty[✉] and Eric Severson

University of California, Davis, CA 95616, USA
{doty,eseverson}@ucdavis.edu

Abstract. We introduce `ppsim` [28], a software package for efficiently simulating population protocols, a widely-studied subclass of chemical reaction networks (CRNs) in which all reactions have two reactants and two products. Each step in the dynamics involves picking a uniform random pair from a population of n molecules to collide and have a (potentially null) reaction. In a recent breakthrough, Berenbrink, Hammer, Kaaser, Meyer, Penschuck, and Tran [6] discovered a population protocol simulation algorithm quadratically faster than the naïve algorithm, simulating $\Theta(\sqrt{n})$ reactions in *constant* time (independently of n, though the time scales with the number of species), while preserving the *exact* stochastic dynamics.

`ppsim` implements this algorithm, with a tightly optimized Cython implementation that can exactly simulate hundreds of billions of reactions in seconds. It dynamically switches to the CRN Gillespie algorithm for efficiency gains when the number of applicable reactions in a configuration becomes small. As a Python library, `ppsim` also includes many useful tools for data visualization in Jupyter notebooks, allowing robust visualization of time dynamics such as histogram plots at time snapshots and averaging repeated trials.

Finally, we give a framework that takes any CRN with only bimolecular (2 reactant, 2 product) or unimolecular (1 reactant, 1 product) reactions, with arbitrary rate constants, and compiles it into a continuous-time population protocol. This lets `ppsim` exactly sample from the chemical master equation (unlike approximate heuristics such as τ-leaping or LNA), while achieving asymptotic gains in running time. In linked Jupyter notebooks, we demonstrate the efficacy of the tool on some protocols of interest in molecular programming, including the approximate majority CRN and CRN models of DNA strand displacement reactions.

Keywords: Population protocol · Chemical reaction network

1 Introduction

A foundational model of chemistry used in natural sciences is that of chemical reaction networks (CRNs) [22]: finite sets of reactions such as $A + B \rightarrow C + D$,

Supported by NSF award 1900931 and CAREER award 1844976.

E. Cinquemani and L. Paulevé (Eds.): CMSB 2021, LNBI 12881, pp. 245–253, 2021.
https://doi.org/10.1007/978-3-030-85633-5_16

representing that molecules A and B, upon colliding, can change into C and D. This gives a continuous time, discrete state, Markov process [22] modelling discrete counts[1] of molecules.

Population protocols [3], a well-studied model of distributed computing with limited agents, are a restricted subset of CRNs (with two reactants and two products in each reaction, and unit rate constants) that nevertheless capture many of the interesting features of CRNs. Different terminology is used: in reaction $A + B \rightarrow C + D$, two *agents* (molecules), whose *states* (species types) are A, B, have an *interaction* (reaction), changing their states respectively to C, D. *Gillespie kinetics for CRNs.* The standard Gillespie algorithm [22] simulates the Markov process mentioned above. Given a fixed volume $v \in \mathbb{R}^+$, the *propensity* of a unimolecular reaction $r : X \xrightarrow{k} \ldots$ is $\rho(r) = k \cdot \#X$, where $\#X$ is the count of X. The propensity of a bimolecular reaction $r : X + Y \xrightarrow{k} \ldots$ is $\rho(r) = k \cdot \frac{\#X \cdot \#Y}{v}$ if $X \neq Y$ and $k \cdot \frac{\#X \cdot (\#X-1)}{2v}$ otherwise. The Gillespie algorithm calculates the sum of the propensities of all reactions: $\rho = \sum_r \rho(r)$. The time until the next reaction is sampled as an exponential random variable T with rate ρ, and a reaction r_{next} is chosen with probability $\rho(r_{\text{next}})/\rho$ to be applied.

Population Protocols. The population protocols model comes with simpler dynamics. At each step, a scheduler chooses a random pair of agents (molecules) to interact in a (potentially null) reaction. The discrete time model counts each interaction as $\frac{1}{n}$ units of time, where n is the population size. A continuous time variant [18] gives each agent a rate-1 Poisson clock, upon which it interacts with a randomly chosen other agent. The expected time until the next interaction is $\frac{1}{n}$, so up to a re-scaling of time, which by straightforward Chernoff bounds is negligible, these two models are equivalent. `ppsim` can use either time model.

There is an important efficiency difference between the algorithms: the Gillespie algorithm automatically skips null reactions. For example, a reaction such as $L + L \rightarrow L + F$, when $\#L = 2$ and $\#F = n - 2$, is much more efficient in the Gillespie algorithm, which simply increments the time until the $L + L \rightarrow L + F$ reaction by an exponential random variable in one step. A naïve population protocol simulation iterates through $\Theta(n)$ expected null interactions ($L+F \rightarrow L+F$ and $F + F \rightarrow F + F$) until the two L's react. To better handle cases like this, `ppsim` dynamically switches to the Gillespie algorithm when the number of null interactions is sufficiently large; see documentation [28] for implementation details.

Other Simulation Algorithms. Variants of the Gillespie algorithm reduce the time to apply a single reaction from $O(|R|)$ to $O(\log |R|)$ [21] or $O(1)$ [31], where $|R|$ is the number of types of reactions. However, the time to apply n reactions still scales with n. A common speedup heuristic for simulating $\omega(1)$ reactions in $O(1)$ time is *τ-leaping* [10,23,24,29,32], which "leaps" ahead by time τ, by assuming reaction propensities will not change and updating counts in a single batch step

[1] Another modelling choice are ODEs that describe real-valued concentrations, the "mean-field" approximation to the discrete behavior in the large scale limit [26].

by sampling according these propensities. Such methods necessarily approximate the kinetics inexactly, though it is possible in some cases to prove bounds on the approximation accuracy [32]. Linear noise approximation (LNA) [11] can be used to approximate the discrete kinetics, by adding stochastic noise to an ODE approximation. A speedup heuristic for population protocol simulation is to sample the number of each interaction that would result from a random matching of size m, and update species counts in a single step. This, too, is an inexact approximation: unlike the true process, it prevents any molecule from participating in more than one of the next m interactions.

The algorithm implemented by `ppsim`, due to Berenbrink, Hammer, Kaaser, Meyer, Penschuck, and Tran [6], builds on this last heuristic. Conditioned on the event that no molecule is picked twice during the next m interactions, these interacting pairs are a random disjoint matching of the molecules. Define the random variable C as the number of interactions until the same molecule is picked twice. Their basic algorithm samples this collision length C according to its exact distribution, then updates counts in batch assuming all pairs of interacting molecules are disjoint until this collision, and finally simulates the interaction involving the collision. By the Birthday Paradox, $\mathrm{E}[C] \approx \sqrt{n}$ in a population of n molecules, giving a quadratic factor speedup over the naïve algorithm. The time to update a batch scales quadratically with q, the total number of states. The "multibatch" variant, used by `ppsim`, samples multiple successive collisions to process an even larger batch, and uses $O\left(q\sqrt{\frac{\log n}{n}}\right)$ time per simulated interaction.

See [6] for details. An advantage of such a fast simulator, specifically for population protocols implementing *algorithms*, is that the very large population sizes it can handle (over 10^{12}) allow one to tell the difference (on a log-scale plot of convergence time) between a protocol converging in time $O(\log n)$ versus, say, $O(\log^2 n)$.

2 Usage of the Ppsim Tool

We direct the reader to [28] for detailed installation, usage instructions, and examples. Here we highlight basic usage examples for specifying protocols.

There are three ways one can specify a population protocol, each best suited for different contexts. The most direct specification of a protocol directly encodes the mapping of input state pairs to output state pairs using a Python `dict` (the following is the well-studied *approximate majority* protocol, which has been studied theoretically [4,13] and implemented experimentally with DNA [12]):

```
a,b,u = 'A','B','U'
approx_majority = {(a,b):(u,u), (a,u):(a,a), (b,u):(b,b)}
```

More complex protocols with many possible species are often specified in pseudocode instead of listing all possible reactions. `ppsim` supports this by allowing the *transition function* mapping input states to output states to be computed by a Python function. The following allows species to be integers and computes an integer average of the two reactants:

```
def discrete_averaging(s: int, r: int):
    return math.floor((s+r)/2), math.ceil((s+r)/2)
```

States and transition rules are converted to integer arrays for internal Cython methods, so there is no efficiency loss for the ease of representing protocol rules, since a Python function defining the transition function is not called during the simulation: producible states are enumerated before starting the simulation.

For complicated protocols, an advantage of ppsim over standard CRN simulators is the ability to represent species/states as Python objects with different fields (as they are often represented in pseudocode), and to plot counts of agents based on their field values.[2]

Finally, protocols can be specified using CRN-like notation for CRNs with reactions that are bimolecular (2-input, 2-output) or unimolecular (1-input, 1-output), with arbitrary rate constants. For instance, this code specifies the CRN

$$A + B \underset{4}{\overset{0.5}{\rightleftharpoons}} 2C, \qquad C \xrightarrow{5} D$$

```
a,b,c,d = species('A B C D')
crn = [(a+b | 2*c).k(0.5).r(4), (c >> d).k(5)]
```

This will then get compiled into a continuous time population protocol that samples the same distribution as Gillespie. See full paper [14] for details.

Any of the three specifications (dict, Python function, or list of CRN reactions) can be passed to the Simulation constructor. The Simulation can be run to generate a history of sampled configurations.

```
init_config = {a: 51, b: 49}
sim = Simulation(init_config, approx_majority)
sim.run(16, 0.1)    # 160 samples up to time 16
sim.history.plot() # Pandas dataframe with counts
```

This would produce the plot shown in Fig. 1a. When the input is a CRN, ppsim defaults to continuous time and produces the exact same distributions as the Gillespie algorithm. Figure 1b shows a test against the package GillesPy2 [25] to confirm they sample the same distribution.

3 Speed Comparison with Other CRN Simulators

We ran speed comparisons of ppsim against both GillesPy2 [25] and StochKit2 [30], the latter being the fastest option we found for Gillespie simulation. Figure 2 shows that ppsim is able to reach significantly larger population sizes. Other tests shown in an example notebook[3] show how each package scales with the number of species and reactions.

[2] Download and run https://github.com/UC-Davis-molecular-computing/ppsim/blob/main/examples/majority.ipynb to visualize such large state protocols.

[3] https://github.com/UC-Davis-molecular-computing/ppsim/blob/main/examples/crn.ipynb shows further plots and explanations.

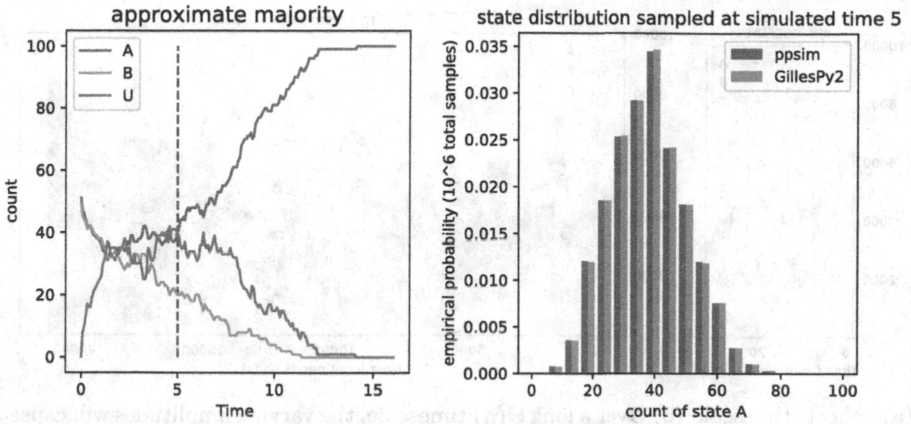

(a) Plot of sim.history. (b) Comparison with Gillespie algorithm.

Fig. 1. Time 5 (dotted line in Fig. 1a) was sampled 10^6 times with **ppsim** and GillesPy2 to verify they both sample the same chemical master equation distribution (Fig. 1b).

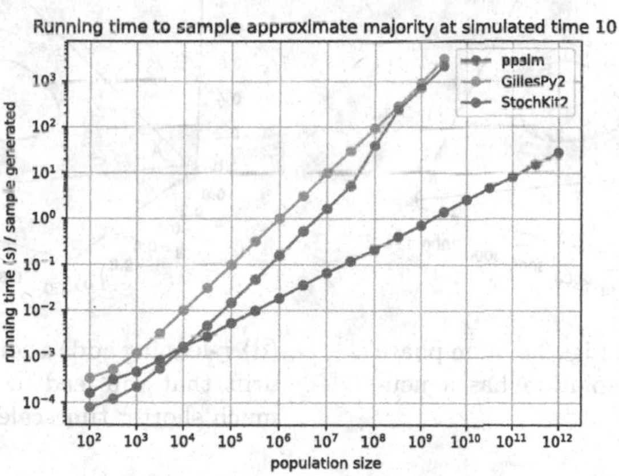

Fig. 2. Comparing runtime with population size n shows $O(n)$ scaling for Gillespie (slope 1 on log-log plot) versus $O(\sqrt{n})$ scaling for **ppsim** (slope 1/2).

4 Issues with Other Speedup Methods

It is reasonable to conjecture that exact stochastic simulation of large-count systems is unnecessary, since Gillespie is fast enough on small-count systems, and faster ODE approximation is "reasonably accurate" for large-count systems. However, there are example large count systems with stochastic effects not observed in ODE simulation, and where τ-leaping introduces systematic inaccuracies that disrupt the fundamental qualitative behavior of the system, demonstrating the need for exact stochastic simulation. A simple such example is the

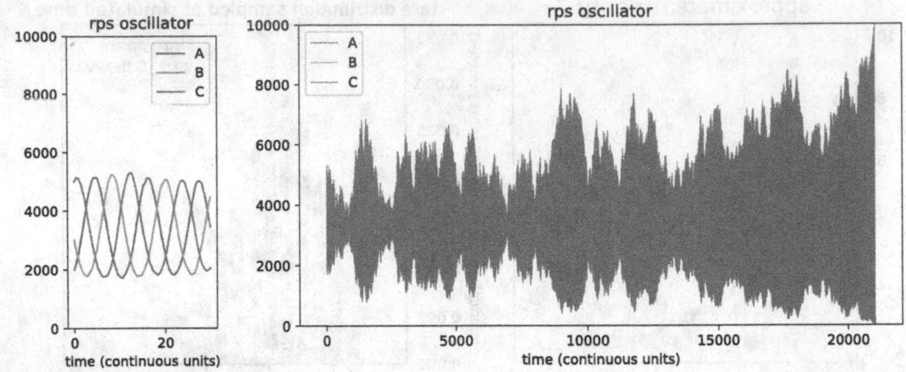

(a) Short timescale oscillations.

(b) Over a long $\Theta(n)$ timescale, the varying amplitudes will cause two species to go extinct.

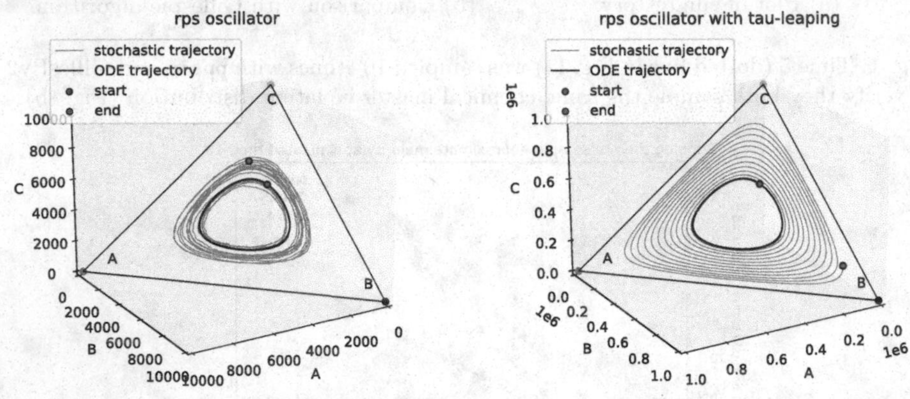

(c) Dynamics from Figs 3a, 3b in phase space. The ODE solution has a neutrally stable orbit.

(d) τ-leaping adds a consistent outward drift that will lead to extinction on a much shorter timescale.

Fig. 3. The rock-paper-scissors oscillator has qualitative dynamics missed by both ODE simulation (never goes extinct) and τ-leaping (too quickly goes extinct).

3-state rock-paper-scissors oscillator: $B + A \to 2B$, $C + B \to 2C$, $A + C \to 2A$. Fig. 3 compares exact simulation of this CRN to τ-leaping and ODEs.

The population protocol literature furnishes more examples, with problems such as leader election [2,5,7–9,15,17,19,20,34,35] and single-molecule detection [1,16],[4] that crucially use small counts in a very large population, a regime not modelled correctly by ODEs. See also [27] for examples of CRNs with

[4] Download and run https://github.com/UC-Davis-molecular-computing/ppsim/blob/main/examples/rps_oscillator.ipynb to see visualizations of the generalized 7-state rps oscillator used for single-molecule detection in [16].

qualitative stochastic behavior not captured by ODEs, yet that behavior appears only in population sizes too large to simulate with Gillespie.

5 Conclusion

Unfortunately, the algorithm of Berenbrink et al. [6] implemented by `ppsim` seems inherently suited to population protocols, not more general CRNs. For instance, reversible dimerization reactions $A+B \rightleftharpoons C$ (used, for example, in [33] to model toehold occlusion reactions in DNA systems) seem beyond the reach of the batching technique of [6]. Although such reactions can be *approximated* by $A+B \rightleftharpoons C+F$ for some anonymous "fuel" species F, the count of F influences the rate of the reverse reaction $F+C \rightarrow A+B$, with a different rate than $C \rightarrow A+B$.

Another area for improvement is the handling of null reactions. There could be a way to more deeply intertwine the logic of the Gillespie and batching algorithms, to gain the simultaneous benefits of each, skipping the null reactions while simulating many non-null reactions in batch.

References

1. Alistarh, D., Dudek, B., Kosowski, A., Soloveichik, D., Uznański, P.: Robust detection in leak-prone population protocols. In: Brijder, R., Qian, L. (eds.) DNA 2017. LNCS, vol. 10467, pp. 155–171. Springer, Cham (2017). https://doi.org/10.1007/978-3-319-66799-7_11
2. Alistarh, D., Gelashvili, R.: Polylogarithmic-time leader election in population protocols. In: Halldórsson, M.M., Iwama, K., Kobayashi, N., Speckmann, B. (eds.) ICALP 2015. LNCS, vol. 9135, pp. 479–491. Springer, Heidelberg (2015). https://doi.org/10.1007/978-3-662-47666-6_38
3. Angluin, D., Aspnes, J., Diamadi, Z., Fischer, M.J., Peralta, R.: Computation in networks of passively mobile finite-state sensors. Distrib. Comput. **18**(4), 235–253 (2006)
4. Angluin, D., Aspnes, J., Eisenstat, D.: A simple population protocol for fast robust approximate majority. Distrib. Comput. **21**(2), 87–102 (2008)
5. Berenbrink, P., Giakkoupis, G., Kling, P.: Optimal time and space leader election in population protocols. In: STOC 2020: Proceedings of the 52nd Annual ACM SIGACT Symposium on Theory of Computing, STOC 2020, pp. 119–129. Association for Computing Machinery, New York (2020). https://doi.org/10.1145/3357713.3384312
6. Berenbrink, P., Hammer, D., Kaaser, D., Meyer, U., Penschuck, M., Tran, H.: Simulating population protocols in sub-constant time per interaction. In: ESA 2020: 28th Annual European Symposium on Algorithms, vol. 173, pp. 16:1–16:22 (2020). https://drops.dagstuhl.de/opus/volltexte/2020/12882
7. Berenbrink, P., Kaaser, D., Kling, P., Otterbach, L.: Simple and efficient leader election. In: 1st Symposium on Simplicity in Algorithms (SOSA 2018), vol. 61, pp. 9:1–9:11 (2018)

8. Bilke, A., Cooper, C., Elsässer, R., Radzik, T.: Brief announcement: population protocols for leader election and exact majority with $O(\log^2 n)$ states and $O(\log^2 n)$ convergence time. In: PODC 2017: Proceedings of the ACM Symposium on Principles of Distributed Computing, pp. 451–453. ACM (2017)
9. Burman, J., et al.: Time-optimal self-stabilizing leader election in population protocols. In: PODC 2021: Proceedings of the 2021 ACM Symposium on Principles of Distributed Computing (2021)
10. Cao, Y., Gillespie, D.T., Petzold, L.R.: Efficient step size selection for the tau-leaping simulation method. J. Chem. Physi. **124**(4), 044109 (2006)
11. Cardelli, L., Kwiatkowska, M., Laurenti, L.: Stochastic analysis of chemical reaction networks using linear noise approximation. Biosystems **149**, 26–33 (2016). https://doi.org/10.1016/j.biosystems.2016.09.004, https://www.sciencedirect.com/science/article/pii/S0303264716302039, selected papers from the Computational Methods in Systems Biology 2015 conference
12. Chen, Y.J., et al.: Programmable chemical controllers made from DNA. Nat. Nanotechnol. **8**(10), 755–762 (2013)
13. Condon, A., Hajiaghayi, M., Kirkpatrick, D., Maňuch, J.: Approximate majority analyses using tri-molecular chemical reaction networks. Nat. Comput. **19**(1), 249–270 (2020)
14. Doty, D., Severson, E.: ppsim: A software package for efficiently simulating and visualizing population protocols. Technical Report 2105.04702, arXiv (2021). arXiv:2105.04702
15. Doty, D., Soloveichik, D.: Stable leader election in population protocols requires linear time. Distrib. Comput. **31**(4), 257–271 (2018), special issue of invited papers from DISC 2015
16. Dudek, B., Kosowski, A.: Universal protocols for information dissemination using emergent signals. In: Proceedings of the 50th Annual ACM SIGACT Symposium on Theory of Computing, pp. 87–99 (2018)
17. Elsässer, R., Radzik, T.: Recent results in population protocols for exact majority and leader election. Bull. EATCS **3**(126) (2018)
18. Fanti, G., Holden, N., Peres, Y., Ranade, G.: Communication cost of consensus for nodes with limited memory. Proc. Natl. Acad. Sci. **117**(11), 5624–5630 (2020)
19. Gąsieniec, L., Stachowiak, G.: Fast space optimal leader election in population protocols. In: SODA 2018: ACM-SIAM Symposium on Discrete Algorithms, pp. 2653–2667. SIAM (2018)
20. Gąsieniec, L., Stachowiak, G., Uznański, P.: Almost logarithmic-time space optimal leader election in population protocols. In: SPAA 2019: 31st ACM Symposium on Parallelism in Algorithms and Architectures, pp. 93–102 (2019)
21. Gibson, M.A., Bruck, J.: Efficient exact stochastic simulation of chemical systems with many species and many channels. J. Phys. Chem. A **104**(9), 1876–1889 (2000)
22. Gillespie, D.T.: Exact stochastic simulation of coupled chemical reactions. J. Phys. Chem. **81**(25), 2340–2361 (1977)
23. Gillespie, D.T.: Approximate accelerated stochastic simulation of chemically reacting systems. J. Phys. Chem. **115**(4), 1716–1733 (2001)
24. Gillespie, D.T.: Stochastic simulation of chemical kinetics. Annu. Rev. Phys. Chem. **58**, 35–55 (2007)
25. GillesPy2. https://github.com/StochSS/GillesPy2
26. Kurtz, T.G.: The relationship between stochastic and deterministic models for chemical reactions. J. Phys. Chem. **57**(7), 2976–2978 (1972)

27. Lathrop, J.I., Lutz, J.H., Lutz, R.R., Potter, H.D., Riley, M.R.: Population-induced phase transitions and the verification of chemical reaction networks. In: Geary, C., Patitz, M.J. (eds.) DNA 26: 26th International Conference on DNA Computing and Molecular Programming. Leibniz International Proceedings in Informatics (LIPIcs), vol. 174, pp. 5:1–5:17. Schloss Dagstuhl–Leibniz-Zentrum für Informatik, Dagstuhl (2020). https://doi.org/10.4230/LIPIcs.DNA.2020.5
28. ppsim Python package. source code (2021). https://github.com/UC-Davis-molecular-computing/ppsim API documentation, https://ppsim.readthedocs.io/ Python package for installation via pip: https://pypi.org/project/ppsim/
29. Rathinam, M., El Samad, H.: Reversible-equivalent-monomolecular tau: a leaping method for "small number and stiff" stochastic chemical systems. J. Comput. Phys. **224**(2), 897–923 (2007)
30. Sanft, K.R., Wu, S., Roh, M., Fu, J., Rone, K.L., Petzold, L.R.: Stochkit2: software for discrete stochastic simulation of biochemical systems with events. Bioinformatics **27**(17), 501–522 (2011). https://academic.oup.com/bioinformatics/article/27/17/2457/224105
31. Slepoy, A., Thompson, A.P., Plimpton, S.J.: A constant-time kinetic monte carlo algorithm for simulation of large biochemical reaction networks. J. Chem. Phys. **128**(20), 05B618 (2008)
32. Soloveichik, D.: Robust stochastic chemical reaction networks and bounded tau-leaping. J. Comput. Biol. **16**(3), 501–522 (2009)
33. Srinivas, N., Parkin, J., Seelig, G., Winfree, E., Soloveichik, D.: Enzyme-free nucleic acid dynamical systems. Science **358**(6369), eaal2052 (2017)
34. Sudo, Y., Masuzawa, T.: Leader election requires logarithmic time in population protocols. Para. Process. Lett. **30**(01), 2050005 (2020)
35. Sudo, Y., Ooshita, F., Izumi, T., Kakugawa, H., Masuzawa, T.: Time-optimal leader election in population protocols. IEEE Trans. Parallel Distrib. Syst. **31**(11), 2620–2632 (2020). https://doi.org/10.1109/TPDS.2020.2991771

Web-Based Structural Identifiability Analyzer

Ilia Ilmer[1(✉)], Alexey Ovchinnikov[2,3], and Gleb Pogudin[4]

[1] Ph.D. Program in Computer Science, CUNY Graduate Center, New York, NY, USA
iilmer@gradcenter.cuny.edu

[2] Ph.D. Programs in Mathematics and Computer Science, CUNY Graduate Center, New York, NY, USA

[3] Department of Mathematics, CUNY Queens College, Queens, NY, USA
aovchinnikov@qc.cuny.edu

[4] LIX, CNRS, École Polytechnique, Institute Polytechnique de Paris, Paris, France
gleb.pogudin@polytechnique.edu

Abstract. Parameter identifiability describes whether, for a given differential model, one can determine parameter values from model equations. Knowing global or local identifiability properties allows construction of better practical experiments to identify parameters from experimental data. In this work, we present a web-based software tool that allows to answer specific identifiability queries. Concretely, our toolbox can determine identifiability of individual parameters of the model and also provide all functions of parameters that are identifiable (also called identifiable combinations) from single or multiple experiments. The program is freely available at https://maple.cloud/app/6509768948056064.

Keywords: Structural identifiability · Identifiability software · Differential algebra

1 Introduction and Related Work

A parameter is said to be structurally *globally* identifiable if, given the input and output of the experiment, one can uniquely recover the parameter's value in the generic case. If the recovered value is not unique but comes from a finite collection, then we say that such a parameter is *locally* identifiable. Otherwise, the parameter is called *non-identifiable.* In the latter case, one wonders if there is a function of that parameter that is identifiable. This is useful in several ways, for instance, it can mitigate the issue of non-identifiability of some parameters [13].

There is a variety of installable packages that deal with parameter identifiability, see, for instance [1,5,10,17,21]. For a more detailed overview of these, see [4,8] and references therein. A general overview of solving parameter identifiability problems was presented, for instance, in [12,14,20]. Among the available

This work was partially supported by the NSF grants CCF-1564132, CCF-1563942, DMS-1760448, DMS-1853650, and DMS-1853482, and by the Paris Ile-de-France Region.

E. Cinquemani and L. Paulevé (Eds.): CMSB 2021, LNBI 12881, pp. 254–265, 2021.
https://doi.org/10.1007/978-3-030-85633-5_17

identifiability software, SIAN [7] written in MAPLE[1] is typically the fastest one for assessing global identifiability of individual parameters (see, e.g., [7, Table 1]). In the case of the lack of identifiability, one may want to find which functions of the parameters are identifiable. For this task, DAISY software [1] implemented in Reduce can be used (under some assumptions, see [15, Remark 3] and [13]). This state of affairs may be inconvenient for the user because

1) the features of interest are scattered among different packages;
2) packages may require proprietary (MAPLE) or less popular (Reduce) software and may not be available for commonly used OS (DAISY is not available for the UNIX-type systems);
3) finally, the packages should be installed.

These issues have been partially addressed by a web-based tool called COMBOS [11] (and its recent refinement COMBOS 2 for linear systems [9]). However, the backend algorithm appears to be less efficient than SIAN [7, Table 1], and it relies on the same assumption on the input model as DAISY.

Our main contribution is a web-based toolbox hosted on Maple Cloud for assessing structural identifiability built upon SIAN and recent software for computing identifiable functions of parameters [13] which uses the Boulier's BLAD software package [2] incorporated into the MAPLE's `Differential Algebra` package. The key features are

1) *efficiency.* We use SIAN for assessing identifiability of individual parameters efficiently. For computing identifiable functions, we use the code from [13] which we speed up by exploiting the results of the computation performed by SIAN.
2) *versatility.* The toolbox allows assessing local and global identifiability of the parameters and initial conditions and compute the identifiable functions in parameter both in the single- and multi-experiment setup. We do not make any assumptions on the input system unlike DAISY or COMBOS.
3) *availability.* The toolbox is a web app, so it can be used in a browser in one click and does not require installing anything.

In Sect. 3, we outline several scenarios in which our application is essential for assessing identifiability of parameters and parameter combinations. We also illustrate the speedup achievable using output from each of its parts. The web-application[2] can be used at https://maple.cloud/app/6509768948056064 and is also available for download.

2 Input-Output Specification

Let us define the specific form of state-space input ODE that our application accepts.

---

[1] For a Julia implementation, see https://github.com/alexeyovchinnikov/SIAN-Julia.

[2] The MAPLE implementations of each underlying algorithm are available on GitHub at https://github.com/pogudingleb/SIAN and https://github.com/pogudingleb/AllIdentifiableFunctions.

Definition 1 (Model in the state-space form). A model in *the state-space form* accepted by the application is a system

$$\Sigma := \begin{cases} \mathbf{x}' &= \mathbf{f}(\mathbf{x}, \boldsymbol{\mu}, \mathbf{u}), \\ \mathbf{y} &= \mathbf{g}(\mathbf{x}, \boldsymbol{\mu}, \mathbf{u}), \\ \mathbf{x}(0) &= \mathbf{x}^*, \end{cases}$$

where $\mathbf{f} = (f_1, \ldots, f_n)$ and $\mathbf{g} = (g_1, \ldots, g_n)$ with $f_i = f_i(\mathbf{x}, \boldsymbol{\mu}, \mathbf{u})$, $g_i = g_i(\mathbf{x}, \boldsymbol{\mu}, \mathbf{u})$ are rational functions over the field of complex numbers $\mathbb{C}$.

The vector $\mathbf{x} = (x_1, \ldots, x_n)$ represents the time-dependent state variables and $\mathbf{x}'$ represents the derivative. The vector-function $\mathbf{u} = (u_1, \ldots, u_s)$ represents the input variable. The m-vector $\mathbf{y} = (y_1, \ldots, y_n)$ represents the output variables. The vector $\boldsymbol{\mu} = (\mu_1, \ldots, \mu_\lambda)$ represents the parameters and $\mathbf{x}^* = (x_1^*, \ldots, x_n^*)$ defines initial conditions of the model.

Below we specify the input format and possible outputs of our toolbox. Note that while used in descriptions below, some outputs, such as number of solutions for each parameter, are not listed here for brevity. The app also provides additional logs for debugging purposes. In Appendix B, we provide more specification examples.

In: A model in state-space form, see Definition 1.

Out:	*Globally* :	Globally identifiable parameters, that is ones uniquely recoverable for a given system.
	Locally not Globally :	Locally but not globally identifiable parameters, with finitely many recoverable values.
	Non-Identifiable :	Non-identifiable parameters, these can have infinitely many values.
	Single-Experiment :	Single-Experiment identifiable functions of parameters, i.e. identifiable from $k \leq 1$ experiments.
	Multi-Experiment :	Multi-Experiment identifiable functions of parameters, i.e. identifiable from $k \leq \beta$ experiments.
	β :	Bound on the number of experiments.

Note that the single- and multi-experiment identifiable combinations returned by the app generate *all* single- and multi-experiment functions of parameters, respectively. We return them in the algebraically simplified form. In addition, the app reports number of solutions per each globally or locally identifiable parameter, which is not explicitly reflected here due to space limitations.

3 Use Cases for Structural Identifiability Toolbox

3.1 Globally Identifiable Example (Two-Species Competition Model)

Let us consider a simple two-species competition model based with logistic growth in homogeneous environment and assume that we are interested in identifiability properties of all parameters and initial conditions:

$$\begin{cases} x_1' = r_1 x_1 \left(1 - \frac{x_1+x_2}{k_1}\right), \\ x_2' = r_2 x_2 \left(1 - \frac{x_1+x_2}{k_2}\right), \\ y_1 = x_1, \quad y_2 = x_2 \end{cases}$$

with population densities x_1, x_2 being time-dependent state variables, and intrinsic growth rates r_1, r_2 and carrying capacities k_1, k_2 being constant. To run the toolbox for this system, we would write the following into the input field:

In:
```
diff(x1(t),t) = r1*x1(t)*(1 - (x1(t) + x2(t))/k1);
diff(x2(t),t) = r2*x2(t)*(1 - (x1(t) + x2(t))/k2);
y1(t) = x1(t);
y2(t) = x2(t)
```

Out: *Globally*: `[x1(0), x2(0), r1, r2, k1, k2]`
Locally not Globally: `[]`
Non-Identifiable: `[]`

To determine the identifiability for this model, we keep default "Check global/local identifiability" and "Print Number of Solutions" options on. After entering the system and running the application, the output field contains the results. In this model, all parameters and initial conditions are globally (and locally) identifiable. One can now proceed to data collection and further experiments.

3.2 Locally Identifiable Model (SIRS Model with Forcing)

Consider an example of a seasonal epidemic model with a periodic forcing term:

$$\begin{cases} s' = \mu - \mu s - b_0(1 + b_1 x_1) i \cdot s + g \cdot r, \\ i' = b_0(1 + b_1 x_1) i \cdot s - (\nu + \mu) i, \\ r' = \nu i - (\mu + g) r, \\ x_1' = -M x_2, \\ x_2' = M x_1, \\ y_1 = i, \quad y_2 = r. \end{cases}$$

The model is taken from [3] and is built into the application as one of the illustrating examples. Assume that we are interested in identifiability of parameters of this model. Without changing default settings, running the application yields the result of $b_1, x_1(0), x_2(0)$ being unidentifiable, and $b_0, g, \mu, \nu, s(0), i(0), r(0)$ as globally identifiable. At the same time, we observe that M which defines oscillation of the term x_1 is the only parameter identifiable locally, not globally. By checking the number of solutions, we see that only two can be found for M with probability $p = 0.99$. Since M represents the oscillation frequency, it is assumed to be positive in practice, hence globally identifiable. Note that we only needed a single section of the app and the result has been obtained in about 7.2 s.

3.3 Identifiable Combination of Non-identifiable Parameters (Tumor Targeting)

In this example, we consider system 3 from [16, Section 3] with unknown initial conditions. The example describes a compartmental model describing tumor targeting with antibodies, see [18]. To arrive at the system below, we suppose equations (B) and (D) are identically zero and that $\frac{5V36}{V3} = 1$. The functions $x_i, i = 1, \ldots, 5$ represent concentrations, $k_i, i = 3, \ldots, 7$ and a, b, d represent rate constants.

$$\begin{cases} x_1' = -(k_3 + k_7)x_1 + k_4 x_2, \\ x_2' = k_3 x_1 - (k_4 + (a + bd)k_5)x_2 + k_6(x_3 + x_4) + k_5 x_2(x_3 + x_4), \\ x_3' = ak_5 x_2 - k_6 x_3 - k_5 x_2 x_3, \\ x_4' = bdk_5 x_2 - k_6 x_4 - k_5 x_2 x_4, \\ x_5' = k_7 x_1, \\ y_1 = x_5. \end{cases}$$

For this model, after computing identifiability properties using SIAN, we observe that everything except parameters $a, b, d, x_3(0), x_4(0)$ is globally or locally identifiable. To investigate further, we consider computation with "Compute Identifiable Combinations" option turned on. Running the program with this additional setting, we see that while parameters a, b, d are not identifiable, their combination $a + bd$ can be identified from at most one experiment. This is especially beneficial since one can connect the meaning of expression $a+bd$ to the overall biological sense of the model's underlying phenomenon. For instance, in the original paper [18], constant a and a product bd may be attributed to total binding sites on normal tissue and number of binding sites on tumor making $a + bd$ the total number of binding sites in the system. Further, one could apply a substitution of the form $\widehat{x}_3 = x_3 + x_4, \widehat{p} = a + bd$ so that in the new system we only have equations for $x_1, x_2, \widehat{x}_3, x_5$ and the parameter combination $a + bd$ will now be globally identifiable as a parameter $\widehat{p}$.

3.4 System with a Non-identifiable Parameter (Lotka-Volterra Model)

Let us consider the following Lotka-Volterra model

$$\begin{cases} x_1' = ax_1 - bx_1x_2 \\ x_2' = -cx_2 + dx_1x_2 \\ y = x_1 \end{cases} \tag{1}$$

By running the application for (1) using only SIAN, we see that parameter b and initial condition $x_2(0)$ are non-identifiable, and the parameters a, b, d and the initial condition $x_1(0)$ are globally identifiable. Furthermore, since a is identifiable and x_1 is observed, from the first equation we conclude that $bx_2(0) = a - x_1'(0)/x_1(0)$ is identifiable. This implies that we have an output-preserving scaling transformation $b \to \lambda b$, $x_2 \to x_2/\lambda$. Therefore, the reparametrization $\hat{x}_2 := bx_2$ makes the model globally identifiable.

3.5 Refining Multi-experiment Identifiability Bound (Slow-Fast Ambiguity in a Chemical Reaction Network)

Consider the following system:

$$\begin{cases} x_A' = -k_1 x_A, \\ x_B' = k_1 x_A - k_2 x_B, \\ x_C' = k_2 x_B, \\ e_A' = e_C' = 0, \\ y_1 = e_A x_A + e_B x_B + e_C x_C, \\ y_2 = x_C, \ y_3 = e_A, \ y_4 = e_C. \end{cases}$$

This model is based on a kinetic reaction $A \xrightarrow{k_1} B \xrightarrow{k_2} C$ from [19] and has an extra output equation y_2. The functions x_A, x_B, x_C are concentrations and e_A, e_B with constant e_C represent molar extinction coefficients. In addition, parameters include unknown rate coefficients k_1, k_2. The application reports global identifiability for $x_C(0)$, $e_A(0)$, $e_C(0)$ and local identifiability for everything else.

It is then of interest to check identifiable parameter combinations. The app reports single-experiment identifiability for k_1k_2, $k_1 + k_2$. This implies that the parameters k_1 and k_2 are identifiable up to a permutation, so it is possible to infer the reaction rates from an experiment but not which rate corresponds to which reaction. Interestingly, the app reports that e_B, k_1, k_2 become globally identifiable if one performs at most 3 experiments. Can we do better? To answer this, we turn on the option "Try to Refine Bound" with default number of refining attempts being 4. As a result, the app reports a new bound for the number of experiments being 2.

Let us illustrate this point in another way. Recall that we can tell SIAN to consider multiple copies of the system when analyzing identifiability. In this mode, SIAN does not output initial conditions for brevity. We observed that the refined bound for parameters e_B, k_1, k_2 was 2. If we set the "Number of experiments (copies of the input system)" to 2, SIAN yields global identifiability of e_B, k_1, k_2, which verifies our earlier finding. Moreover, turning off "Attempt Bypass using SIAN" option in the search for combinations, we observe that the application still returns e_B, k_1, k_2 as identifiable with 3 experiments, however, single experiment check overwrites this result, yielding bound of 1.

Acknowledgements. We are grateful to Joseph DiStefano III, Jürgen Gerhard, John May, Maria Pia Saccomani, and Eduardo Sontag for fruitful discussions, useful feedback, and technical assistance.

A Details on the Underlying Algorithms

The application solves two problems: identifiability properties of individual parameters and that of combinations (functions) of parameters. Note that we return generators of the field of *all* identifiable functions.

The input for both problems follows the same structure where we pass a collection of ODEs and output functions. For querying identifiability of individual parameters and initial conditions we use SIAN [7]. In short, it expresses the Taylor coefficients of output functions in terms of the initial conditions and parameters and checks whether the parameters or initial conditions of interest can be expressed via these coefficients. For better efficiency, this is checked for a randomly sampled solution of the system. The probability that such a solution will exhibit the generic behavior is quantified in [8]. Therefore, the overall algorithm is randomized Monte Carlo, that is the result is guaranteed correct with user-specified probability p.

To answer the question on identifiability of parameter functions we take advantage of work [13]. Note that our application distinguishes single- and multi-experiment identifiable combinations as opposed to existing methods for identifiable combination queries. The latter is equivalent to having multiple copies of original ODE system sharing the parameters, outputs, and inputs. We also provide a bound on the number of experiments which can be refined by changing ordering of variables in the underlying algorithm.

We compute the input-output equations, that is, differential equations relating inputs, outputs, and parameters of the differential model. Identifiable functions of parameters are then extracted from the coefficients of these equations using methods of differential algebra and computational algebraic geometry, including Gröbner basis computation. To minimize the computational overhead, we take advantage of the Gröbner walk procedure, by changing the order from total degree reverse to pure lexicographic. This algorithm is deterministic or a Monte Carlo probabilistic, depending on how/which of the Gröbner basis implementation is used.

To achieve maximal speed of computation without compromising the functionality of the application, we take advantage of the fact that SIAN is typically faster than the algorithm from [13] and its output can be sometimes used to obtain the output of [13] without further computation. More precisely, if all parameters are reported as globally identifiable with probability p, then, with the same probability, we report these parameters as their own identifiable combinations and an example of this is presented in Appendix B.5.

With the current implementation, the application does not support specifying initial conditions, however this functionality is planned for future versions.

B Systems in Structural Identifiability Toolbox Input Form

Below we present input and output form for examples discussed in this paper (see also Figs. 1 and 2). The input is shown in the MAPLE syntax form. The toolbox also supports a different input format:

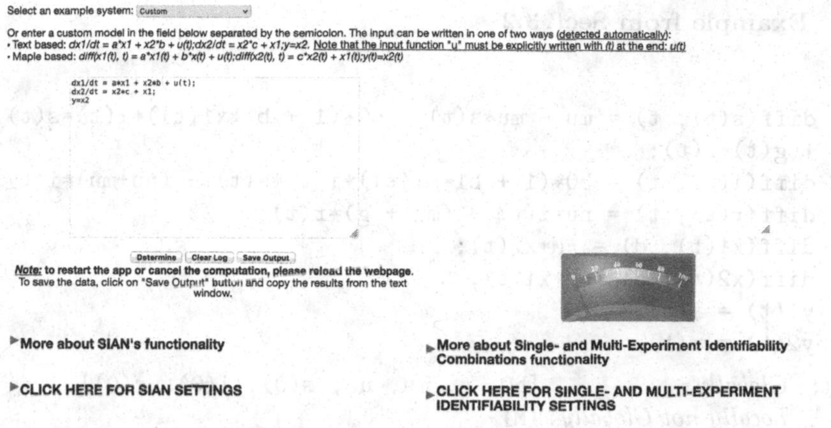

Fig. 1. Main view of the application. The dial on the right side indicates whether an app is running. The arrows are clickable and show additional settings for each section of the program as well as documentation.

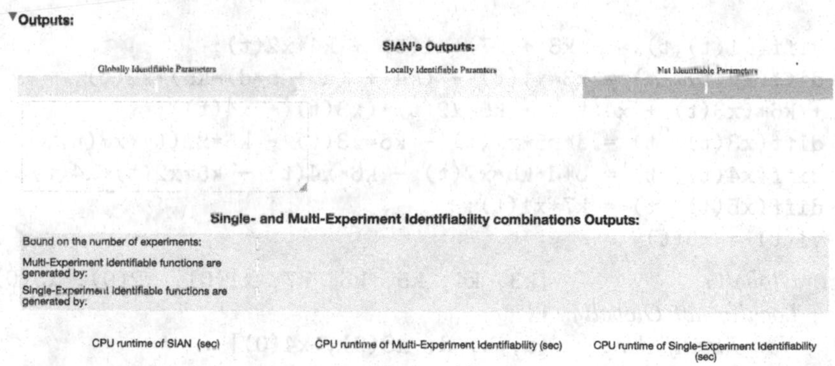

Fig. 2. Output fields of the application. We present individual parameters' results and identifiable combinations with the bound separately. The white field on the left displays number of solutions per identifiable parameter.

```
dx1/dt = a*x1 + x2*b + u(t);
dx2/dt = x2*c + x1;
y = x2
```

where inputs `u(t)` are required to have argument written explicitly.

B.1 Example from Sect. 3.2

In:
```
diff(s(t), t) = mu - mu*s(t) - b0*(1 + b1*x1(t))*i(t)*s(t)
+ g(t)*r(t);
diff(i(t), t) = b0*(1 + b1*x1(t))*i(t)*s(t) - (nu+mu)*i(t);
diff(r(t), t) = nu*i(t) - (mu + g)*r(t);
diff(x1(t), d) = -M*x2(t);
diff(x2(t), d) = M*x1(t);
y1(t) = i(t);
y2(t) = r(t)
```

Out:
Globally: `[b0, g, mu, nu, s(0), i(0), r(0)]`
Locally not Globally: `[M]`
Non-Identifiable: `[b1, x1(0), x2(0)]`

B.2 Example from Sect. 3.3

In:
```
diff(x1(t),t) = -(k3 + k7)*x1(t) + k4*x2(t);
diff(x2(t), t) = k3*x1(t) - (k4 + (a + b*d)*k5)*x2(t)
+ k6*(x3(t) + x4(t)) + k5*x2(t)*(x3(t) + x4(t));
diff(x3(t), t) = a*k5*x2(t) - k6*x3(t) - k5*x2(t)*x3(t);
diff(x4(t), t) = b*d*k5*x2(t) - k6*x4(t) - k5*x2(t)*x4(t);
diff(x5(t), t) = k7*x1(t);
y1(t) = x5(t)
```

Out:
Globally: `[k3, k4, k5, k6, k7, x1(0), x2(0), x5(0)]`
Locally not Globally: `[]`
Non-Identifiable: `[a, b, d, x3(0), x4(0)]`
Single-Experiment: `[k3, k4, k6, k7, k5/k7, bd+a]`
Multi-Experiment: `[k3, k4, k6, k7, k5/k7, bd+a]`
$\beta =$ 1

B.3 Example from Sect. 3.4

In:
```
diff(x1(t), t) = a*x1(t) - b*x1(t)*x2(t);
diff(x2(t), t) = -c*x2(t) + d*x1(t)*x2(t);
y(t) = x1(t)
```

Out:
Globally: `[a, c, d, x1(0)]`
Locally not Globally: `[]`
Non-Identifiable: `[b, x2(0)]`
Single-Experiment: `[a, c, d]`
Multi-Experiment: `[a, c, d]`
$\beta =$ 1

B.4 Example from Sect. 3.5

In:
```
diff(xA(t), t) = -k1*xA(t);
diff(xB(t), t) = k1*xA(t) - k2*xB(t);
diff(xC(t), t) = k2*xB(t);
diff(eA(t), t) = 0;
diff(eC(t), t) = 0;
y1(t) = eA(t)*xA(t) + eB*xB(t) + eC(t)*xC(t);
y2(t) = xC(t);
y3(t) = eA(t);
y4(t) = eC(t)
```

Out:

Globally:	`[xC(0), eA(0), eC(0)]`
Locally not Globally:	`[eB, k1, k2, xA(0), xB(0)]`
Non-Identifiable:	`[]`
Single-Experiment =	`[k1k2, k1+k2]`
Multi-Experiment =	`[eB, k1, k2]`
$\beta =$	3

B.5 Example of Speedup with Bypasses

Consider the system from [6]

$$\begin{cases} x_1' = -k_1 x_1 x_2 + k_2 x_4 + k_4 x_6, \\ x_2' = k_1 x_1 x_2 + k_2 x_4 + k_3 x_4, \\ x_3' = k_3 x_4 + k_5 x_6 - k_6 x_3 x_5, \\ x_4' = k_1 x_1 x_2 - k_2 x_4 - k_3 x_4, \\ x_5' = k_4 x_6 + k_5 x_6 - k_6 x_3 x_5, \\ x_6' = -k_4 x_6 - k_5 x_6 + k_6 x_3 x_5, \\ y_1 = x_3, \\ y2 = x_2. \end{cases}$$

This is an example of a mixed-mechanism network, where the state functions $x_i(t), i = 1, \dots, 6$ are concentrations and the parameters $k_i, i = 1, \dots, 6$ are rate constants. The app returns global and local identifiability for all parameter in under 4 s. This is used to conclude that multi-experiment identifiable combinations with the bound of 1 are parameters themselves. If we turn off the "Attempt Bypass" function, the multi-experiment identifiable combinations $k_1, k_3, k_5, k_6, \frac{-k_2 k_4 + k_3 k_5}{k_2 + k_3}, k_2 - k_3$ with bound 1 are returned in 433 s. The input form for this example is presented below

In:
```
diff(x1(t), t) =-k1*x1(t)*x2(t) + k2*x4(t) + k4*x6(t);
diff(x2(t), t) = k1*x1(t)*x2(t) + k2*x4(t) + k3*x4(t);
diff(x3(t), t) = k3*x4(t) + k5*x6(t) - k6*x3(t)*x5(t);
diff(x4(t), t) = k1*x1(t)*x2(t) - k2*x4(t) - k3*x4(t);
diff(x5(t), t) = k4*x6(t) + k5*x6(t) - k6*x3(t)*x5(t);
diff(x6(t), t) =-k4*x6(t) - k5*x6 (t)+ k6*x3(t)*x5(t);
y1(t) = x3(t);
y2(t) = x2(t)
```

Out:

Globally:	[k1, k2, k3, k4, k5, k6, x1(0), x2(0), x3(0), x4(0), x5(0), x6(0)]
Locally not Globally:	[]
Non-Identifiable:	[]
Single-Experiment:	[k1, k2, k3, k4, k5, k6]
Multi-Experiment:	[k1, k2, k3, k4, k5, k6]
$\beta =$	1

References

1. Bellu, G., Saccomani, M.P., Audoly, S., D'Angiò, L.: DAISY: A new software tool to test global identifiability of biological and physiological systems. Comput. Methods Programs Biomed. **88**(1), 52–61 (2007). https://doi.org/10.1016/j.cmpb.2007.07.002
2. Boulier, F.: BLAD–bibliothèques lilloises d'algèbre différentielle (2014). https://cristal.univ-lille.fr/~boulier/pmwiki/pmwiki.php/Main/BLAD
3. Capistrán, M.A., Moreles, M.A., Lara, B.: Parameter estimation of some epidemic models. the case of recurrent epidemics caused by respiratory syncytial virus. Bull. Math. Biol. **71**(8), 4890 (2009). https://doi.org/10.1007/s11538-009-9429-3
4. Chiş, O.-T., Banga, J.R., Balsa-Canto, E.: Structural identifiability of systems biology models: A critical comparison of methods. PLoS ONE **6**(11), e27755 (2011). https://doi.org/10.1371/journal.pone.0027755
5. Chiş, O.-T., Banga, J.R., Balsa-Canto, E.: GenSSI: a software toolbox for structural identifiability analysis of biological models. Bioinformatics **27**(18), 2610–2611 (2011). https://doi.org/10.1093/bioinformatics/btr431
6. Conradi, C., Shiu, A.: Dynamics of posttranslational modification systems: Recent progress and future directions. Biophys. J. **114**(3), 507–515, 2018. ISSN 0006-3495. https://doi.org/10.1016/j.bpj.2017.11.3787
7. Hong, H., Ovchinnikov, A., Pogudin, G., Yap, C.: SIAN: software for structural identifiability analysis of ODE models. Bioinformatics **35**(16), 2873–2874 (2019). https://doi.org/10.1093/bioinformatics/bty1069
8. Hong, H., Ovchinnikov, A., Pogudin, G., Yap, C.: Global identifiability of differential models. Commun. Pure Appl. Math. **73**(9), 1831–1879 (2020). https://doi.org/10.1002/cpa.21921
9. Yazdi, A.K., Nadjafikhah, M., Distefano, J.III.: COMBOS2: an algorithm to the input-output equations of dynamic biosystems via gaussian elimination. J. Taibah University Sci. **14**(1), 896–907 (2020). https://doi.org/10.1080/16583655.2020.1776466

10. Ligon, T.S., et al.: GenSSI 2.0: multi-experiment structural identifiability analysis of SBML models. Bioinformatics **34**(8), 1421–1423 (2018). https://doi.org/10.1093/bioinformatics/btx735
11. Meshkat, N., Kuo, C.E., DiStefano, J.III.: On finding and using identifiable parameter combinations in nonlinear dynamic systems biology models and COMBOS: a novel web implementation. PLoS One **9**(10), e110261 (2014). https://doi.org/10.1371/journal.pone.0110261
12. Miao, H., Xia, X., Perelson, A.S., Hulin, W.: On identifiability of nonlinear ODE models and applications in viral dynamics. SIAM Rev. **53(1)**, 3–39 (2011). https://doi.org/10.1137/090757009
13. Ovchinnikov, A., Pillay, A., Pogudin, G., Scanlon, T.: Computing all identifiable functions for ODE models. arXiv preprint arXiv:2004.07774 (2020). https://arxiv.org/abs/2004.07774
14. Raue, A., Karlsson, J., Saccomani, M.P., Jirstrand, M., Timmer, J.: Comparison of approaches for parameter identifiability analysis of biological systems. Bioinformatics **30**(10), 1440–1448 (2014). https://doi.org/10.1093/bioinformatics/btu006
15. Saccomani, M., Audoly, S., D'Angiò, L.: Parameter identifiability of nonlinear systems: the role of initial conditions. Automatica **39**, 619–632 (2003). https://doi.org/10.1016/S0005-1098(02)00302-3
16. Saccomani, M.P., Audoly, S., Bellu, G., D'Angiò, L.: Examples of testing global identifiability of biological and biomedical models with the DAISY software. Comput. Biol. Med. **40**(4), 402–407 (2010). https://doi.org/10.1016/j.compbiomed.2010.02.004
17. Saccomani, M.P., Bellu, G., Audoly, S., d'Angió, L.: A new version of DAISY to test structural identifiability of biological models. In: Bortolussi, L., Sanguinetti, G. (eds.) CMSB 2019. LNCS, vol. 11773, pp. 329–334. Springer, Cham (2019). https://doi.org/10.1007/978-3-030-31304-3_21
18. Thomas, G.D., et al.: Effect of dose, molecular size, affinity, and protein binding on tumor uptake of antibody or ligand: a biomathematical model. Cancer Res. **49**(12), 3290–3296 (1989). http://www.ncbi.nlm.nih.gov/pubmed/2720683
19. Vajda, S., Rabitz, H.: Identifiability and distinguishability of first-order reaction systems. J. Phys. Chem. **92**(3), 701–707 (1988). https://doi.org/10.1021/j100314a024
20. Villaverde, A.F.:. Observability and structural identifiability of nonlinear biological systems. Complexity (2019). https://doi.org/10.1155/2019/8497093
21. Villaverde, A.F., Barreiro, A., Papachristodoulou, A.: Structural identifiability of dynamic systems biology models. PLoS Comput. Biol. **12**(10) (2016). https://doi.org/10.1371/journal.pcbi.1005153

BioFVM-X: An MPI+OpenMP 3-D Simulator for Biological Systems

Gaurav Saxena[1(✉)], Miguel Ponce-de-Leon[1(✉)], Arnau Montagud[1(✉)], David Vicente Dorca[1(✉)], and Alfonso Valencia[1,2(✉)]

[1] Barcelona Supercomputing Center (BSC), Barcelona, Spain
{gaurav.saxena,miguel.ponce,arnau.montagud,david.vicente, alfonso.valencia}@bsc.es

[2] Institució Catalana de Recerca i Estudis Avançats (ICREA), Passeig de Lluís Companys 23, 08010 Barcelona, Spain

Abstract. Multi-scale simulations require parallelization to address large-scale problems, such as real-sized tumor simulations. BioFVM is a software package that solves diffusive transport Partial Differential Equations for 3-D biological simulations successfully applied to tissue and cancer biology problems. Currently, BioFVM is only shared-memory parallelized using OpenMP, greatly limiting the execution of large-scale jobs in HPC clusters. We present BioFVM-X: an enhanced version of BioFVM capable of running on multiple nodes. BioFVM-X uses MPI+OpenMP to parallelize the generic core kernels of BioFVM and shows promising scalability in large 3-D problems with several hundreds diffusible substrates and ≈0.5 billion voxels. The BioFVM-X source code, examples and documentation, are available under the BSD 3-Clause license at https://gitlab.bsc.es/gsaxena/biofvm_x.

Keywords: Multi-scale modeling · Lattice-free modeling · OpenMP · MPI · Shared-memory · Distributed-memory · Parallelization

1 Introduction

Advances in understanding complex biological systems such as tumors require multi-scale simulations that integrate intracellular processes, cellular dynamics, and their interaction with the environment. Computational biologists use a wide range of approaches to simulate how single cells affect multi-cellular systems' dynamics [17,24]. Nevertheless, large-scale multi-scale modeling still needs tools to accurately simulate the environment in an efficient manner.

BioFVM [8] is a Finite Volume Method (FVM) [20] based simulation software for solving Partial Differential Equations (PDEs) [29] that model complex processes like the uptake, release and diffusion of substrates for multi-cellular systems such as tissues, tumors or microbial communities. Apart from being a self-contained callable library that can be used to implement and simulate biological models, BioFVM forms the core component of PhysiCell [9] - a flexible, lattice-free, agent-based multi-cellular framework capable of simulating cell

E. Cinquemani and L. Paulevé (Eds.): CMSB 2021, LNBI 12881, pp. 266–279, 2021.
https://doi.org/10.1007/978-3-030-85633-5_18

mechanics, such as cell movement, cell-cell interaction and different cell phenotypes, as well as the micro-environment consisting of diffusing substrates, signaling factors, drugs, etc. BioFVM is capable of handling multiple substrates and can simulate chemical and biological processes using both cell and bulk sources. The following diffusive PDE on a computational domain Ω (and boundary $\partial\Omega$) is solved for a substrate density vector ρ:

$$\frac{\partial \rho}{\partial t} = \nabla \cdot (\mathbf{D} \circ \nabla \rho) - \lambda \circ \rho + \mathbf{f}, \tag{1}$$

with the boundary condition $(\mathbf{D} \circ \nabla \rho) \cdot n = 0$ on $\partial\Omega$ and the initial condition $\rho(\mathbf{x}, t_0) = g$ in Ω. In (1) above, $\mathbf{D}$ is the matrix of (constant) diffusion coefficients, λ is the decay rate, $\mathbf{f}$ is the net source term and $\circ$ is the termwise product of vectors [8]. Without loss of generality, the substrate density ρ can represent any kind of molecule such as a nutrient, a by-product, a signal molecule or a drug. As a consequence, modeling complex environments requires simulating many densities, posing a challenging scaling problem. Simulating the environment requires the numerical solution of the linear system obtained by a Finite Volume Discretization of the PDE given by Eq. (1), which BioFVM solves using the Thomas algorithm [31] - a fast, direct solver for tridiagonal systems. BioFVM's biggest scalability limitation is that it cannot execute on multiple nodes of an HPC cluster to solve a single, coherent problem and thus the problem *must* fit into the memory of a single node.

We present BioFVM-X[1]: an enhanced distributed version that uses MPI (Message-Passing Interface [21]) to parallelize the core kernels of BioFVM - enabling one to solve very large problems which were not previously solvable using the shared-memory only version. This contribution represents the first and the most critical step on the road to a distributed implementation of PhysiCell.

2 Related Work

Different agent-based approaches have been proposed to model and simulate multi-cellular systems, including on-lattice cellular automata, the Cellular-Potts model [10] and overlapping spheres, among others [23]. BioFVM [8,9] was created with the goal of achieving simplicity of usage, flexibility in expressing cell models, and optimizing execution speed while minimising dependencies on external libraries but is only shared-memory parallelized using OpenMP [22].

For realistic, complex simulations, the need is to simulate billions of cells and dynamic, complex 3-D environments, only achievable by optimal, full scale utilization of parallel systems [12,14]. Biocellion [14] is a flexible, discrete agent-based simulation framework that uses MPI for inter-node communication, as well as other dependencies, such as PNNL Global Arrays [25], CHOMBO [3], the Intel TBB [11] and the iterative Multigrid solver [2,32]. Nevertheless, Biocellion has fixed routines to describe system behaviors, is dependent on external libraries

[1] Available at: https://gitlab.bsc.es/gsaxena/biofvm_x under BSD 3-Clause license.

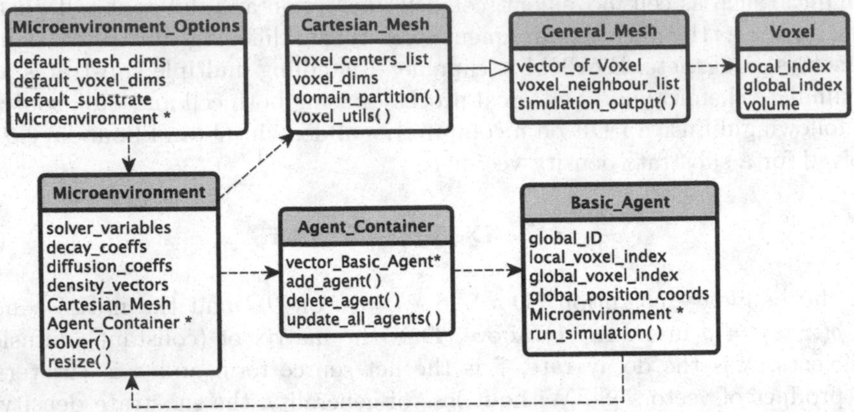

Fig. 1. Key classes in BioFVM, along with their member data and functions. Functions are distinguishable by a leading parenthesis i.e. ( ). Names are arbitrary but convey semantic information. Solid, thick arrow with an un-shaded triangle represents inheritance and dashed arrows denote a pointer or class relationship - the class (or its pointer) being pointed to by the arrow is a data member of the class from which the arrow originates.

and is closed source, which might deter potential users. Chaste is an open-source, general purpose simulation package for modeling soft tissues and discrete cell populations [18] that can be used with MPI using PETSc [1] but which itself suffers from multiple dependencies. Timothy [4,5] is another open-source, MPI based tool but with several dependencies, such as Zoltan [6], Hypre [7] and SPRNG [19].

3 Internal Design and Domain Partitioning

The simplicity, flexibility, minimal dependence on external libraries, execution speed and openness of BioFVM make it an ideal experimental candidate for distributed parallelization. In BioFVM, the 3-D simulation domain is divided into Voxels (Volumetric pixels). The principal classes depicting the internal architecture and their relationship in BioFVM is shown in Fig. 1.

The top-level biological entities along with related classes (see Fig. 1) are: (1) Biological Environment (`Microenvironment` and `Microenvironment_Options`), (2) Physical Domain represented as 2-D/3-D Mesh (`General_Mesh`, `Cartesian_Mesh` and `Voxel`), and (3) Cells (`Basic_Agent` and `Agent_Container`). The data members of some classes are either the objects or the pointers of another class type (see dashed arrows in Fig. 1). The `Microenvironment` class sets the micro-environment name, the diffusion/decay rates of substrates, defines constants for the Thomas algorithm, contains an object of `Cartesian_Mesh`, a pointer to the `Agent_Container` class and performs I/O.

A group of resizing functions that determine the global/local voxels are members of the `Cartesian_Mesh` class. The `Microenvironment_Options` class helps to set oxygen as the first default substrate and the default dimensions of the domain/voxel. The `Cartesian_Mesh` class is publicly derived from `General_Mesh` (thick arrow in Fig. 1). The `Basic_Agent` class forms an abstraction of a cell. An object of the `Basic_Agent` class can either act as a source or sink of/for substrates. Each agent has a unique ID, a type, and maintains the local/global index of its current voxel.

We initialize MPI with the `MPI_THREAD_FUNNELED` thread support level and after domain partitioning [27,28], assign the sub-domains to individual MPI processes. Our implementation as of now supports only a 1-D x-decomposition (see Appendix A). The randomly generated positions of basic agents are mapped to respective processes (see Appendix B) after which they are created individually and in parallel on the MPI processes. Each MPI process initializes an object of the `Microenvironment` class, maintains the local and global number of voxels, local (`mesh_index`) and global voxel indices (`global_mesh_index`) and the center of each local voxel's global coordinates. A 1-D x-decomposition permits us to employ the optimal *serial* Thomas algorithm [30,31] in the undivided y and z dimensions. This enables all threads within a node to simultaneously act on elements belonging to different linear systems.

The Thomas algorithm is used to solve a tridiagonal system of linear equations in serial and consists of two steps, namely, Forward Elimination (FE) step followed by a Backward Substitution (BS) step. Unfortunately, both the steps involve serial and dependent operations and thus, the solver is inherently serial and cannot be fully (trivially) parallelized. Although we decompose data in the $x-$direction, the solver still runs *serially* i.e. MPI process rank i must finish the FE before this step can begin on MPI process rank $i+1$. Thus, the performance of this multi-node but serial Thomas solver is expected to be worse than a single-node Thomas solver due to the overhead of communication. The performance penalty is least in the $x-$direction as the data is contiguous in the memory as compared to the y and z direction where the data in the voxels' vector is non-contiguous. Thus, we decompose data only in the $x-$direction and avoid decomposition in the other directions. We expect to replace this non-optimized implementation by a modified, MPI+OpenMP version of the modified Thomas algorithm [15] in future versions.

4 Experiments

We used the MareNostrum 4 (MN4) supercomputer at the Barcelona Supercomputing Center (BSC) for all our experiments. Each node has two 24-core Intel Xeon Platinum 8160 processors and a total memory of 96 GB. BioFVM-X only requires a C++ compiler and an MPI implementation for compilation. We used `GCC 8.1` and `OpenMPI 3.1.1` running atop the SUSE Linux Enterprise Server 12 SP2 OS. The parallel file system is the IBM General Parallel File System and the compute nodes are interconnected with the Intel Omni-Path technology with a bandwidth of 100 Gbits/s. We pinned the threads to individual cores and bind each MPI process to a single processor (socket). We set the OpenMP environment variables `OMP_PROC_BIND=spread`, `OMP_PLACES=threads` [26] and used the `--map-by ppr:1:socket:pe=24` notation to allocate resources (see https://gitlab.bsc.es/gsaxena/biofvm_x).

We used a cubic physical domain and cubic voxels for all our tests. Our implementation assumed that the total number of voxels in the BioFVM's x-direction are completely divisible by the total number of MPI processes. The example that we used to demonstrate the benefits of Hybrid parallelism is *tutorial1* in the *BioFVM/examples* directory. This example: (1) Initializes and resizes the micro-environment (μ-environment, MC kernel) (2) Creates a Gaussian profile (GPG kernel) of the substrate concentration (3) Writes the initial and final concentrations to a `.mat` file (I/O kernel) (4) Creates Basic Agents (Sources and Sinks, BAG kernel) and (5) Simulates Sources/Sinks and Diffusion (Solver kernel).

Figure 2 presents timing results for the MC, GPG, BAG, I/O and Solver kernels on physical domains of sizes $1000^3, 1920^3$ and 3840^3. Cubic voxels had a volume of 10^3 with 5×10^2 sources and 5×10^2 sinks in this example. We denote the Hybrid implementation as "Hyb (n = a)", where "a" denotes the total number of nodes. For example, with Hyb (n = 2), we obtain a total of 2 (nodes) × 2 (MPI processes) × 24 (OpenMP threads) = 96 OpenMP threads, as we always run 2 MPI processes per node and 24 OpenMP threads per MPI process. Instead of 8 MPI processes for the domain of size 1000^3, we used 10 MPI processes due to a divisibility problem. Figure 3 shows the initial and final concentration of the diffusing substrate (oxygen) for a domain of size 1000^3. The simulation plots were obtained with Hyb (n = 1) by executing the `cross_section_surface.m` Matlab script bundled with BioFVM.

In summary for Hyb (n = 1), both MC and BAG kernels took advantage of the multiple MPI processes as initialization of the `Microenvironment` and `Basic_Agent` class objects were simultaneously carried out on separate processes in BioFVM-X as opposed to a single thread in BioFVM. The (MPI) I/O kernel showed significant performance gains over serial I/O for the tests considered (Fig. 2). Nevertheless, the Solver kernel execution run-times did not reflect a significant gain in the Hybrid version. An extended analysis of these results can be found in Appendix C. Note that it is generally very difficult for an MPI+OpenMP implementation to outperform the pure OpenMP implementation on a *single* node, as is the case of Fig. 2, due to the additional memory footprint of MPI and the cost of message-passing/synchronization. Our aim in the current work was to tackle very large problems that cannot fit into the memory a single node and to reduce their time to solution in a multi-node scenario.

After testing with increased voxels and basic agents, we run a performance test to evaluate the scalability in the number of substrates. We found that the pure OpenMP BioFVM version is incapable of executing a simulation of 400 substrates on a domain of 1500^3 due to memory limitations. Nonetheless, we successfully run a Hybrid simulation using 400, and even 800 substrates, on a domain of 1500^3 by distributing the computation between 2 nodes.

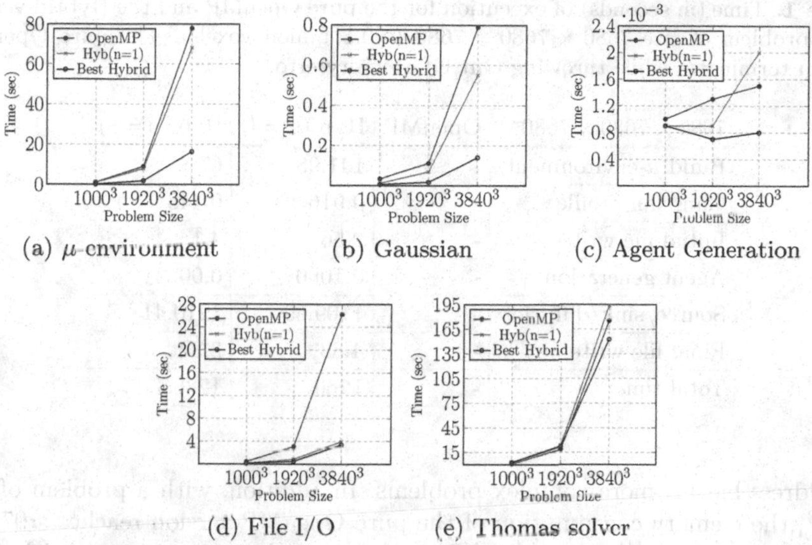

(a) μ-environment (b) Gaussian (c) Agent Generation

(d) File I/O (e) Thomas solver

Fig. 2. Pure OpenMP Vs Hybrid MPI execution times for increasing problem sizes. Hyb (n = 1) represents the time when a single node with 2 MPI processes and 24 threads is used. The Best Hybrid represents the least time for that kernel for *any* number of experimental nodes considered.

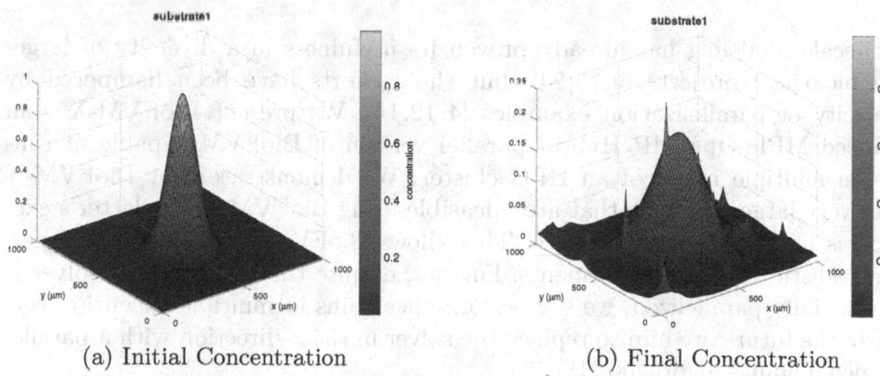

(a) Initial Concentration (b) Final Concentration

Fig. 3. 3-D concentration density of oxygen simulated using Hyb (n = 1) for a domain of size 1000^3 and 1000 Basic Agents.

To further showcase BioFVM-X capabilities, we run a parallelized version of the model of tumor growth in a heterogeneous micro-environment from BioFVM [8]. We verified that the BioFVM-X distributed-memory 3-D tumor example yielded the exact same results as the shared-memory one (see Appendix D and Fig. 8). This is further proof that BioFVM-X correctly distributes the original BioFVM models with a boost in performance due to the load distribution and the potential of scaling simulations to a cluster of nodes, thus enabling researchers

Table 1. Time (in seconds) of execution for the pure OpenMP and the Hybrid version for a problem of size 7680 × 7680 × 7680 (≈0.5 billion voxels). The pure OpenMP version terminates while throwing Out Of Memory error.

7680 × 7680 × 7680	OpenMP	Hyb (n = 4)	Hyb (n = 8)
Build μ-environment	-	141.98	67.81
Gaussian profile	-	0.916	0.448
Initial file write	-	2.56	4.1
Agent generation	-	0.1060	0.0023
Source/sink/diffusion	-	1109.69	1210.41
Final file write	-	4.83	3.32
Total time	-	1260	1286.1

to address bigger, more complex problems. In addition, with a problem of size 7680^3, the memory consumption of the pure OpenMP version reaches ≈97% of the total memory of the node (96 GB) and the simulation terminates with a bus error. For the same problem size, the Hybrid code on 4 (with 192 threads) and 8 nodes (with 384 threads) executes successfully (Table 1).

5 Conclusion and Future Work

Multi-scale modeling has already proven its usefulness in a diversity of large-scale biological projects [9,16,24], but these efforts have been hampered by a scarcity of parallelization examples [4,12,14]. We present BioFVM-X - an enhanced MPI+OpenMP Hybrid parallel version of BioFVM capable of running on multiple nodes of an HPC cluster. We demonstrate that BioFVM-X solves very large problems that are infeasible using BioFVM as the latter's execution is limited to a single node. This allows BioFVM-X to simulate bigger, more realistic *in-silico* experiments. Further, despite the fact that our solver is only partially parallelized, we see performance gains in multiple execution kernels. In the future, we aim to replace the solver in the x-direction with a parallel modified Thomas algorithm [15].

BioFVM-X is open source under the BSD 3-Clause license and freely available at https://gitlab.bsc.es/gsaxena/biofvm_x. Even though it can be used to easily implement and simulate biological models in a self-contained manner, BioFVM-X also forms the lower layer of our ongoing efforts to have a parallel large-scale and multi-scale modeling framework termed PhysiCell-X, based on PhysiCell [9] - a framework that is under active development and has multiple stable releases.

Acknowledgements. The research leading to these results has received funding from EU H2020 Programme under the PerMedCoE project, grant agreement number 951773 and the INFORE project, grant agreement number 825070. The authors would like to thank Paul Macklin and Randy Heiland from Indiana University for their constant support and advice regarding BioFVM.

Appendix A 1-D Pure x-Domain Decomposition

Figure 4 shows a 1-D x-direction domain partition of a 3-D domain (x-direction is the unit-stride dimension).

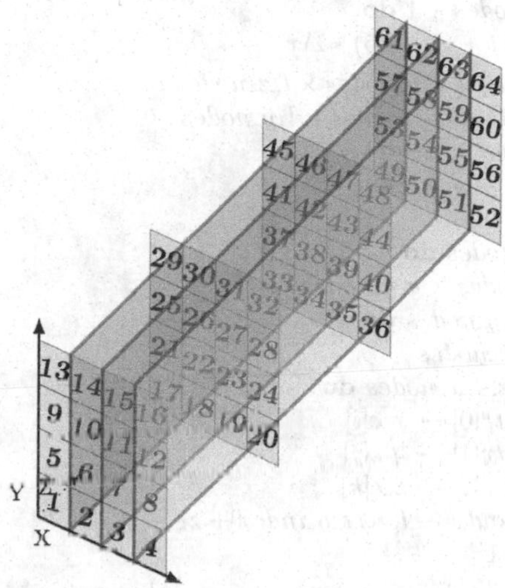

Fig. 4. A 3-D domain of dimensions $4 \times 4 \times 4$ visualized as four 2-D plates (shaded gray) of dimension 4×4 arranged one after the other. A 1-D domain partition (shown with blue, red and green planes) of 4 MPI processes in the x-direction divides the voxels numbered 1 to 64 into 4 parts. Rank 0, Rank 1, Rank 2, and Rank 3 processes contain voxel IDs numbered $4n + 1$, $4n + 2$, $4n + 3$, and $4n + 4$ respectively, where $n = 0, 1, 2, \ldots, 15$. Data is contiguous in the x-direction and the distance between 2 consecutive elements in the y and z directions is 4 and 16, respectively. (Color figure online)

Figure 5 shows the algorithm for domain partitioning where voxels are assigned to each MPI process. First, the domain dimensions (e.g. $xmin, xmax$) and the voxel dimensions (Δx) are used to decide the total number of global voxels (*g_x_nodes*). Given the total number of MPI processes (P), the voxels per MPI process (*l_x_nodes*) in the x-direction are computed next. This is followed by the computation of the global coordinates for the centers of voxels (for brevity, lines 1–6 in Fig. 5 show this for the $x-$direction *only*, with the treatment of remaining directions being analogous to the $x-$direction). Further, since each MPI process must maintain the local and corresponding global voxel index, the global mesh index of the first voxel (*l_strt_g_index*) is computed on each process - used subsequently to assign the global mesh index to each voxel on that process (see the triply nested loop in Fig. 5). In addition to the assignment of a local/global voxel index on each process, a list of the immediate directional-neighbours of each voxel is also maintained (*not* shown in Fig. 5). In parallel,

Require: $xmin, xmax$, $d[]$ (Topology Dimensions), $c[]$ (Process Coordinates), Δx (Voxel x-length)

$g\_x\_nodes \leftarrow \frac{(xmax-xmin)}{\Delta x}$ ▷ y and z analogous
$l\_x\_nodes \leftarrow \frac{g\_x\_nodes}{d[1]}$
$l\_x\_start \leftarrow xmin + (c[1] \times l\_x\_nodes \times \Delta x)$
$i \leftarrow 0$
while $i{+}{+} \leq l\_x\_nodes - 1$ **do**
    $x\_c[i] \leftarrow l\_x\_start + (i + 0.5) * \Delta x$
$z_l \leftarrow c[2] \times g\_x\_nodes \times g\_y\_nodes \times l\_z\_nodes$
$y_l \leftarrow (d[0] - c[0] - 1) \times g\_x\_nodes \times l\_y\_nodes$
$x_l \leftarrow c[1] \times l\_x\_nodes$
$l\_strt\_g\_index \leftarrow x_l + y_l + z_l$
$n, i, j, k \leftarrow 0$
while $k{+}{+} < l\_z\_nodes$ **do**
    $z_k \leftarrow k \times g\_x\_nodes \times g\_y\_nodes$
    while $j{+}{+} < l\_y\_nodes$ **do**
        $y_j \leftarrow j \times g\_x\_nodes$
        while $i{+}{+} < l\_x\_nodes$ **do**
            $vxl[n].cntr[0] \leftarrow x\_c[i]$
            $vxl[n].cntr[1] \leftarrow y\_c[j]$
            $vxl[n].cntr[2] \leftarrow z\_c[k]$
            $vxl[n].g\_indx \leftarrow l\_strt\_g\_index + z_k + y_j + i$
            $n \leftarrow n + 1$

Fig. 5. Assignment of voxels to MPI processes in 1-D x-Domain Decomposition. Only partitioning of x-dimension is shown (same for y and z-directions). Prefixes $l\_$ and $g\_$ stand for "local" and "global", respectively. Array `d[]` contains the topology dimensions and array `c[]` contains MPI process coordinates [21]. The triply nested loop sets the global voxel (vxl) centers ($cntr$) and the global voxel index ($indx$).

such a scheme must accommodate for the cases when there is no local x, y or z neighbour but a global neighbour exists on the neighbouring process or when the process is aligned to the physical boundary of the domain. In BioFVM, a list for the Moore neighbourhood is also built for each voxel. The Moore neighbourhood equates to a 9-pt stencil in 2-D and a 27-pt stencil in 3-D [13].

Appendix B Mapping Basic Agents to a Voxel

A mapping that relates the position coordinates of the Basic Agent to the local index of a process-specific voxel is illustrated with the help of an algorithm in Fig. 6. Given the positions vector (denoted by $p[]$ in Fig. 6) of a Basic Agent, first the MPI Cartesian coordinates of the MPI process that contains the Basic Agent are computed (denoted by x_p, y_p and z_p). This is followed by the computation of the global x, y and z index (denoted by $first_x$, $first_y$ and $first_z$) of the first voxel of the MPI process that contains the Basic Agent. After calculating the directional i.e. x, y and z global indices of the voxel (denoted by vox_x, vox_y and

Require: $xmin, ymin, zmin,$ $d[]$ (Topology Dimensions), $d[]$ (MPI Cartesian dimensions),$p[]$ (Agent position coordinates), $\Delta x, \Delta y, \Delta z$ (Voxel x/y/z-length)

$x_p \leftarrow d[0] - 1 - \lfloor (p[1] - ymin)/(l\_y\_nodes * \Delta y) \rfloor$
$y_p \leftarrow (p[0] - xmin)/(l\_x\_nodes * \Delta x)$
$z_p \leftarrow (p[2] - zmin)/(l\_z\_nodes * \Delta z)$
$first_x \leftarrow y_p * l\_x\_nodes$
$first_y \leftarrow (d[0] - 1 - x_p) * l\_y\_nodes$
$first_z \leftarrow z_p * l\_z\_nodes$
$vox_x \leftarrow \lfloor (p[0] - xmin)/\Delta x \rfloor$
$vox_y \leftarrow \lfloor (p[1] - ymin)/\Delta y \rfloor$
$vox_z \leftarrow \lfloor (p[2] - zmin)/\Delta z \rfloor$
$d_x \leftarrow vox_x - first_x$
$d_y \leftarrow vox_y - first_y$
$d_z \leftarrow vox_z - first_z$
$l\_index \leftarrow (d_z * l\_y\_nodes + d_y) * l\_x\_nodes + d_x$

Fig. 6. Mapping the position of a Basic Agent (in array $p[]$) to the process-local index (l_index) of a voxel that contains $p[]$. Prefix $l\_$ stands for "local". Array $d[]$ contains the MPI Cartesian topology dimensions. $l\_x/y/z\_nodes$ give the number of process local voxels and $\Delta x, \Delta y, \Delta z$ denote voxel dimensions (generally $\Delta x = \Delta y = \Delta z$).

vox_z) that contains the Basic Agent, indices of the "first" voxel of the MPI process computed above is subtracted from the directional indices to obtain a local offset (denoted by d_x, d_y and d_z) of voxel indices in each direction. Subsequently, to obtain the local index of the process-specific voxel (l_index), the directional local offsets are appropriately multiplied by the number of process-local voxels.

Appendix C Extended Results

For an 8x increase in the number of voxels, the OpenMP MC, GPG, BAG and I/O kernels show a 7.86 – 8.67x, 3.29 – 7.05x, 1.15 – 1.3x and 6.78 – 8.51x increase, respectively (Fig. 2). The increase in the corresponding kernels for the best overall Hybrid version are: 8.7–9.4x, 3–7.78x, 0.77–1.14x and 3.14–6.68x, respectively (Fig. 2). Both MC and BAG kernels can take advantage of the multiple MPI processes as initialization of the `Microenvironment` and `Basic_Agent` class objects are simultaneously carried out on separate processes in BioFVM-X as opposed to a single thread in BioFVM. The (MPI) I/O kernel shows significant performance gains over serial I/O for the tests considered. For an 8x increase in the mesh resolution, the 6.78 – 8.11x increase for Hybrid version in the Solver kernel looks promising as compared to the 9.24–15.93x pure OpenMP increase, but the Hybrid version's absolute execution run-times do not reflect a significant gain. To help solve this, future versions of BioFVM-X will use the parallel modified Thomas solver [15] in the x-direction.

Appendix D Correctness Checking

To verify the correctness of the simulation, we run a simulation on a domain of size 1000^3 but increase the number of Basic Agents to 2×10^6 (Fig. 7 and Table 2).

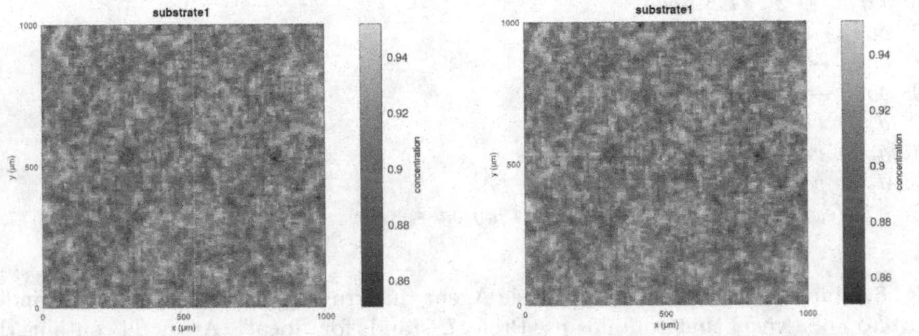

(a) OpenMP (48 threads) (b) Hybrid(n=1)

Fig. 7. 2-D cross-section of the final concentration density of a given substrate with 2×10^6 agents on a domain of size 1000^3 using (a) Pure OpenMP (b) Hybrid MPI + OpenMP

Table 2. Time (in seconds) of execution of simulation for the OpenMP version and the Hybrid version in a domain of size $1000 \times 1000 \times 1000$ with 2×10^6 Basic Agents.

$1000 \times 1000 \times 1000$	OpenMP	Hyb (n = 1)
Build μ-environment	1.14	1.03
Gaussian profile	0.0157	0.0117
Initial file write	0.219	0.084
Agent generation	2.46	1.45
Source/sink/diffusion	7.48	5.88
Final file write	0.22	0.063
Total time	11.56	8.54

To further underline the correctness of BioFVM-X, we compared the results of a tumor growth model in a heterogeneous environment from BioFVM [8] available at this link. In this model a 2-D tumor growth is driven by a substrate supplied by a continuum vascular system and cells die when it is insufficient. Additionally, the tumor cells have motility and can degrade the vascular system. We first expanded this example to a 3-D example (instead of the original 2-D) and specified the domain as $80 \times 80 \times 80$ voxels for a total of 512 000 voxels. We choose two different configurations:

– a shared-memory configuration (OpenMP) of 48 threads
– a hybrid shared and distributed-memory configuration (MPI+OpenMP) of 2 MPI processes running 24 threads each on a single node.

The comparison of the shared-memory and distributed-memory simulations yields identical results as shown in Fig. 8, further confirming that BioFVM-X provides the same results as BioFVM. The code to reproduce the figure is available on the BioFVM-X code repository.

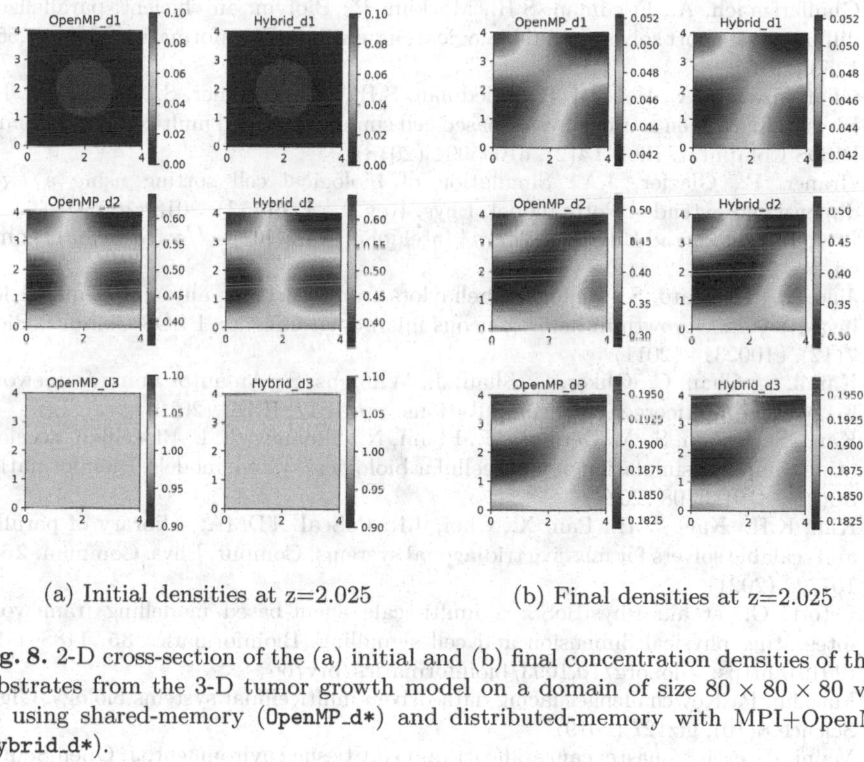

(a) Initial densities at z=2.025 (b) Final densities at z=2.025

Fig. 8. 2-D cross-section of the (a) initial and (b) final concentration densities of three substrates from the 3-D tumor growth model on a domain of size 80 × 80 × 80 voxels using shared-memory (`OpenMP_d*`) and distributed-memory with MPI+OpenMP (`Hybrid_d*`).

References

1. Balay, S., et al.: PETSc Users Manual. No. ANL-95/11-Revision 3.15 (2021). https://www.mcs.anl.gov/petsc
2. Briggs, W.L., McCormick, S.F., et al.: A Multigrid Tutorial, vol. 72. SIAM (2000)
3. Adams, M., et al.: Chombo Software Package for AMR Applications - Design Document. Lawrence Berkeley National Laboratory Technical Report LBNL-6616E
4. Cytowski, M., Szymanska, Z.: Large-scale parallel simulations of 3d cell colony dynamics. Comput. Sci. Eng. **16**(5), 86–95 (2014). https://doi.org/10.1109/MCSE.2014.2

5. Cytowski, M., Szymanska, Z.: Large-scale parallel simulations of 3d cell colony dynamics: the cellular environment. Comput. Sci. Eng. **17**(5), 44–48 (2015). https://doi.org/10.1109/MCSE.2015.66
6. Devine, K.D., Boman, E.G., Leung, V.J., Riesen, L.A., Catalyurek, U.V.: Dynamic load balancing and partitioning using the Zoltan toolkit (2007). https://www.osti.gov/biblio/1147186
7. Falgout, R.D., Yang, U.M.: *hypre*: a library of high performance preconditioners. In: Sloot, P.M.A., Hoekstra, A.G., Tan, C.J.K., Dongarra, J.J. (eds.) ICCS 2002. LNCS, vol. 2331, pp. 632–641. Springer, Heidelberg (2002). https://doi.org/10.1007/3-540-47789-6_66
8. Ghaffarizadeh, A., Friedman, S.H., Macklin, P.: Biofvm: an efficient, parallelized diffusive transport solver for 3-d biological simulations. Bioinformatics **32**(8), 1256–1258 (2015)
9. Ghaffarizadeh, A., Heiland, R., Friedman, S.H., Mumenthaler, S.M., Macklin, P.: PhysiCell: an open source physics-based cell simulator for 3-d multicellular systems. PLOS Computat. Biol. **14**(2), e1005991 (2018)
10. Graner, F., Glazier, J.A.: Simulation of biological cell sorting using a two-dimensional extended Potts model. Phys. Rev. Lett. **69**(13), 2013 (1992)
11. Intel: Intel®Thread Building Blocks | Intel®Software. https://software.intel.com/en-us/tbb
12. Jiao, Y., Torquato, S.: Emergent behaviors from a cellular automaton model for invasive tumor growth in heterogeneous microenvironments. PLOS Comput. Biol. **7**(12), e1002314 (2011)
13. Kamil, S., Chan, C., Oliker, L., Shalf, J., Williams, S.: An auto-tuning framework for parallel multicore stencil computations, pp. 1–12. IEEE (2010)
14. Kang, S., Kahan, S., McDermott, J., Flann, N., Shmulevich, I.: Biocellion: accelerating computer simulation of multicellular biological system models. Bioinformatics **30**(21), 3101–3108 (2014)
15. Kim, K.H., Kang, J.H., Pan, X., Choi, J.I.: PaScaL_TDMA: a library of parallel and scalable solvers for massive tridiagonal systems. Comput. Phys. Commun. **260**, 107722 (2021)
16. Letort, G., et al.: PhysiBoSS: a multi-scale agent-based modelling framework integrating physical dimension and cell signalling. Bioinformatics **35**, 1188–1196 (2019). https://doi.org/10.1093/bioinformatics/bty766
17. Macklin, P.: Key challenges facing data-driven multicellular systems biology. GigaScience **8**(10), giz127 (2019)
18. Maini, P., et al.: Chaste: cancer, heart and soft tissue environment. J. Open Source Softw. **5**(47), 1848 (2020)
19. Mascagni, M., Srinivasan, A.: Algorithm 806: SPRNG: a scalable library for pseudorandom number generation. ACM Trans. Math. Softw. (TOMS) **26**(3), 436–461 (2000)
20. Mazumder, S.: Numerical Methods for Partial Differential Equations: Finite Difference and Finite Volume Methods. Academic Press (2015)
21. Message Passing Interface Forum: MPI: A message-passing interface standard version 3.1 (June 2015). https://www.mpi-forum.org/docs/mpi-3.1/mpi31-report.pdf
22. OpenMP Architecture Review Board: OpenMP application program interface version 5.0 (November 2018). https://www.openmp.org/wp-content/uploads/OpenMP-API-Specification-5.0.pdf
23. Osborne, J.M., Fletcher, A.G., Pitt-Francis, J.M., Maini, P.K., Gavaghan, D.J.: Comparing individual-based approaches to modelling the self-organization of multicellular tissues. PLOS Comput. Biol. **13**(2), e1005387 (2017)

24. Ozik, J., Collier, N., Heiland, R., An, G., Macklin, P.: Learning-accelerated discovery of immune-tumour interactions. Mol. Syst. Des. Eng. **4**(4), 747–760 (2019)
25. Pacific Northwest National Laboratory: PNNL: Global Arrays Toolkit. https://hpc.pnl.gov/globalarrays/
26. Van der Pas, R., Stotzer, E., Terboven, C.: Using OpenMP–The Next Step: Affinity, Accelerators, Tasking, and SIMD. MIT Press (2017)
27. Saxena, G., Jimack, P.K., Walkley, M.A.: A cache-aware approach to domain decomposition for stencil-based codes, pp. 875–885. IEEE (2016)
28. Saxena, G., Jimack, P.K., Walkley, M.A.: A quasi-cache-aware model for optimal domain partitioning in parallel geometric multigrid. Concurrency Comput. Pract. Exp. **30**(9), e4328 (2018)
29. Strauss, W.A.: Partial Differential Equations: An Introduction. Wiley (2007)
30. Süli, E., Mayers, D.F.: An Introduction to Numerical Analysis. Cambridge University Press (2003)
31. Thomas, L.: Elliptic Problems in Linear Differential Equations Over a Network. Watson Scientific Computing Laboratory. Columbia University, NY (1949)
32. Trottenberg, U., Oosterlee, C.W., Schuller, A.: Multigrid. Elsevier (2000)

Author Index

Printed in the United States
by Baker & Taylor Publisher Services

Printed in the United States
by Baker & Taylor Publisher Services